AF602093

ISNM
International Series of Numerical Mathematics
Vol. 146

Modeling, Simulation, and Optimization of Integrated Circuits

Proceedings of a Conference held at the
Mathematisches Forschungsinstitut, Oberwolfach,
November 25–December 1, 2001

K. Antreich
R. Bulirsch
A. Gilg
P. Rentrop
Editors

Springer Basel AG

Editors:

Kurt Antreich
Technische Universität München
Lehrstuhl für Entwurfsautomatisierung
Arcisstr. 21
D-80333 München
Germany

Albert Gilg
Siemens AG
CT PP 2
Otto-Hahn-Ring 6
D-81730 München
Germany

Roland Bulirsch
Technische Universität München
Zentrum Mathematik - M2
Boltzmannstr. 3
D-85748 Garching b. München
Germany

Peter Rentrop
Technische Universität München
Zentrum Mathematik – M2
Boltzmannstr. 3
D-85748 Garching b. München
Germany

2000 Mathematics Subject Classification 34-06, 34B60, 35Q60, 65-06, 65Fxx, 65Lxx, 65Mxx, 65Nxx, 65Z05, 78Mxx, 80A20, 94Cxx

A CIP catalogue record for this book is available from the Library of Congress, Washington D.C., USA

Bibliographic information published by Die Deutsche Bibliothek
Die Deutsche Bibliothek lists this publication in the Deutsche Nationalbibliografie; detailed bibliographic data is available in the Internet at <http://dnb.ddb.de>.

ISBN 978-3-7643-2192-5 ISBN 978-3-0348-8065-7 (eBook)
DOI 10.1007/978-3-0348-8065-7

Originally published by Birkhäuser Verlag, Basel, Switzerland in 2003
Printed on acid-free paper produced of chlorine-free pulp. TCF ∞

ISBN 978-3-7643-2192-5

9 8 7 6 5 4 3 2 1

www.birkhasuer-science.com

Contents

Preface

The simulation chain: process simulation – device simulation – circuit simulation – equipment simulation is a central focus in Chip Industry. Since simulation plays a key role in industrial development its definition is fixed by the German VDI Norm 3633: Simulation ist das Nachbilden eines Systems mit seinen dynamischen Prozessen in einem experimentierfähigen Modell, um zu Erkenntnissen zu gelangen, die auf die Wirklichkeit übertragbar sind. Instead of real objects, physical or mathematical models are studied. In short terms, simulation can be expressed as experiment with a model. Thus numerical simulation in Chip Industry replaces time consuming and expensive experiments and saves resources.

During the last decade the Mathematical Research Centre at Oberwolfach, Germany, has hosted three Conferences on the topic Mathematical Models and Numerical Simulation in Electronic Industry. The two preceeding conferences in 1988 and 1992 – the Proceedings have been published in this series ISNM 93 (1990) und ISNM 117 (1994) – concentrated more on device and process simulation. In this conference two new topics were introduced. Co-Simulation, i.e., the coupling of different simulation tools with strongly varying time scales and electromagnetic effects have been of central interest. Additionally few talks were added from different areas, in order to introduce concepts, which are of potential interest for Chip design.

The research facilities in Oberwolfach have hosted 43 participants from mathematical departments, from departments of electrical engineering and from industry. Their contributions try to bridge the gap between basic and applied mathematics, research in electrical engineering and the needs of industry.

The valuable assistance and LaTeX expertise of Dr. R. Pulch and Dr. U. Rettig is gratefully acknowledged.

The editors

List of Participants

Prof. Dr. Angelo Marcello Anile
anile@dmi.unict.it
Dipartimento di Mathematica
Città Universitaria
Viale A. Doria, 6 - 1
I - 95125 Catania

Prof. Dr. Kurt Antreich
kurt.antreich@ei.tum.de
Lehrstuhl für Entwurfsautomatisierung
Technische Universität München
Arcisstr. 21 D - 80290 München

Andreas Bartel
bartel@iwrmm.math.uni-karlsruhe.de
IWRMM
Universität Karlsruhe (TH)
D - 76128 Karlsruhe

Prof. Dr. Dr. h.c. Roland Bulirsch
bulirsch@mathematik.tu-muenchen.de
Zentrum Mathematik
Technische Universität München
D - 80290 München

Prof. Dr. Angelika Bunse-Gerstner
angelika@mathematik.uni-bremen.de
Fachbereich 3
Mathematik und Informatik
Universität Bremen
Postfach 330440
D - 28334 Bremen

Dr. Maria Paz Calvo
maripaz@mac.cie.uva.es
Departamento de Mathematica
Aplicada y Computacion
Facultad de Ciencias
Universidad de Valladolid
Prado de la Magdalena s/n
E - 47005 Valladolid

Dr. Georg Denk
Georg.Denk@infineon.com
Infineon Technologies
MP TI CS ATS
Balanstr. 73
D - 81541 München

Prof. Dr. Olaf Dössel
doessel@ibt.etec.uni-karlsruhe.de
Fakultät für Elektro- und
Informationstechnik
Universität Karlsruhe (TH)
D - 76128 Karlsruhe

Dr. Uwe Feldmann
uwe.feldmann@infineon.com
Infineon Technologies
MP TI CS ATS
Balanstr. 73
D - 81541 München

Dr. Roland W. Freund
freund@research.bell-labs.com
Bell Laboratories
Room 2C - 525
700 Mountain Avenue
Murray Hill, NJ 07974-0636
USA

Dr. Albert Gilg
Siemens AG
CT PP2
Otto-Hahn-Ring 6
D - 81739 München

Dr. Helmut Gräb
helmut.graeb@ei.tum.de
Lehrstuhl für Entwurfsautomatisierung
Technische Universität München
Arcisstr. 21
D - 80333 München

Prof. Dr. Friedrich Grund
grund@wias-berlin.de
Weierstraß-Institut für
Angewandte Analysis und Stochastik
im Forschungsverbund Berlin e.V.
Mohrenstr. 39
D - 10117 Berlin

Prof. Dr. Michael Günther
guenther@math.uni-wuppertal.de
Arbeitsgruppe Angewandte
Mathematik/Numerik
Fachbereich Mathematik
Bergische Universität Wuppertal
Gaußstr. 20
D - 42119 Wuppertal

Prof. Dr. Ralf Hiptmair
hiptmair@sam.math.ethz.ch
Seminar of Applied Mathematics
ETHZ
CH - 8092 Zürich

Prof. Dr. Ronald H.W. Hoppe
hoppe@math.uni-augsburg.de
Lehrstuhl für Angewandte
Mathematik I
Universität Augsburg
Universitätsstr. 14
D - 86159 Augsburg

Dr. Stephan Houben
stephanh@win.tue.nl
Philips Research Laboratories
Electronic Design Tools
Prof. Holstlaan 4
Building Way - 3.73
NL - 5656 AA Eindhoven

Ansgar Jüngel
juengel@fmi.uni-konstanz.de
FB Mathematik und Statistik
Universität Konstanz
Fach D 193
D - 78457 Konstanz

Prof. Dr. Franz X. Kärtner
kaertner@mit.edu
Fakultät für Elektro- und
Informationstechnik
Universität Karlsruhe (TH)
D - 76128 Karlsruhe

Dr. Martin Kahlert
martin.kahlert@gmx.de
Sonnenstr. 16
D-85229 Markt Indersdorf

Dr. Anne Kvaerno
anne@math.ntnu.no
Department of Mathematical Sciences
The Norwegian University of Science and Technology
N - 7034 Trondheim

Prof. Dr. Ulrich Langer
ulanger@numa.uni-linz.ac.at
Institut für Numerische Mathematik
Universität Linz
Altenbergstr. 69
A - 4040 Linz

Prof. Dr. Roswitha März
maerz@mathematik.hu-berlin.de
Institut für Mathematik
Humboldt-Universität Berlin
D - 10099 Berlin

Dr. E. Jan W. ter Maten
jan.ter.maten@philips.com
Philips Research Laboratories
Electr. Design & Tools,
Analogue Simulation
Prof. Holstlaan 4 (WAY 3)
NL - 5656 AA Eindhoven

Prof. Dr. Wolfgang Mathis
mathis@tet.uni-hannover.de
Institut für Theoretische Elektrotechnik
Universität Hannover
Appelstr. 9A
D - 30167 Hannover

Prof. Dr. Volker Mehrmann
mehrmann@math.tu-berlin.de
Institut für Mathematik
Technische Universität Berlin
Straße des 17. Juni 136
D - 10623 Berlin

Prof. Dr. Alexander Ostermann
alexander.ostermann@uibk.ac.at
Institut für Technische Mathematik, Geometrie und Bauinformatik
Universität Innsbruck
Technikerstr. 13
A - 6020 Innsbruck

Michael Pronath
michael.pronath@ei.tum.de
Lehrstuhl für Entwurfsautomatisierung
Technische Universität München
Arcisstr. 21
D - 80333 München

Roland Pulch
pulch@mathematik.tu-muenchen.de
Zentrum Mathematik - M2
Technische Universität München
Boltzmannstr. 3
D - 85748 Garching bei München

Dr. Gunther Reißig
gr@ieee.org
Massachusetts Institute of Technology
Room 66-363
77 Massachusetts Avenue
Cambridge, MA 02139
USA

Prof. Dr. Peter Rentrop
toth-pinter@ma.tum.de
Zentrum Mathematik - M2
Technische Universität München
Boltzmannstr. 3
D - 85748 Garching bei München

Prof. Dr. Ursula van Rienen
ursula.van-rienen@ ...
... etechnik.uni-rostock.de
Fachbereich Elektrotechnik und
Informationstechnik
Universität Rostock
Albert-Einstein-Str. 2
D - 18051 Rostock

Prof. Dr. Werner Römisch
romisch@mathematik.hu-berlin.de
Institut für Mathematik
Humboldt-Universität Berlin
D - 10099 Berlin

Prof. Dr. Willy H.A. Schilders
wil.schilders@philips.com
Philips Research Laboratories
Building WAY 4.077
Prof. Holstlaan 4
NL - 5656 AA Eindhoven

Dr. Peter Schwarz
schwarz@ieas.iis.fhg.de
Frauenhofer Institut für
Integrierte Schaltungen
Außenstelle Entwurfsautomatisierung
Zeunerstr. 38
D - 01069 Dresden

Guido Stehr
guido.stehr@eda.ei.tum.de
Lehrstuhl für Entwurfsautomatisierung
Technische Universität München
Arcisstr. 21
D - 80333 München

Prof. Dr. Klaus Taubert
taubert@math.uni-hamburg.de
Fachbereich Mathematik
Universität Hamburg
Bundesstr. 55 D - 20146 Hamburg

Dr. Caren Tischendorf
caren@mathematik.hu-berlin.de
Institut für Mathematik
Humboldt-Universität Berlin
D - 10099 Berlin

Prof. Dr. Ljiljana Trajkovic
School of Engineering Science
ASB 9811
Simon Fraser University
Burnaby, BC V5A 1S6
Canada

Prof. Dr. Donato Trigiante
trigiant@cesit1.unifi.it
Dipartimento di Energetica
Università di Firenze
Via C. Lombroso 6/17
I - 50134 Firenze

Prof. Dr. Jan G. Verwer
janv@cwi.nl
Centrum voor Wiskunde en
Informatica
Kruislaan 413
NL - 1098 SJ Amsterdam

Prof. Dr. Gerhard Wachutka
wachutka@tep.ei.tum.de
Lehrstuhl für Entwurfsautomatisierung
Technische Universität München
Arcisstr. 21
D - 80333 München

Prof. Dr. Robert Weigel
r.weigel@ieee.org
Institut für Nachrichtentechnik/
Informationstechnik
Johannes Kepler Universität Linz
Altenbergerstr. 69
A - 4040 Linz

Dr. Renate Winkler
winkler@mathematik.hu-berlin.de
Institut für Mathematik
Humboldt-Universität Berlin
D - 10099 Berlin

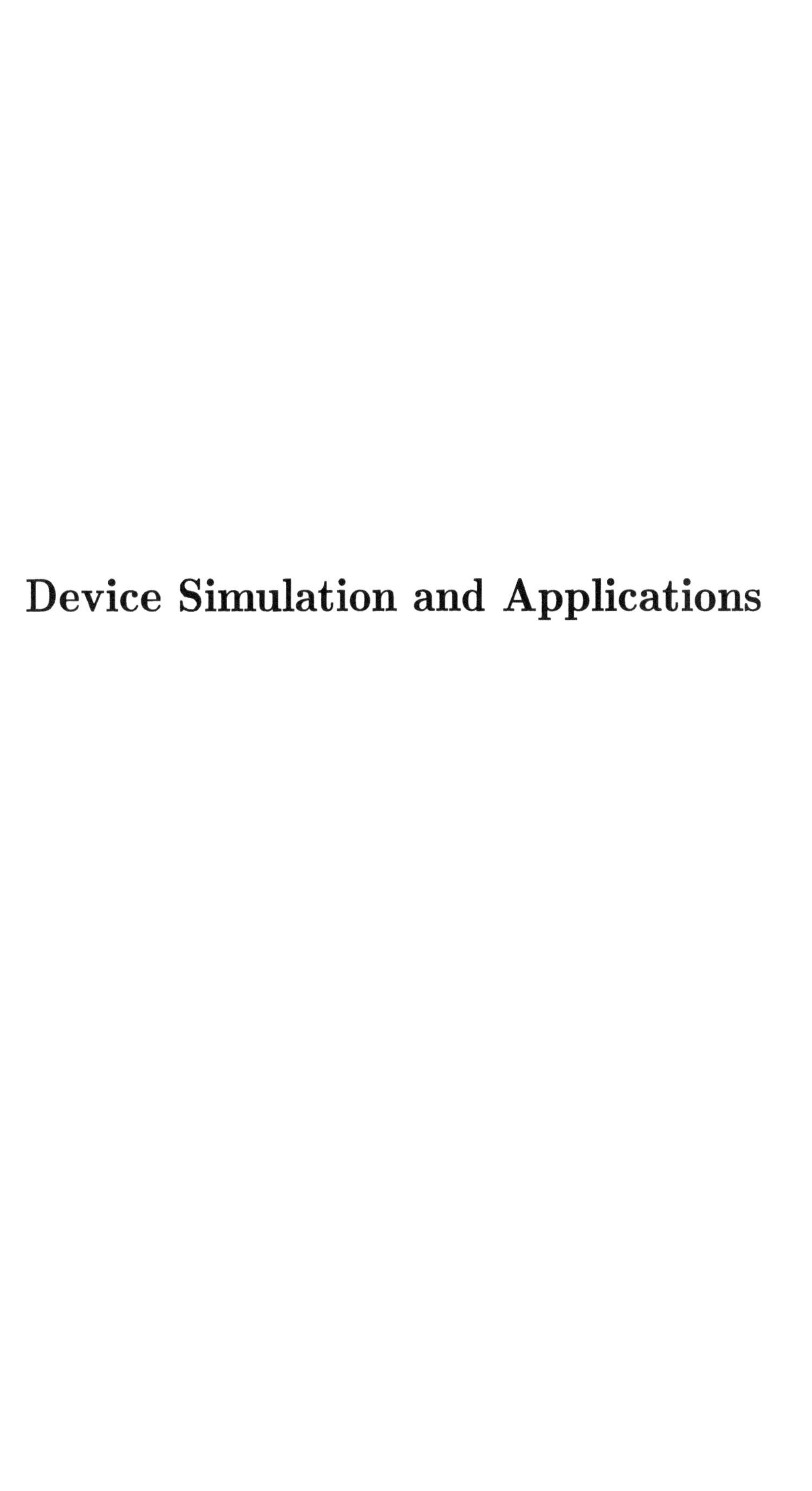

Device Simulation and Applications

Modeling, Simulation and Optimization of Integrated Circuits
(Eds.) K. Antreich, R. Bulirsch, A. Gilg and P. Rentrop

International Series of Numerical Mathematics Vol. 146, 3–15

Hydrodynamical Model for GaAs Semiconductors Based on the Maximum Entropy Principle with Application to Electronic Devices

A.M. Anile, G. Mascali, and V. Romano

Abstract. A hydrodynamical model for electron transport in GaAs semiconductors, which is free of any fitting parameter, has been formulated in [1] on the basis of the maximum entropy principle, by including both Γ and L conduction band valleys described by the Kane dispersion relation.

Here we test the model in the bulk case [1] and by simulating an $n^+ - n - n^+$ GaAs diode and a Gunn oscillator [2].

1. The model

Hydrodynamical models are obtained from the infinite hierarchy of the moment equations of the Boltzmann transport equation by suitable truncation procedures. However most of them suffer from serious theoretical drawbacks due to the *ad hoc* treatment of the closure problem [3]. Recently, in the case of silicon semiconductors, a moment approach has been introduced [4, 5] (see also [6] for a complete review) in which the closure procedure is based on the maximum entropy principle, while the conduction bands are described by the Kane dispersion relation. In [1, 2] the same approach has been employed for GaAs by including the Γ-valley and the four equivalent L-valleys. Electrons have been considered as a mixture of two fluids, one representing electrons in the Γ-valley and the other electrons in the four equivalent L-valleys. The model comprises the balance equations of electron density, energy density, velocity and energy flux for both populations, coupled to the Poisson equation for the electric potential. Apart from the Poisson equation, the system is hyperbolic in the physically relevant region of the field variables.

In this section we give only a brief sketch of the model. For more details the interested reader is referred to [1].

One assumes that the conduction band [7] is described in the neighborhood of each minimum, the so-called valleys, by the Kane dispersion approximation.

Received by the editors December 1, 2001.
2000 *Mathematics Subject Classification.* Primary 82D37; Secondary 82C70.
Key words and phrases. Semiconductors, GaAs, Gunn diode.
This work was completed with the support of our TEX-pert.

At a kinetic level, the system is described by two Boltzmann equations, one for the Γ-valley and the other for one L-valley. The macroscopic balance equations are deduced by taking the moments of the Boltzmann transport equations, that is by multiplying each transport equation by suitable weight functions and integrating over the wave vector space which in the Kane approximation is identified with $\mathbb{R}^3$. We will consider the set of weight functions necessary to get the macroscopic balance equations for densities, momenta, energies and energy-fluxes of both Γ and L valley electrons, these equations read

$$\frac{\partial n_A}{\partial t} + \frac{\partial (n_A V_A^i)}{\partial x^i} = n_A C_{n_A}, \tag{1.1}$$

$$\frac{\partial (n_A P_A^i)}{\partial t} + \frac{\partial (n_A U_A^{ij})}{\partial x^j} + n_A q E^i = n_A C_{P_A^i}, \tag{1.2}$$

$$\frac{\partial (n_A W_A)}{\partial t} + \frac{\partial (n_A S_A^j)}{\partial x^j} + n_A q V_A^j E_j = n_A C_{W_A}, \tag{1.3}$$

$$\frac{\partial (n_A S_A^i)}{\partial t} + \frac{\partial (n_A F_A^{ij})}{\partial x^j} + n_A q E_j G_A^{ij} = n_A C_{S_A^i}, \tag{1.4}$$

where $\mathbf{E}$ represents the self-consistent electric field, q the elementary charge and

$$n_A = \int_{\mathbb{R}^3} f_A d\mathbf{k}_A \quad \text{is the electron density,}$$

$$V_A^i = \frac{1}{n_A} \int_{\mathbb{R}^3} v_A^i f_A d\mathbf{k}_A \quad \text{the average electron velocity,}$$

$$W_A = \frac{1}{n_A} \int_{\mathbb{R}^3} \mathcal{E}_A(k_A) f_A d\mathbf{k}_A \quad \text{the average electron energy,}$$

$$S_A^i = \frac{1}{n_A} \int_{\mathbb{R}^3} v_A^i \mathcal{E}_A(k_A) f_A d\mathbf{k}_A \quad \text{the average energy flux,}$$

$$P_A^i = \frac{1}{n_A} \int_{\mathbb{R}^3} \hbar k_A^i f_A d\mathbf{k}_A = m_A^* \left(V_A^i + 2\alpha_A S_A^i \right)$$

the average crystal momentum,

$$U_A^{ij} = \frac{1}{n_A} \int_{\mathbb{R}^3} v_A^i \hbar k_A^j f_A d\mathbf{k}_A \quad \text{the average crystal momentum flux,}$$

$$G_A^{ij} = \frac{1}{n_A} \int_{\mathbb{R}^3} \left[\frac{\partial}{\partial k_A^j} \left(\frac{1}{\hbar} v_A^i \, \mathcal{E}_A(\mathbf{k}) \right) \right] f_A d\mathbf{k}_A,$$

$$
\begin{aligned}
F_A^{ij} &= \frac{1}{n_A}\int_{\mathbb{R}^3} v_A^i v_A^j \mathcal{E}_A(k_A) f_A d\mathbf{k}_A \quad \text{the average flux of energy flux,}\\
C_{n_A} &= \frac{1}{n_A}\int_{\mathbb{R}^3} \mathcal{C}_A[f] d\mathbf{k}_A \quad \text{the density production,}\\
C_{P_A^i} &= \frac{1}{n_A}\int_{\mathbb{R}^3} \hbar k_A^i \mathcal{C}_A[f] d\mathbf{k}_A \quad \text{the crystal momentum production,}\\
C_{W_A} &= \frac{1}{n_A}\int_{\mathbb{R}^3} \mathcal{E}_A(k_A)\mathcal{C}_A[f] d\mathbf{k}_A \quad \text{the energy production,}\\
C_{S_A^i} &= \frac{1}{n_A}\int_{\mathbb{R}^3} v_A^i \mathcal{E}_A(k_A)\mathcal{C}_A[f] d\mathbf{k}_A \quad \text{the energy flux production.}
\end{aligned}
$$

All these quantities refer to electrons in the A-valley, $A = \Gamma,\ L$.

$\mathcal{C}_A[f]$ are the scattering operators which appear in the electron transport equations. They take into account the most relevant interactions in GaAs, that is those between electrons and intervalley non-polar optical phonons, intravalley polar and acoustic phonons and ionized impurities. The electron-electron scattering will not be considered here.

The above-written system is coupled to the Poisson equation

$$
\mathbf{E} = -\nabla\Phi, \quad \epsilon\Delta\Phi = -q(N_+ - N_- - n_\Gamma - 4n_L) \tag{1.5}
$$

where Φ is the electric potential, N_+ and N_- the donor and acceptor densities respectively, and ϵ the dielectric constant. These moment equations do not constitute a set of closed relations because of the fluxes and production terms. Therefore constitutive assumptions must be prescribed.

If we assume as fundamental variables n_A, V_A^i, W_A and S_A^i, $A = \Gamma, L$, which have a direct physical interpretation, the closure problem consists of expressing U_A^{ij}, F_A^{ij} and G_A^{ij} and the moments of the collision terms C_{n_A}, $C_{P_A}^i$, C_{W_A} and $C_{S_A^i}$ as functions of n_A, V_A^i, W_A and S_A^i.

The Maximum Entropy Principle (hereafter MEP) leads to a systematic way of obtaining constitutive relations on the basis of information theory. According to the MEP, the distribution functions f_A^{ME} which can be used to evaluate the unknown moments of f_A are stationary points of the electron entropy functional under the constraint of fixed fundamental variables.

This procedure, which fits in the framework of extended thermodynamics [8, 9] or equivalently Levermore's moment theory [10], has been used in [1] upon the ansatz of small anisotropy for the f_A^{ME}. Formally a *small* anisotropy parameter δ has been introduced and explicit constitutive equations have been obtained for fluxes and production terms up to the first order in δ. Here we do not report the constitutive relations. The interested reader is referred to [1].

One can prove that the system (1.1)–(1.4), closed with the maximum entropy principle, is hyperbolic in the physically relevant region of the dependent variables [2].

For the numerical integration we use the extension of the scheme developed in [11, 12] for homogeneous hyperbolic systems, which has been adapted in [13, 14, 15, 16, 17, 18] for balance laws with (possibly *stiff*) source terms.

The complete method is based on a second-order splitting technique which separately solves the system with the source put equal to zero (convection step) and the one with the flux vector put equal to zero (relaxation step). For more details see [2].

2. Applications to the case of bulk GaAs

In this section we test the model in the case of a uniformly doped GaAs semiconductor. Two different impurity concentrations, $N_+ = 10^{14}\text{cm}^{-3}$, 10^{17}cm^{-3} are considered, and all the relevant scattering mechanisms are taken into account.

Since the problem is homogeneous, in the evolution equations we can drop the spatial dependence and the balance equations reduce to the following set of ordinary differential equations

$$\frac{d}{dt} n_\Gamma = n_L\, C_{n\ \Gamma L}^{(np)+}(W_L) - n_\Gamma\, C_{n\ \Gamma L}^{(np)-}(W_\Gamma), \tag{2.1}$$

$$m_\Gamma^* \frac{d}{dt} n_\Gamma (V_\Gamma^i + 2\alpha_\Gamma S_\Gamma^i) = -q n_\Gamma E^i + c_{11}^\Gamma\, n_\Gamma V_\Gamma^i + c_{12}^\Gamma\, n_\Gamma S_\Gamma^i, \tag{2.2}$$

$$\frac{d}{dt} n_\Gamma W_\Gamma = -q n_\Gamma V_\Gamma^k E_k + n_L C_{W\ \Gamma L}^{(np)+}(W_L)$$

$$+ \qquad n_\Gamma \left[C_{W_\Gamma}^{(p)}(W_\Gamma) - C_{W\ \Gamma L}^{(np)-}(W_\Gamma) \right], \tag{2.3}$$

$$\frac{d}{dt} n_\Gamma S_\Gamma^i = -q\, n_\Gamma E_k G_\Gamma^{ik} + c_{21}^\Gamma n_\Gamma V_\Gamma^i + c_{22}^\Gamma n_\Gamma S_\Gamma^i, \tag{2.4}$$

$$\frac{d}{dt} n_L = -n_L\, C_{n\ L\Gamma}^{(np)-}(W_L) + n_\Gamma\, C_{n\ L\Gamma}^{(np)+}(W_\Gamma), \tag{2.5}$$

$$m_L^* \frac{d}{dt} n_L (V_L^i + 2\alpha_L S_L^i) = -q n_L E^i + c_{11}^L\, n_L V_L^i + c_{12}^L\, n_L S_L^i, \tag{2.6}$$

$$\frac{d}{dt} n_L W_L = -q n_L V_L^k E_k + n_\Gamma C_{W\ L\Gamma}^{(np)+}(W_\Gamma) + n_L \left[C_{W_L}^{(p)}(W_L) + C_{W\ LL}^{(np)}(W_L) \right.$$

$$\left. - C_{W\ L\Gamma}^{(np)-}(W_L) \right], \tag{2.7}$$

$$\frac{d}{dt} n_L S_L^i = -q\, n_L E_k G_L^{ik} + c_{21}^L n_L V_L^i + c_{22}^L n_L S_L^i, \tag{2.8}$$

the meaning of the various productions terms which appear in the equations can be found in [1].

From (2.1), (2.5) and the expressions of the production terms, one has that $n = n_\Gamma + 4\, n_L = \text{const}$, so that the total electron number is conserved, as it must be. In cases when a constant bias voltage is applied to the semiconductor, the Poisson equation is satisfied with n equal to the value of the doping concentration and **E** constant. The motion is along the direction of the electric field and, if we take this as the x-direction, the system (2.1)–(2.8) reads

$$\frac{d}{dt}n_\Gamma = n_L\, C^{(np)+}_{n\;\Gamma L}(W_L) - n_\Gamma\, C^{(np)-}_{n\;\Gamma L}(W_\Gamma), \tag{2.9}$$

$$\frac{d}{dt}n_\Gamma V_\Gamma = \left(2\alpha_\Gamma G_\Gamma - \frac{1}{m^*_\Gamma}\right) q\,E\,n_\Gamma + \left(\frac{c^\Gamma_{11}}{m^*_\Gamma} - 2\,\alpha_\Gamma\, c^\Gamma_{21}\right) n_\Gamma V_\Gamma + + \left(\frac{c^\Gamma_{12}}{m^*_\Gamma} - 2\,\alpha_\Gamma\, c^\Gamma_{22}\right) n_\Gamma S_\Gamma, \tag{2.10}$$

$$\frac{d}{dt}n_\Gamma W_\Gamma = -q n_\Gamma V_\Gamma E + n_L C^{(np)+}_{W\;\Gamma L}(W_L) + n_\Gamma\left[C^{(p)}_{W_\Gamma}(W_\Gamma) - C^{(np)-}_{W\;\Gamma L}(W_\Gamma)\right], \tag{2.11}$$

$$\frac{d}{dt}n_\Gamma S_\Gamma = -q\,n_\Gamma E\,G_\Gamma + c^\Gamma_{21} n_\Gamma V_\Gamma + c^\Gamma_{22} n_\Gamma S_\Gamma, \tag{2.12}$$

$$\frac{d}{dt}\left(n_\Gamma + 4\,n_L\right) = 0, \tag{2.13}$$

$$\frac{d}{dt}n_L V_L = \left(2\alpha_L G_L - \frac{1}{m^*_L}\right) q\,E\,n_L + \left(\frac{c^L_{11}}{m^*_L} - 2\,\alpha_L\, c^L_{21}\right) n_L V_L + + \left(\frac{c^L_{12}}{m^*_L} - 2\,\alpha_L\, c^L_{22}\right) n_L S_L, \tag{2.14}$$

$$\frac{d}{dt}n_L W_L = -q n_L V_L E + n_\Gamma C^{(np)+}_{W\;L\Gamma}(W_\Gamma) + n_L\left[C^{(p)}_{W_L}(W_L) + C^{(np)}_{W\;L`L}(W_L) - C^{(np)-}_{W\;L\Gamma}(W_L)\right], \tag{2.15}$$

$$\frac{d}{dt}n_L S_L = -q\,n_L E\,G_L + c^L_{21} n_L V_L + c^L_{22} n_L S_L\,, \tag{2.16}$$

where V_A and S_A are the x-components of $\mathbf{V}_A$ and $\mathbf{S}_A$ and G_A is the xx-component of G^{ij}_A, $A = \Gamma,\ L$. In the evolution equations the x-component E of the electric field enters as a parameter.

As initial conditions for (2.9)–(2.16), we take, in suitable units,

$$\begin{aligned}
n_\Gamma(0) + 4\,n_L(0) &= 1,\\
\frac{n_\Gamma(0)}{n_L(0)} &= \frac{\int_{\Re^3} f^{(eq)}_\Gamma d\mathbf{k}_\Gamma}{\int_{\Re^3} f^{(eq)}_L d\mathbf{k}_L} = \left(\frac{m^*_\Gamma}{m^*_L}\right)^{3/2} \frac{d^\Gamma_0\left(\frac{1}{k_B T_L}\right)}{d^L_0\left(\frac{1}{k_B T_L}\right)},\\
V_A(0) &= 0,\\
W_A(0) &= g_A\left(\frac{1}{k_B T_L}\right),\\
S_A(0) &= 0, \qquad\qquad A = \Gamma\,, L\,.
\end{aligned}$$

The crystal temperature, T_L, is assumed to be 300 K. The expressions of d^Γ_0, d^L_0 and g_A are given in [1].

The solutions of (2.9)–(2.16) for electric fields respectively equal to 0.2, 0.5, 1, 2 and $6\frac{V}{\mu m}$, are reported in Fig. 1. The stationary regime is reached in a few

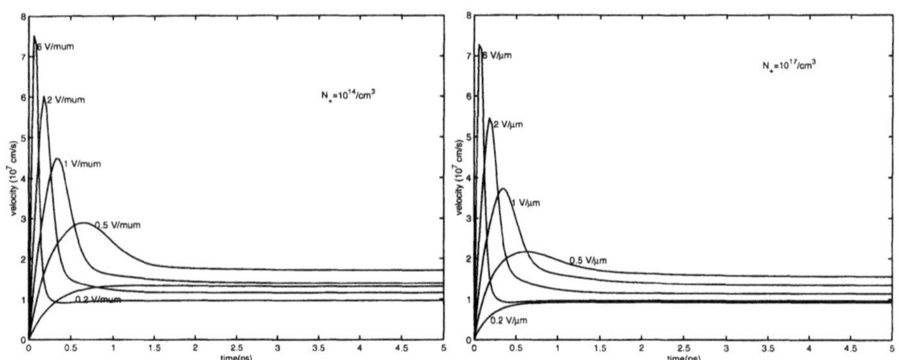

FIGURE 1. The time evolution of the electron average velocity for different values of the electric field and for $N_+ = 10^{14}/\text{cm}^3$ and $10^{17}/\text{cm}^3$ respectively.

picoseconds. The typical phenomena of overshoot and saturation of the velocity are both qualitatively and quantitatively well described.

We also report the curves representing the electron valley occupancy, $\frac{n_\Gamma}{n_\Gamma+4n_L}$ and $\frac{4n_L}{n_\Gamma+4n_L}$, and average velocity as functions of the electric field (see Figs. 2 and 3) for the above impurity concentrations. Fig. 2 clearly shows the electron transfer

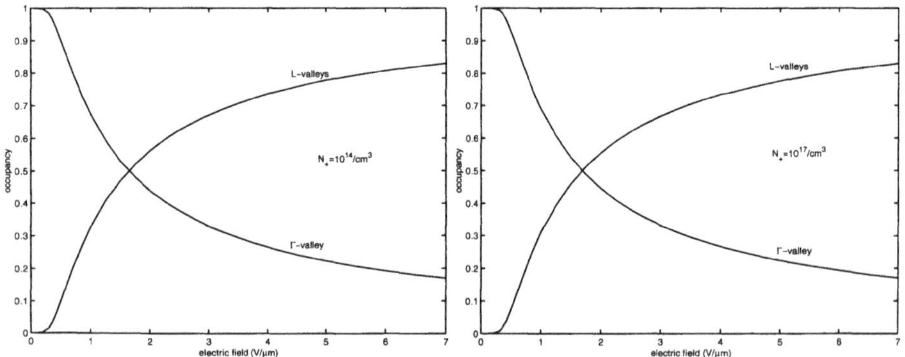

FIGURE 2. Electron occupancy in the Γ- and L-valleys for $N_+ = 10^{14}/\text{cm}^3$ and $10^{17}/\text{cm}^3$ respectively

from the Γ-valley to the L-valleys. The population inversion is observed at electric fields of about $1.8V/\mu m$.

Fig. 3 shows that the low field mobility and the electron peak velocity decrease with the increase of the impurity concentration.

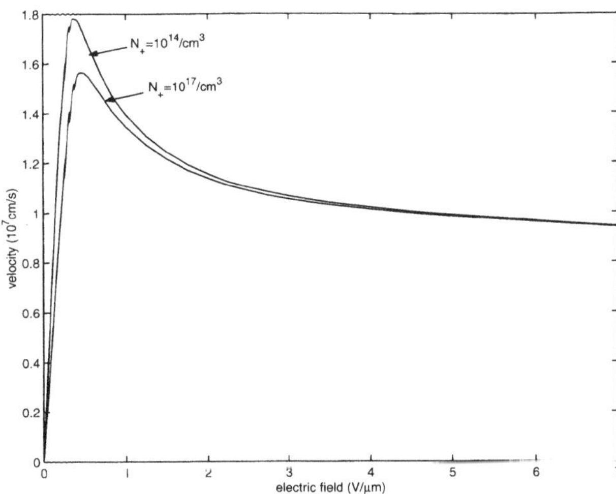

FIGURE 3. Electron velocity versus electric field characteristics for $N_+ = 10^{14}/\mathrm{cm}^3$ and $10^{17}/\mathrm{cm}^3$ respectively

3. Simulation a GaAs $n^+ - n - n^+$ diode

In this section we consider the case of a $n^+ - n - n^+$ diode which models the channel of a MOSFET. The device is made of GaAs and its temperature is 300 K. The n^+ regions are 0.1 μm long, while the channel length is 0.4 μm. The doping

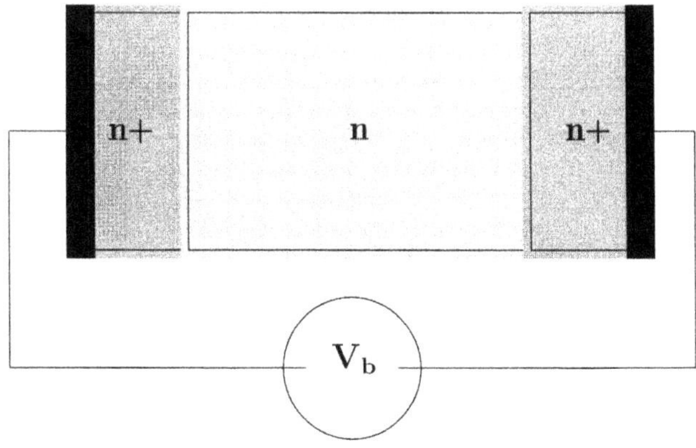

FIGURE 4. Schematic representation of a $n^+ - n - n^+$ diode

profile is

$$N_+ = \begin{cases} 2 \times 10^{18}\text{cm}^{-3} & \text{in the } n^+ \text{ regions} \\ 10^{16}\text{cm}^{-3} & \text{in the } n \text{ region} \end{cases}$$

An external voltage of 2 V is applied. To avoid initiating too much complicated transient behaviour in the device, the voltage is slowly raised to this value. The following initial and boundary conditions are considered

$$n_\Gamma(x,0) = \frac{r^{(0)}}{4 + r^{(0)}} N_+(x), \qquad n_L(x,0) = \frac{N_+(x) - n_\Gamma(x,0)}{4},$$
$$W_A(x) = W_A^{(0)}, \qquad V_A(x,0) = S_A(x,0) = 0,$$

$$n_\Gamma(a,t) = \frac{r^{(0)}}{4 + r^{(0)}} N_+(a), \qquad n_L(a) = \frac{N_+(a) - n_\Gamma(a)}{4},$$
$$W_A(a) = W_A^{(0)}, \qquad \frac{\partial V_A}{\partial x}|_a = \frac{\partial S_A}{\partial x}|_a = 0, \qquad A = \Gamma, L,$$
$$a = 0, 0.6\,\mu\text{m}.$$

where $r^{(0)}$ is the equilibrium ratio between the Γ and the L-electrons and $W_\Gamma^{(0)}$ and $W_L^{(0)}$ are the equilibrium energies at the lattice temperature. Boundary conditions for Poisson's equation are imposed by specifying the electric potential at the device contacts:

$$\Phi(0) = 0 \qquad \Phi(0.6) = 2.$$

Simulations are run to a final time of $t = 75$ps, at which the solutions are judged to reach a steady state. The results relative to the electric field, the total electron average velocity and energy are represented in Figs. 5, 6, where a comparison with analogous results for a Si $n^+ - n - n^+$ diode is also shown. The great difference in the velocity, which can be seen in Fig. 5-b is essentially due to the difference between the effective masses of Silicon electrons and Γ-electrons in GaAs. Indeed

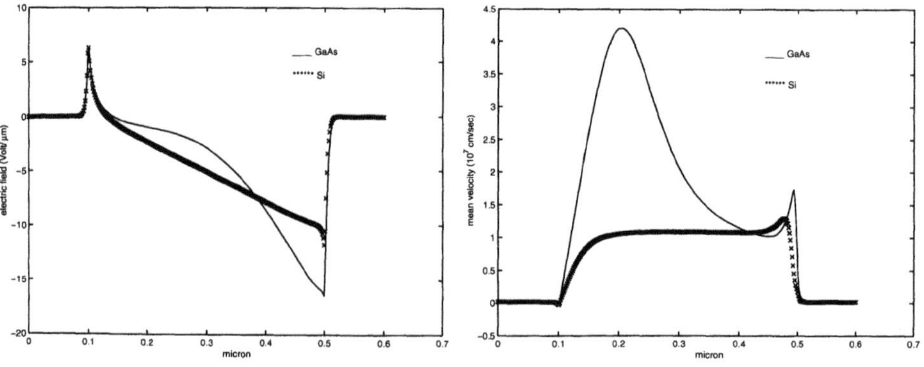

FIGURE 5. A comparison between a Si diode and a GaAs one, electric field and velocity vs position.

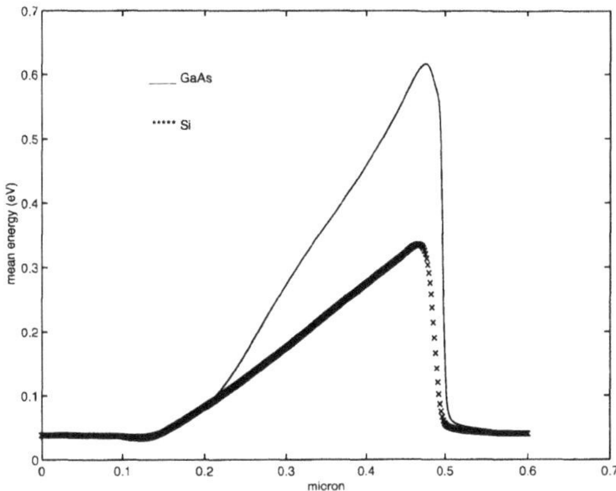

FIGURE 6. A comparison between a Si diode and a GaAs one, energy vs position.

near the first junction the mean energy is low and electrons are mainly in the Γ-valley where they have a small effective mass and consequently high velocities. Instead close to the second junction, where the electric field reaches its maximum value, the L-valley is almost exclusively populated. Since here the effective mass is much greater than that in the Γ-valley and comparable with the effective mass in silicon, the mean velocity is lower.

4. Gunn oscillations

In this section we consider a GaAs diode with the same doping profile as that in [19] and [20]. This is also coupled to an RLC tank circuit which stimulates Gunn oscillatory effects. The one-dimensional diode has length $L_d = 2\,\mu\text{m}$ and its doping profile is

$$N_+(x) = \begin{cases} 10^{17} & \text{for} \quad x < 0.125\,\mu\text{m}\,, \\ 10^{16} & \text{for} \quad 0.125\,\mu\text{m} < x < 0.15\,\mu\text{m}\,, \\ 0.5\times 10^{16} & \text{for} \quad 0.15\,\mu\text{m} < x < 0.1875\,\mu\text{m}\,, \\ 10^{16} & \text{for} \quad 0.1875\,\mu\text{m} < x < 1.875\,\mu\text{m}\,, \\ 10^{17} & \text{for} \quad 1.875\,\mu\text{m} < x\,. \end{cases} \quad (\text{donors/cm}^3) \tag{4.1}$$

The transitions in the doping profile at the device junctions are discontinuous (in contrast to [19]).The same initial and boundary conditions as before have been used except for the potential. Now $\Phi(L_d)$ may be either equal to 2V or is determined by coupling the device to a system of ODE which models the circuit. These equations read

$$\frac{dV_d}{dt} = \frac{1}{C}\left(I - I_d - \frac{V_d}{R}\right), \quad \frac{dI}{dt} = \frac{1}{\Lambda}\left(V_B - V_d\right), \tag{4.2}$$

where V_d is the voltage through the device, $V_B = 2V$ the bias voltage of the circuit and I_d, the particle current in the device, is calculated as

$$I_d = -\frac{q\,A}{L_d}\int_0^{L_d}(n_\Gamma v_\Gamma + 4\,n_L v_L)\,dx\,. \tag{4.3}$$

Simple finite difference versions of equations (4.2) and (4.3) allow the diode voltage to be updated at each simulation time step. The values used for the capacitance, C, resistance, R, and inductance, Λ, of the circuit are

$$C = \left(\epsilon A/L_d + 0.82\times10^{-12}\right)F\,,\quad R = 25\,\text{ohm}\,,\quad \Lambda = 3.5\times10^{-12}\,\text{henry}\,,$$

where the cross-sectional area, A, of the diode is $A = 1.0\times10^{-3}\,\text{cm}^2$.

The oscillator equations (4.2) are given the initial state

$$V_d(t_0) = 2\text{V}\,,\quad I(t_0) = 0\,. \tag{4.4}$$

The circuit is engaged at time $t_0 = 75$ps when the GaAs diode is judged to have reached the steady state illustrated in Figs. 7, 8, 9.

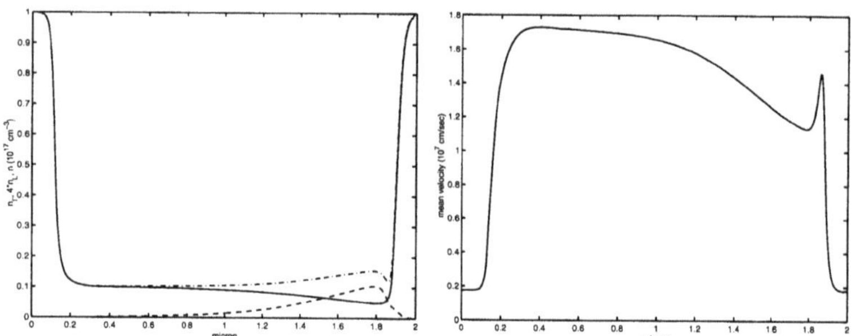

FIGURE 7. Eletron density and velocity vs position. Fig.7a : Continuus line: Γ-electron density, dotted line: L-electron density, dotted-dashed line: total electron density

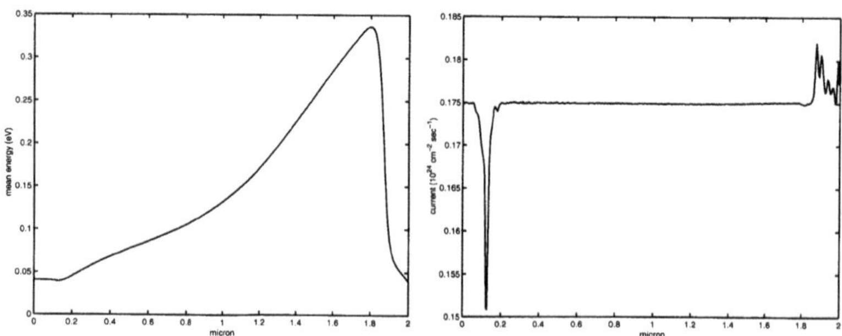

FIGURE 8. Eletron energy and current vs position.

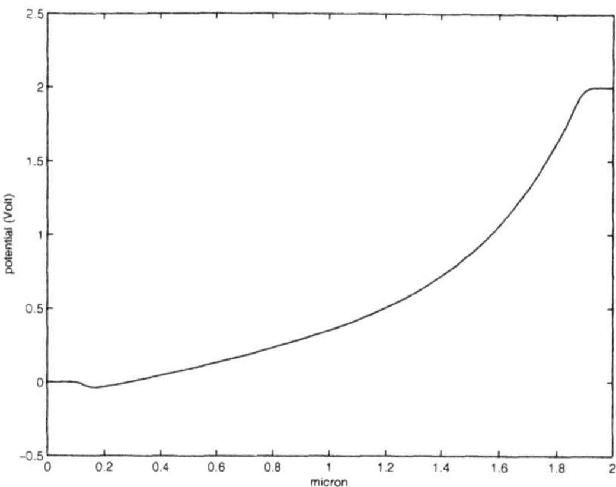

FIGURE 9. Potential vs position.

One observes, Fig. 10, that there are some initial oscillations that smooth out and become negligible after about 200 ps. The qualitative behaviour, at variance with other hydrodynamical models, as those presented in [20], is very similar to the MC simulation reported in [20], even though this latter has been obtained in the parabolic band approximation.

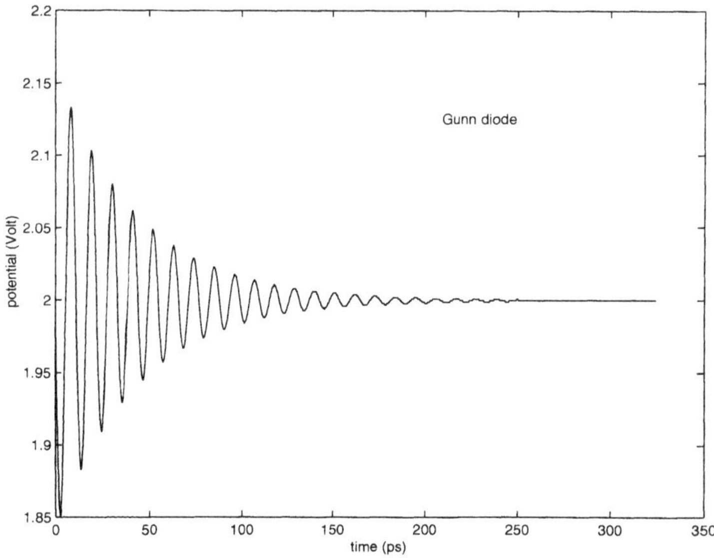

FIGURE 10. The potential V_d versus time for the Gunn diode.

References

[1] G. Mascali and V. Romano, *Hydrodynamical model of charge transport in GaAs based on the maximum entropy principle*, to appear in Cont. Meh. Thermodyn. (TMR preprint archive *Asymptotic Methods in Kinetic theory* http://www.math.tu-berlin.de/~tmr/preprint).

[2] G. Mascali and V. Romano, *Simulation of Gunn oscillations with a non-parabolic hydrodynamical model based on the maximum entropy principle*, preprint (2002) (TMR preprint archive *Asymptotic Methods in Kinetic theory* http://www.math.tu-berlin.de/~tmr/preprint).

[3] A.M. Anile and O. Muscato, *Improved hydrodynamical model for carrier transport in semiconductors*, Phys. Rev. B 51 (1995), pp. 16728–16740.

[4] A.M. Anile and V. Romano, *Non parabolic band transport in semiconductors: closure of the moment equations*, Cont. Mech. Thermodyn., 11 (1999) pp. 307–325.

[5] V. Romano, *Non parabolic band transport in semiconductors: closure of the production terms in the moment equations*, Cont. Mech. Thermodyn., 12 (2000) pp. 31–51.

[6] A.M. Anile and V. Romano, *Hydrodynamical modeling of charge transport in semiconductors*, Meccanica 35 (2000) pp. 249–296.

[7] N.W. Ashcroft and N.D. Mermin, *Solid State Physics*, Philadelphia, Sounders College Publishing International Edition, 1976.

[8] I. Müller and T. Ruggeri, *Rational Extended Thermodynamics*, Berlin, Springer-Verlag 1998.

[9] D. Jou, J. Casas-Vazquez and G. Lebon, *Extended irreversible thermodynamics*, Berlin, Springer-Verlag 1993.

[10] C.D. Levermore, *Moment Closure Hierarchies for Kinetic Theories*, J. Stat. Phys 83 (1996) pp. 331–407.

[11] H. Nessyahu and E. Tadmor, *Non-oscillatory central differencing for hyperbolic conservation law*, J. Comp. Physics 87 (1990), pp. 408–463.

[12] G-S. Jiang and E. Tadmor, *Nonoscillatory central schemes for multidimensional hyperbolic conservation laws*, SIAM J. Sci. COMPUT., 19 (1998), pp. 1892–1917.

[13] V. Romano and G. Russo, *Numerical solutions for hydrodynamical models of semiconductors*, Math. Models Method. Appl. Sciences 7 (2000) pp. 1099–1120.

[14] A.M. Anile, V. Romano and G. Russo, *Extended hydrodynamical model of carrier transport in semiconductors*, SIAM J. Appl. Math., 61 (2000), pp. 74–101.

[15] F. Liotta, V. Romano and G. Russo, *Central schemes for systems of balance laws*, International Series of Numerical Mathematics, 130 (1999) pp. 651–660.

[16] F. Liotta, V. Romano and G. Russo, *Central schemes for balance laws of relaxation type*, SIAM J. Num. Analysis 38 (2000), 1337–1356.

[17] V. Romano, *Nonparabolic band hydrodynamical model of silicon semiconductors and simulation of electron devices*, Math. Meth. Appl. Sciences 24 (2001) 439–471 .

[18] V. Romano, *2D simulation of a silicon MESFET with a nonparabolic hydrodynamical model based on the maximum entropy principle*, J. Comp. Phys. 176 (2002) 70–92.

[19] G.-Q. Chen, J.W. Jerome, C.W. Shu and D. Wang, *Two Carrier Semiconductor Device Models with Geometric Structure and Symmetry Properties*, in Modelling

and Computation for Applications in Mathematics, Science, and Engineering (ed. J. Jerome), Oxford University Press, 1998, pp. 103–140.

[20] A.M. Anile and S. Hern, *Two-valley hydrodynamical model for electron reansport in Gallium Arsenide: simulation of Gunn oscillations*, preprint (2001) submitted to VLSI Design.

Acknowledgments

This work has been partially supported by MURST and by CNR grants n. 98.01041.CT01, n. 99.01714.01, n. 00.00128.ST74 and n. CNRG00DB7 (program Agenzia 2000).

A. M. Anile
Dipartimento di Matematica e Informatica
Università di Catania
viale A. Doria 6
I - 95125 Catania, Italy
email: anile@dmi.unict.it

G. Mascali
Dipartimento di Matematica
Università della Calabria
Via Ponte Bucci
cubo 30 b
I - 87030 Arcavacata di Rende (Cs), Italia
email: g.mascali@unical.it

V. Romano
Dipartimento di Matematica e Informatica
Università di Catania
viale A. Doria 6
I - 95125 Catania, Italy
email: romano@dmi.unict.it

Modeling, Simulation and Optimization of Integrated Circuits
(Eds.) K. Antreich, R. Bulirsch, A. Gilg and P. Rentrop

International Series of Numerical Mathematics Vol. 146, 17–31

Modeling and Simulation of Electro-Quasistatic Fields

Ursula van Rienen, Jürgen Flehr, Ute Schreiber, Victor Motrescu

Abstract. We consider time-dependent electromagnetic fields with negligible propagation effect. These fields are called quasistatic. While the case with negligible displacement current (mainly inductive effects; magneto-quasistatics) is usually treated in classical theory, the case of fields free of eddy currents (mainly capacitive; electro-quasistatics) is still not too common. Yet, the electro-quasistatic model is applicable in many different constellations, especially for microelectronic devices, too.

In this paper, we derive the electro-quasistatic (EQS) equations and deal with their discretization using the Finite Integration Technique. We treat the time-harmonic and the transient equations as well as anisotropic materials. For some application from high-voltage engineering we study Krylov-subspace methods with algebraic multigrid preconditioning. As special application, neuronal microelectrode arrays are chosen affording the coupling of the transient EQS equations with the Hodgkin-Huxley equations in order to simulate the so-called action potential of the nerves.

1. Mathematical model for electro-quasistatic fields

Maxwell's equations describe all macroscopic electromagnetic phenomena reflecting the relations between the characteristic quantities of electromagnetic fields, the electric and magnetic field strength $\mathbf{E}(\mathbf{r},t)$ and $\mathbf{H}(\mathbf{r},t)$. These are related to the other quantities, i.e., the electric flux density $\mathbf{D}(\mathbf{r},t)$, the magnetic flux density $\mathbf{B}(\mathbf{r},t)$ and the current density $\mathbf{J}(\mathbf{r},t)$ by the constitutive relations (1.1)–(1.3). $\mathbf{J}(\mathbf{r},t)$ is composed of the conduction current density $\mathbf{J}_L(\mathbf{r},t) = \sigma(\mathbf{r},t)\mathbf{E}(\mathbf{r},t)$ arising in materials with electric conductivity $\sigma(\mathbf{r},t)$ from the subsisting field strength, some impressed current density $\mathbf{J}_E(\mathbf{r},t)$ independent of all field forces and the convection current density $\mathbf{J}_K(\mathbf{r},t) = \delta \operatorname{grad} \rho(\mathbf{r},t)$ being the density of a current of

---

2000 *Mathematics Subject Classification.* Primary 78A25, 78M25, 92C20, 65N22, 65F10, 65N55; Secondary 31C20, 65L05, 78A30, 92C30, 92B20,96C99.

Key words and phrases. Electro-Quasistatics, Finite Integration Technique, Anistropic Material Operator, Hodgkin-Huxley Equations, Action Potential, Neural Microelectrode Arrays, Krylov-Subspace Methods, Algebraic Multigrid.

This work was supported by the DFG (Deutsche Forschungsgemeinschaft) under the contracts RI 814/2, GRK-45, RI 814/10, RI 814/12 and by CST GmbH.

free electrical charges with the electric charge densities $\rho(\mathbf{r},t)$ and the diffusion constant δ.

$$\begin{aligned}
\mathbf{D}(\mathbf{r},t) &= \epsilon(\mathbf{r},t)\mathbf{E}(\mathbf{r},t) = \epsilon_0\mathbf{E}(\mathbf{r},t) + \mathbf{P}(\mathbf{r},t) && (1.1)\\
\mathbf{B}(\mathbf{r},t) &= \mu(\mathbf{r},t)\mathbf{H}(\mathbf{r},t) = \mu_0(\mathbf{H}(\mathbf{r},t) + \mathbf{M}(\mathbf{r},t)) && (1.2)\\
\mathbf{J}(\mathbf{r},t) &= \mathbf{J}_L(\mathbf{r},t) + \mathbf{J}_E(\mathbf{r},t) + \mathbf{J}_K(\mathbf{r},t) &&\\
&= \sigma(\mathbf{r},t)\mathbf{E}(\mathbf{r},t) + \mathbf{J}_E(\mathbf{r},t) + \delta \,\mathrm{grad}\,\rho(\mathbf{r},t) && (1.3)
\end{aligned}$$

with permittivity $\epsilon(\mathbf{r},t)$, dielectric constant ϵ_0, electric polarization density $\mathbf{P}(\mathbf{r},t)$, permeability $\mu(\mathbf{r},t)$, vacuum permeability μ_0, magnetization $\mathbf{M}(\mathbf{r},t)$. Please note that the material parameter functions may also depend on other physical quantities such as temperature or field strength. In free space, permittivity and permeability are constant and the condition $c = 1/\sqrt{\epsilon_0\mu_0}$ holds where c is the speed of light. Maxwell's equations can either be written in integral form

$$\begin{aligned}
\int_{\partial A} \mathbf{E}(\mathbf{r},t)\cdot d\mathbf{s} &= -\iint_A \frac{\partial \mathbf{B}(\mathbf{r},t)}{\partial t}\cdot d\mathbf{A}\\
\int_{\partial A} \mathbf{H}(\mathbf{r},t)\cdot d\mathbf{s} &= \iint_A \left(\frac{\partial \mathbf{D}(\mathbf{r},t)}{\partial t} + \mathbf{J}(\mathbf{r},t)\right)\cdot d\mathbf{A}\\
\iint_{\partial V} \mathbf{D}(\mathbf{r},t)\cdot d\mathbf{A} &= \iiint_V \rho(\mathbf{r},t)\cdot dV\\
\iint_{\partial V} \mathbf{B}(\mathbf{r},t)\cdot d\mathbf{A} &= 0
\end{aligned} \qquad (1.4)$$

where A is any given surface and V any given volume, or in differential form

$$\begin{aligned}
\mathrm{curl}\ \mathbf{E}(\mathbf{r},t) &= -\frac{\partial \mathbf{B}(\mathbf{r},t)}{\partial t} & \mathrm{div}\,\mathbf{D}(\mathbf{r},t) &= \rho(\mathbf{r},t)\\
\mathrm{curl}\ \mathbf{H}(\mathbf{r},t) &= \frac{\partial \mathbf{D}(\mathbf{r},t)}{\partial t} + \mathbf{J}(\mathbf{r},t) & \mathrm{div}\,\mathbf{B}(\mathbf{r},t) &= 0.
\end{aligned} \qquad (1.5)$$

They present a hyperbolic system of partial differential equations.

1.1. Electro-quasistatic fields

In applications where the wavelength of the studied electromagnetic fields is large compared to the extension of the studied object the wave propagation phenomena may be neglected, i.e., either electric or magnetic fields may be regarded. Theses "slowly-varying" fields are denoted as electro- and magneto-quasistatic, respectively.

Electro-quasistatics gives a reasonable approximation for low frequency fields which can be thought to be free of eddy currents (i.e., the time derivative of the magnetic flux density may be neglected and thus $\mathrm{curl}\,\mathbf{E}(\mathbf{r},t) \approx 0$), while the effects of displacement current, which is the time derivative $\frac{\partial \mathbf{D}(\mathbf{r},t)}{\partial t}$ of the electric flux density, are dominant. Furthermore, no current of free electrical charges shall be present, i.e., $\mathbf{J}_K(\mathbf{r},t) = 0$. So, we assume

$$\begin{aligned}
\text{curl } \mathbf{E}(\mathbf{r},t) &\approx 0 \\
\text{curl } \mathbf{H}(\mathbf{r},t) &= \frac{\partial}{\partial t}\epsilon(\mathbf{r},t)\mathbf{E}(\mathbf{r},t) + \sigma(\mathbf{r},t)\mathbf{E}(\mathbf{r},t) + \mathbf{J}_E(\mathbf{r},t) \\
\rho(\mathbf{r},t) &= \text{div } (\epsilon(\mathbf{r},t)\mathbf{E}(\mathbf{r},t)) \\
0 &= \text{div } (\mu(\mathbf{r},t)\mathbf{H}(\mathbf{r},t))
\end{aligned} \tag{1.6}$$

The derivation of estimations about the regime in which the quasistatic approximation is valid can be found in [1], compare also [2]:

$$|kR| \ll 1 \qquad \text{with the wave number} \qquad k = \omega\sqrt{\mu\epsilon\left(1 - i\frac{\sigma}{\omega\epsilon}\right)}$$

and some characteristic dimension R of the studied system.

According to (1.6) the electric field $\mathbf{E}$ is free from eddy currents, so the field can be described uniquely by a scalar potential function (shortly potential):

$$\mathbf{E}(\mathbf{r},t) = -\text{grad}\,\varphi(\mathbf{r},t) \tag{1.7}$$

since curl grad $\equiv 0$. Then, replacing the electric field strength $\mathbf{E}$ in (1.6) by (1.7), Ampère-Maxwell's law reads as

$$\text{curl } \mathbf{H}(\mathbf{r},t) = -\frac{\partial}{\partial t}\left(\epsilon(\mathbf{r},t)\text{grad}\,\varphi(\mathbf{r},t)\right) - \sigma(\mathbf{r},t)\text{grad}\,\varphi(\mathbf{r},t) + \mathbf{J}_E(\mathbf{r},t). \tag{1.8}$$

Taking the divergence of both sides, reminding that div curl $\equiv 0$, we get

$$\text{div}\left[\frac{\partial}{\partial t}\left(\epsilon(\mathbf{r},t)\text{grad}\,\varphi(\mathbf{r},t)\right)\right] + \text{div }\left(\sigma(\mathbf{r},t)\text{grad}\,\varphi(\mathbf{r},t)\right) = \text{div}\,\mathbf{J}_E(\mathbf{r},t). \tag{1.9}$$

The exchange theorem for partial derivatives allows us to interchange the divergence with the time derivative:

$$\frac{\partial}{\partial t}\left[\text{div }\left(\epsilon(\mathbf{r},t)\text{grad}\,\varphi(\mathbf{r},t)\right)\right] + \text{div }\left(\sigma(\mathbf{r},t)\text{grad}\,\varphi(\mathbf{r},t)\right) = \text{div}\,\mathbf{J}_E(\mathbf{r},t). \tag{1.10}$$

This equation is nothing else than the continuity equation

$$\text{div}\,\mathbf{J}(\mathbf{r},t) + \frac{\partial\rho(\mathbf{r},t)}{\partial t} = 0$$

since, according to (1.7), the constitutive equations and Coulomb's law,

$$\text{div }\left(\epsilon(\mathbf{r},t)\text{grad}\varphi(\mathbf{r},t)\right) = -\text{div}\,\mathbf{D}(\mathbf{r},t) = -\rho(\mathbf{r},t)$$

and, with $\mathbf{J}_K(\mathbf{r},t) = 0$,

$$\text{div}\,\mathbf{J}(\mathbf{r},t) = \text{div}\,\mathbf{J}_E(\mathbf{r},t) + \text{div }\left(\sigma(\mathbf{r},t)\text{grad}\,\varphi(\mathbf{r},t)\right).$$

For the rest of the paper, we will assume time invariant, locally constant material behaviour, i.e., permittivity and conductivity are described by tensors $\underline{\underline{\epsilon}}$ and $\underline{\underline{\sigma}}$.

Then, we can use the exchange theorem to interchange the time derivative with the gradient in (1.9). This yields

$$\Delta\left[\underline{\underline{\epsilon}}\,\frac{\partial\varphi(\mathbf{r},t)}{\partial t}+\underline{\underline{\sigma}}\,\varphi(\mathbf{r},t)\right]=\operatorname{div}\mathbf{J}_E(\mathbf{r},t) \tag{1.11}$$

as governing partial differential equation for electro-quasistatics with time invariant, locally constant material.

1.2. Time-harmonic electro-quasistatic fields

For generally time-dependent fields we may use the method of variable separation in (1.6) and then solve for the Fourier transforms of the electromagnetic fields:

$$\mathbf{E}(\mathbf{r},t)\rightarrow\operatorname{Re}(\underline{\mathbf{E}}(\mathbf{r})e^{i\omega t}),\qquad \mathbf{H}(\mathbf{r},t)\rightarrow\operatorname{Re}(\underline{\mathbf{H}}(\mathbf{r})e^{i\omega t}) \tag{1.12}$$

with the complex amplitudes $\underline{\mathbf{E}}(\mathbf{r})=\mathbf{E}(\mathbf{r})e^{i\phi}$ and $\underline{\mathbf{H}}(\mathbf{r})$ where ϕ is the phase angle of the cosine function. This field representation may also be used if a problem with time-harmonic excitation is studied. Again, we assume time invariant, locally constant, and linear material behaviour.
Differentiating and taking out the term $e^{i\omega t}$, we get

$$\begin{aligned}
\operatorname{curl}\underline{\mathbf{E}}(\mathbf{r}) &\approx 0 && (1.13)\\
\operatorname{curl}\underline{\mathbf{H}}(\mathbf{r}) &= i\omega\underline{\mathbf{D}}(\mathbf{r})+\underline{\underline{\sigma}}\,\underline{\mathbf{E}}(\mathbf{r})+\underline{\mathbf{J}}_E(\mathbf{r}) && (1.14)\\
\operatorname{div}\underline{\mathbf{D}}(\mathbf{r}) &= \underline{\rho}(\mathbf{r}) && (1.15)\\
\operatorname{div}\underline{\mathbf{B}}(\mathbf{r}) &= 0. && (1.16)
\end{aligned}$$

Equations (1.13) and (1.15) are sufficient to determine $\underline{\mathbf{E}}(\mathbf{r})$ uniquely and are, therefore, the fundamental equations of time-harmonic electro-quasistatics.

Time-harmonic problems treated in this paper comprise the study of contaminated high-voltage insulators. Equations (1.13)–(1.16) are also the underlying mathematical model for many more applications: for example, they are the correct model in electrical impedance tomography [3] used in medical imaging, geophysics and non destructive testing or to study effects of low-frequent fields on the human body [15].

Now, we proceed in full analogy to the general case. First, we describe the electric phasor $\underline{\mathbf{E}}(\mathbf{r})$ by a scalar potential function:

$$\underline{\mathbf{E}}(\mathbf{r})=-\operatorname{grad}\underline{\varphi}(\mathbf{r}) \tag{1.17}$$

where the complex potential $\underline{\varphi}(\mathbf{r})$ equals the amplitude of the real, time-harmonic potential: $\varphi(\mathbf{r},t)=\operatorname{Re}(\underline{\varphi}(\mathbf{r})e^{i\omega t})$. Then, we follow the same steps as in the derivation of Eqs. (1.8)–(1.11) which finally yields

$$\Delta\left[i\omega\underline{\underline{\epsilon}}\,\underline{\varphi}(\mathbf{r})+\underline{\underline{\sigma}}\,\underline{\varphi}(\mathbf{r})\right]=\operatorname{div}\underline{\mathbf{J}}_E(\mathbf{r}). \tag{1.18}$$

Hence, we have to solve a complex Poisson equation, i.e., an elliptic partial differential equation, in order to compute the phasor $\underline{\varphi}(\mathbf{r})$ of the time-harmonic electro-quasistatic potential.

2. Electromagnetic simulation model

For the numerical solution of the differential equations described above we apply some discretization scheme which is called Finite Integration Technique (shortly FIT) and may be regarded as a kind of Finite Volume scheme. Historically, it was developed by Weiland [4], [5] as a generalization of the well-known FDTD scheme (Finite Difference Time Domain) by Yee [6]. The Finite Integration Technique solves Maxwell's equations on a domain Ω which holds the actual solution domain Ω_s as a sub-domain: $\Omega = \Omega_s \cup \Omega_o$.[1] FIT yields an *exact* representation of Maxwell's equations in integral form on a grid doublet $(G, \tilde{G})$, denoted as *Maxwell-Grid-Equations*:

$$\begin{aligned} \mathbf{C}\widehat{\mathbf{e}} &= -\tfrac{\partial}{\partial t}\widehat{\widehat{\mathbf{b}}} & \qquad \widetilde{\mathbf{S}}\widehat{\widehat{\mathbf{d}}} &= \mathbf{q} \\ \widetilde{\mathbf{C}}\widehat{\mathbf{h}} &= \tfrac{\partial}{\partial t}\widehat{\widehat{\mathbf{d}}} + \widehat{\widehat{\mathbf{j}}} & \qquad \widetilde{\mathbf{S}}\widehat{\widehat{\mathbf{b}}} &= 0 \end{aligned} \tag{2.1}$$

The FIT grid doublet $(G, \tilde{G})$ is not coordinate-bounded, not necessarily orthogonal, not necessarily regular. Often the solution domain Ω_s possesses symmetries or some geometrical invariance such that the 3D problem may be reduced to a 2D problem by appropriate variable separation. One example for a non coordinate 2D grid is the triangular-hexagonal grid doublet of Delaunay-Voronoi type introduced in [7] (see also [8] or [9]).

The operators $\mathbf{C}$, $\widetilde{\mathbf{C}}$, $\mathbf{S}$ and $\widetilde{\mathbf{S}}$ can be interpreted as discrete curl operators $\mathbf{C}, \widetilde{\mathbf{C}}$, discrete divergence $\mathbf{S}, \widetilde{\mathbf{S}}$ and discrete gradient operators $\mathbf{G} = -\widetilde{\mathbf{S}}^T, \widetilde{\mathbf{G}} = -\mathbf{S}^T$. The vectors $\widehat{\mathbf{e}}, \widehat{\mathbf{h}}$, etc. hold scalar state variables defined as field integrals along edges $L_i, \tilde{L}_i$ and across facets $A_i, \tilde{A}_i$, yielding the so-called *grid voltages* $\widehat{\mathbf{e}}, \widehat{\mathbf{h}}$ and *grid flux* vectors $\widehat{\widehat{\mathbf{b}}}, \widehat{\widehat{\mathbf{d}}}, \widehat{\widehat{\mathbf{j}}}$ on the primary grid G and the dual grid $\tilde{G}$, respectively:

$$\begin{aligned} \widehat{e}_i &= \int_{L_i} \mathbf{E}\cdot\mathbf{ds}, & \widehat{\widehat{b}}_i &= \iint_{A_i} \mathbf{B}\cdot\mathbf{dA}, & & \\ \widehat{h}_i &= \int_{\tilde{L}_i} \mathbf{H}\cdot\mathbf{ds}, & \widehat{\widehat{d}}_i &= \iint_{\tilde{A}_i} \mathbf{D}\cdot\mathbf{dA}, & \widehat{\widehat{j}}_i &= \iint_{\tilde{A}_i} \mathbf{J}\cdot\mathbf{dA}. \end{aligned} \tag{2.2}$$

In case of scalar potential problems the discrete potential values are allocated in the primary grid points like in the classical Finite Difference scheme. With these definitions it is straightforward to derive the discrete form of Maxwell's equations (compare the integral form given in (1.4)) and all derived equations like Poisson's equation, the wave equation and so on. For more details on FIT and mathematical properties of this discretization method we refer to the literature cited above and further references therein.

2.1. Discrete material operators

While the derivation of the Maxwell-Grid-Equations needed no approximations yet because of the specific definition of grid voltages and flux vectors the set-up of

[1] Discretization on Ω rather than on Ω_s has the advantage of allowing for higher topological regularity leading to matrices with regular (band) pattern. Usually Ω_o, the overhead, is relatively small.

discrete constitutive laws (material equations) involves introducing an approximation: Regard a dual facet $\tilde{A}_i$ of a 3D orthogonal grid with lexicographic numbering as, e.g., a Cartesian grid. It intersects four primary facets A_j. As shown in Fig. 1 the electric grid voltage $\widehat{e}_i$ and the electric grid flux $\widehat{\widehat{d}}_i$ as well as the current flux $\widehat{\widehat{j}}_i$ are all three allocated in the same position with same orientation (the normal direction to $\tilde{A}_i$). On each primary facet A_j, the magnetic grid voltage $\widehat{h}_j$ and the magnetic grid flux $\widehat{\widehat{b}}_j$ also have the same position and orientation, each. In order to find a discrete analogue of the constitutive relations, Eqs. (1.1)–(1.3), we need to relate each of these voltages and flux vectors with coinciding location. In the conformal FIT (CFIT) [10], [11], applied here, the exact boundary shape is taken into account in the primary facets A_j, ensuring second order accuracy to be compared with classical FIT which already allowed triangular filling of the cells leading to somewhat better convergence order than the classical stair case finite difference approaches limited by first order accuracy.

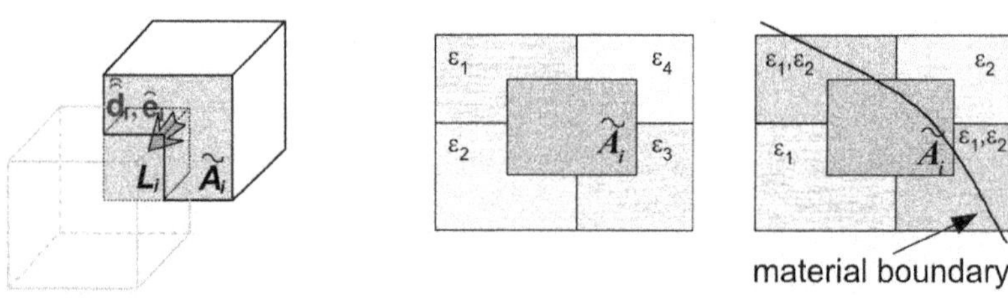

FIGURE 1. Left: Facet $\tilde{A}_i$ of a 3D Cartesian grid with the electric grid voltage $\widehat{e}_i$, the electric grid flux $\widehat{\widehat{d}}_i$ and the current flux $\widehat{\widehat{j}}_i$. Right: The dual facet $\tilde{A}_i$ intersects four primary cells which might be filled with different material, each. Conformal FIT allows for arbitrary partial filling of the primary cells.

2.1.1. ISOTROPIC MATERIAL For isotropic material we replace **E** in the integrals given in (2.2) defining $\widehat{\widehat{d}}_i$ and $\widehat{e}_i$ by a Fourier series around its midpoint value. Then, we draw this in front of the integrals and receive after division by the midpoint value

$$\frac{\widehat{\widehat{d}}_w}{\widehat{e}_w} = \frac{\iint \varepsilon du dv}{\int dw} + O(\Delta^k) \approx \overline{\varepsilon}\,\frac{\iint du dv}{\int dw} + O(\Delta^k), \tag{2.3}$$

where u, v, w is a tuple out of three Cartesian coordinate directions, $\overline{\varepsilon}$ is the weighted average over the four permittivity values in the primary cells which are intersected by the dual facet $\tilde{A}_i$ and k takes values between $k = 2$ for varying permittivities or non-uniform step size and $k = 3$ otherwise. Equation (2.3) refers to primary cells completely filled with only one material. For partial filling as taken into account in CFIT a subvolume approach is incorporated. Introducing diagonal matrices $\widetilde{\mathbf{D}}_A$, $\mathbf{D}_s$ and $\mathbf{D}_\epsilon$ holding the areas of dual facets, lengths of primary edges and permittivities of the primary cells, we can write

$$\widehat{\widehat{\mathbf{d}}} = \widetilde{\mathbf{D}}_A \mathbf{D}_\epsilon \mathbf{D}_s^{-1} \widehat{\mathbf{e}} =: \mathbf{M}_\epsilon \widehat{\mathbf{e}}. \tag{2.4}$$

In complete analogy we can define the material operators $\mathbf{M}_\sigma$ and $\mathbf{M}_\mu$. Obviously, the material operators are diagonal. This is an important property necessary to ensure stability of explicit time domain schemes [12].

2.1.2. ANISOTROPIC MATERIAL For anisotropic material, we have to deal with rank two tensors for permittivity, permeability and/or conductivity. Often, these tensors are diagonal in some local coordinate system but have to be transformed to a global coordinate system leading to full tensors. An example in our context are the muscles in the human body for which permittivity and conductivity possess a transversal isotropic anisotropy, cf. [13] or [15] for more details. In Eq. (2.3) we could easily erase all field-related entries since the electric flux density $\mathbf{D}$ and the electric field $\mathbf{E}$ are related component-wise for isotropic material. For anisotropic material however, the flux density component $\mathbf{D}_v$ is related to all three field components $\mathbf{E}_u, \mathbf{E}_v, \mathbf{E}_w$. The same holds true for the discrete state variables on the FIT grid. Regard a grid point (i,j,k) with the coordinates $(u(i), v(j), w(k))$. As only $\widehat{e}_v$ is allocated at $\widehat{\widehat{d}}_v$'s location, this implies the necessity of interpolating $\widehat{e}_u$ and $\widehat{e}_w$:

$$\begin{aligned}
\widehat{\widehat{d}}_v(i,j,k) &= \iint_{\tilde{A}_{uw}(i,j,k)} \mathbf{D}(u(i), v(j), w(k)) \cdot \mathbf{e}_v \, dudw \\
&= \iint_{\tilde{A}_{uw}(i,j,k)} \underline{\underline{\varepsilon}}(u(i), v(j), w(k)) \mathbf{E}(u(i), v(j), w(k)) \cdot \mathbf{e}_v \, dudw \\
&= \iint_{\tilde{A}_{uw}(i,j,k)} (\varepsilon_{uv}(i,j,k) E_u(u(i), v(j), w(k)) \\
&\qquad + \varepsilon_{vv}(i,j,k) E_v(u(i), v(j), w(k)) \\
&\qquad + \varepsilon_{wv}(i,j,k) E_w(u(i), v(j), w(k))) \, dudw.
\end{aligned} \tag{2.5}$$

The interpolation [16] is carried out in two steps: First, the left- and right-handed pair of $\widehat{e}_u$ are interpolated with weight $\frac{1}{2}$ onto the intermediate dual edge $\tilde{L}_u(i,j,k)$. Next, these two values are interpolated, again with weight $\frac{1}{2}$, onto the position of $\widehat{\widehat{d}}_w(i,j,k)$:

$$\begin{aligned}
&\widehat{e}_u(i - \frac{1}{2}, j + \frac{1}{2}, k) \\
&= \frac{1}{4} \left(\widehat{e}_u(i-1,j,k) + \widehat{e}_u(i-1,j+1,k) + \widehat{e}_u(i,j,k) + \widehat{e}_u(i,j+1,k) \right).
\end{aligned} \tag{2.6}$$

This procedure is illustrated in Fig. 2. The interpolation leads to non-symmetric off-diagonal entries in the material matrix $\mathbf{M}_\epsilon$. But, symmetry may be reached using the approximation $|L_u| = |\tilde{L}_u| + O(\Delta)$ leading to:

$$\frac{\tilde{A}_{uw}}{L_u} = \frac{\tilde{L}_u \tilde{L}_w}{L_u} \approx \tilde{L}_w \approx \frac{\tilde{L}_v \tilde{L}_w}{L_v} = \frac{\tilde{A}_{vw}}{L_v} \tag{2.7}$$

for all points (i, j, k) in the grid. Thus, for anisotropic materials a symmetric, but no longer diagonal material operator may be used.

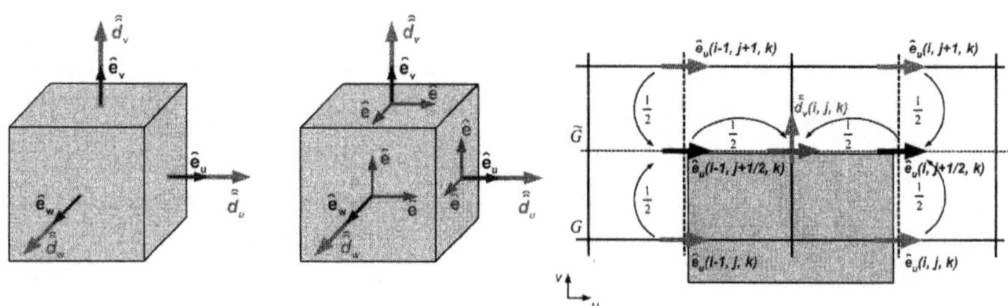

FIGURE 2. Left: Allocation of the electric voltage $\widehat{\mathbf{e}}$ and flux $\widehat{\widehat{\mathbf{d}}}$ with respect to a dual grid cell. Middle: Original and interpolated state variables. Right: Interpolation of $\widehat{e}_w$ to the location of $\widehat{\widehat{d}}_w$.

2.2. Discrete Electro-Quasistatic Equations

For the electro-quasistatic case we derived the time-dependent equation (1.11) and the Poisson equation (1.18) for the time-harmonic case or Fourier transform, respectively. The electric field $\mathbf{E}$ was chosen as $\mathbf{E}(\mathbf{r}, t) = -\operatorname{grad} \varphi(\mathbf{r}, t)$ and $\underline{\mathbf{E}}(\mathbf{r}) = -\operatorname{grad} \underline{\varphi}(\mathbf{r})$, respectively. The vectors $\mathbf{\Phi}(t)$ and $\underline{\mathbf{\Phi}}$ hold the discrete potential values. Then, we receive for the grid voltage $\widehat{\mathbf{e}}$ along edge L_i:

$$\widehat{e}_i = \int_{L_i} \mathbf{E}(\mathbf{r}, t) \cdot d\mathbf{s} = - \int_{L_i} \operatorname{grad} \varphi(\mathbf{r}, t) \cdot d\mathbf{s} = \mathbf{\Phi}_m(t) - \mathbf{\Phi}_n(t) \qquad (2.8)$$

where we assumed that $\mathbf{\Phi}_n(t)$ lies on the end of the integration path and $\mathbf{\Phi}_m(t)$ at the beginning. Now all relevant quantities on the FIT grid are introduced and we can directly write down the discrete FIT equations corresponding to (1.11) and (1.18):

$$\widetilde{\mathbf{S}}\mathbf{M}_\epsilon \widetilde{\mathbf{S}}^T \frac{d}{dt} \mathbf{\Phi}(t) + \widetilde{\mathbf{S}}\mathbf{M}_\sigma \widetilde{\mathbf{S}}^T \mathbf{\Phi}(t) = \widetilde{\mathbf{S}} \widehat{\widehat{\mathbf{j}}}(t) \qquad (2.9)$$

for the transient computation of the discrete electro-quasistatic potential $\mathbf{\Phi}(t)$ and

$$\left(i\omega \widetilde{\mathbf{S}}\mathbf{M}_\epsilon \widetilde{\mathbf{S}}^T + \widetilde{\mathbf{S}}\mathbf{M}_\sigma \widetilde{\mathbf{S}}^T \right) \underline{\mathbf{\Phi}} = \widetilde{\mathbf{S}} \left(\mathbf{M}_\sigma + i\omega \mathbf{M}_\epsilon \right) \widetilde{\mathbf{S}}^T \underline{\mathbf{\Phi}} = \widetilde{\mathbf{S}} \widehat{\widehat{\mathbf{j}}} \qquad (2.10)$$

for the time-harmonic simulation. Please remind that in general $\mathbf{M}_\epsilon$ and $\mathbf{M}_\sigma$ are determined by the permittivity $\underline{\underline{\varepsilon}}$ and conductivity $\underline{\underline{\sigma}}$, i.e., nonlinear, time-dependent rank two tensors which are assumed to be piecewise constant functions with $\underline{\underline{\varepsilon}} > 0$ and $\underline{\underline{\sigma}} \geq 0$ if not stated differently.

The symmetric material operators allow for an explicit time-stepping scheme to solve (2.9). Since we solve a coupled system of transient EQS plus Hodgkin-Huxley equations, details on the chosen strategy will be given below.

In the time-harmonic case, we can start the simulation based on (2.10). The linear system $\widetilde{\mathbf{S}} \left(\mathbf{M}_\sigma + i\omega \mathbf{M}_\epsilon \right) \widetilde{\mathbf{S}}^T \underline{\mathbf{\Phi}} = \widetilde{\mathbf{S}} \widehat{\widehat{\mathbf{j}}}$ has to be solved in order to determine the electro-quasistatic potential $\underline{\mathbf{\Phi}}$ with the symmetric material matrices $\mathbf{M}_\sigma$ and $\mathbf{M}_\epsilon$

as described above. Obviously, the system matrix is complex symmetric. Extensive numerical studies have been carried out to find most robust and fast solution methods leading to the choice of the Krylov-subspace method BiCGCR with algebraic multigrid (AMG) as preconditioner, see, e.g., [14], [9] and [17] and references therein for more details. Some recent results will be shown below.

3. Neurophysiological model for neuron activity

Several groups around the world are studying the integration of neural cells into microelectrode sensor arrays and sensor-actor-systems. Networks of living neurons are cultivated on a chip [18]. A wide variety of possible applications is conceivable like, e.g., pollution monitoring in fluids. One possible realization follows ideas of Gross [19] and uses gold-plated indium-tin-oxide microelectrodes on a glass substrate as shown schematically in Fig. 3. In order to simulate the electrical behaviour of such neural microelectrode arrays we can use equivalent circuits as shown in the same figure. The parameters concerning the neural cells shall be determined by field simulation. For design optimization of the microelectrode array, like, e.g., increasing the signal output, field simulations should be used for larger parts of the array including the electrodes. This gave reason for the study described below. We will see that the transient electro-quasistatic model coupled to the Hodgkin-Huxley Equations yields the appropriate simulation model to simulate a neural network.

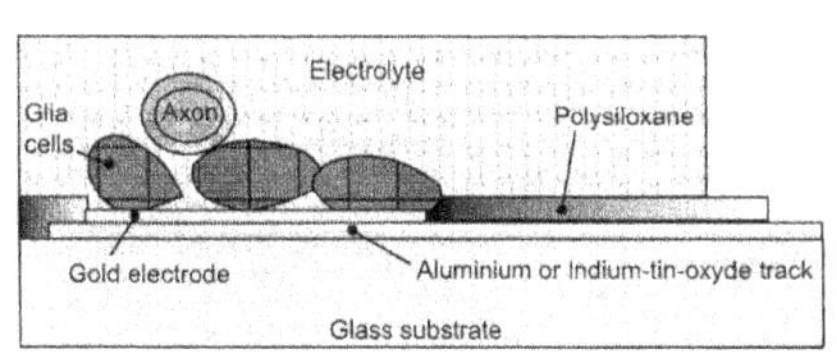

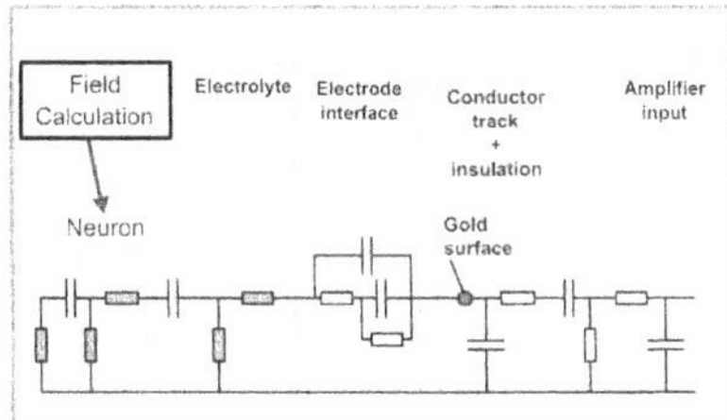

FIGURE 3. Left: Principle of potential derivation via gold electrodes and transparent strip lines. Right: Equivalent circuit for the neuron-gold pad-transition according to Gross [19].

3.1. Some basics on neural cells

First, we will recall very briefly some basics on nerves and their electric activity. The interested reader can find more information in [20]; animations and graphs may be found in [21]. Figure 4 shows a model of the neural cell body with its nucleus together with some of its dendrites and the axon. The information flow starts near the cell body and runs along the axon to its ends. Information is transferred as electric signal thus allowing for high speed. In stationary state, the neuron has a negative potential in its inside relative to the outside. The equilibrium rest potential equals the potential difference across the cell membrane and is typically of the order of -70 mV. As soon as the nerve is excited, a momentary polarity reversal, called depolarization, is generated in a confined zone of the axon's membrane. The polarity reversal propagates by itself along the axon. This functional behaviour is depicted in Fig. 4. The potential difference between the resting potential and the peak value during depolarization, but also the propagation of the depolarization along the nerve are both denoted as action potential. The triggering happens chemically by neuro-transmitters.

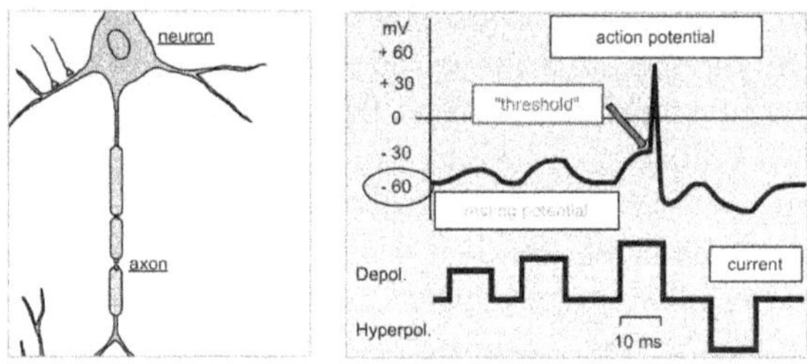

FIGURE 4. Left: Model of a nerve cell with its axon (adopted from [21]). Right: Typical functional behaviour of current and action potential.

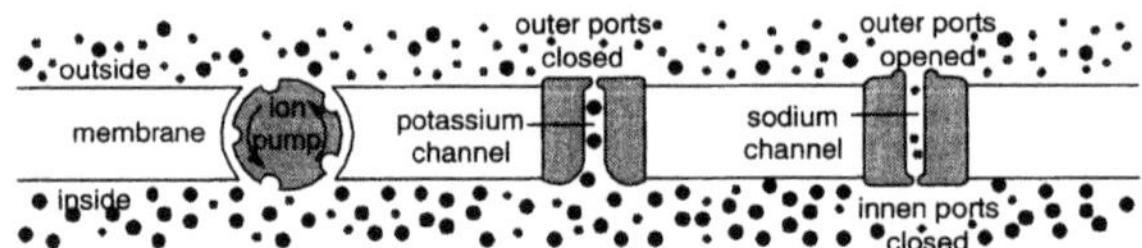

FIGURE 5. Schematic view of the axon's membrane with typical concentration of potassium and sodium ions in- and outside, ion pump and ion channels (adopted from [21]).

Figure 5 shows a schematic view of the axon's membrane. In stationary state, potassium ions diffuse out of the neuron and ion pumps steadily pump 3 Na^+ ions out of the neuron and 2 K^+ ions into the cell thus working against the concentration gradient. The neural cell membrane holds voltage-controlled ports (channels) for sodium (Na^+) and potassium (K^+) which are closed at the stage of equilibrium rest potential (resting potential). When the depolarization exceeds a certain threshold, first the sodium ports open such that sodium flows inside the neuron, then the potassium ports open and potassium flows out of the neuron. The sodium inflow locally depolarizes the membrane even stronger causing a self-reinforcement. The sodium ions propagate along the membrane attracting negative ions at the outer side. As soon as their concentration is high enough, they depolarize the neighbourhood of the originally excited zone such that sodium ports can open there, too. Ports close after a few milliseconds. Previously excited zones can not be excited again for several milliseconds thus preventing reversibility of the propagation direction (refractory period).

3.2. The Hodgkin-Huxley equations

In 1952 Hodgkin and Huxley formulated a differential equation system of reaction-diffusion type modeling very precisely the action potential and its propagation:

$$\begin{aligned}
\frac{\partial V}{\partial t} &= \frac{-1}{C_m}\left[\overline{g}_{Na} m^3 h)\left(V - V_{Na}^*\right) + \overline{g}_K n(^4\left(V - V_K^*\right) + \overline{g}_l\left(V - V_l^*\right)\right] \quad (3.1)\\
\frac{\partial n}{\partial t} &= \alpha_n\left(1-n\right) - \beta_n n\\
\frac{\partial m}{\partial t} &= \alpha_m\left(1-m)\right) - \beta_m m\\
\frac{\partial h}{\partial t} &= \alpha_h)\left(1-h\right) - \beta_h h
\end{aligned}$$

with the membrane voltage $V(t)$, the resting potential (Nernst voltage) V_{Na}^*, V_K^*, V_l^* for sodium, potassium and other ions, the membrane capacity C_m, the sodium conductance $g_{Na}(V) = \overline{g}_{Na} m(V)^3 h(V)$ with some constant $\overline{g}_{Na}$, the sodium activation $m(V)$ and the sodium inactivation $h(V)$, the potassium conductance $g_K(V) = \overline{g}_K n(V)^4$ with some constant $\overline{g}_K$ and the potassium activation $n(V)$ and the remaining constant conductance $\overline{g}_l$ for other ions. The transfer rates $\alpha_i(V)$ and $\beta_i(V)$ determining the activation and inactivation functions $m(V), n(V), h(V)$ gained by Hodgkin and Huxley by fits to experimental data may be found in the original article of Hodgkin and Huxley [22] or in [20].

3.3. Coupling of Hodgkin-Huxley equations and electro-quasistatic equations

Using the Hodgkin-Huxley equations (3.1), we can determine the nonlinear parameters $g_{Na}(V)$ and $g_k(V)$, i.e., the sodium and potassium conductance. At moment, this is done with MATLAB [23]. In a second step, these conductance values are written to some input file for the static solver of the CAE tool MAFIA [24] which is based on the FIT equations (2.9).

The conductivity-dependent term in (1.11) and (2.9) is then replaced by the current of sodium, potassium and the rest which we collect in the grid flux vector $\widehat{\widehat{\mathbf{j}}}_m$ and put the corresponding term on the right-hand side. $\mathbf{J}_E(\mathbf{r},t)$ is taken equal to zero. Then, with FIT we get

$$\widetilde{\mathbf{S}}\mathbf{M}_\epsilon\widetilde{\mathbf{S}}^T\frac{d}{dt}\mathbf{\Phi}(t) = \widetilde{\mathbf{S}}\widehat{\widehat{\mathbf{j}}}_m(t). \quad (3.2)$$

Time discretization is introduced by $\frac{\partial\mathbf{\Phi}(t)}{\partial t} \approx \frac{\mathbf{\Phi}^{(i+1)} - \mathbf{\Phi}^{(i)}}{\Delta t}$. Now we can reformulate (3.2) as explicit time-stepping scheme for the electro-quasistatic potential $\mathbf{\Phi}$ according to

$$\widetilde{\mathbf{S}}\mathbf{M}_\epsilon\widetilde{\mathbf{S}}^T\mathbf{\Phi}^{(i+1)} = \widetilde{\mathbf{S}}\mathbf{M}_\epsilon\widetilde{\mathbf{S}}^T\mathbf{\Phi}^{(i)} + \Delta t\cdot\widetilde{\mathbf{S}}\widehat{\widehat{\mathbf{j}}}_m^{(i)}. \quad (3.3)$$

The initial value $\mathbf{\Phi}^{(0)} := \mathbf{\Phi}(t_0)$ is determined in an electrostatic simulation.

First results have been obtained for simple models of the axon. Figure 6 shows a comparison of the ascending branch of the action potential determined by the coupled model and the solution of the Hodgkin-Huxley equations. Please note, that not only the membrane potential is computed in the coupled model but also the potential and thus the electric field inside and outside of the nerve. This is a crucial point since the final goal is the optimization of electrodes and their arrangement in the neural sensor with respect to the output signal.

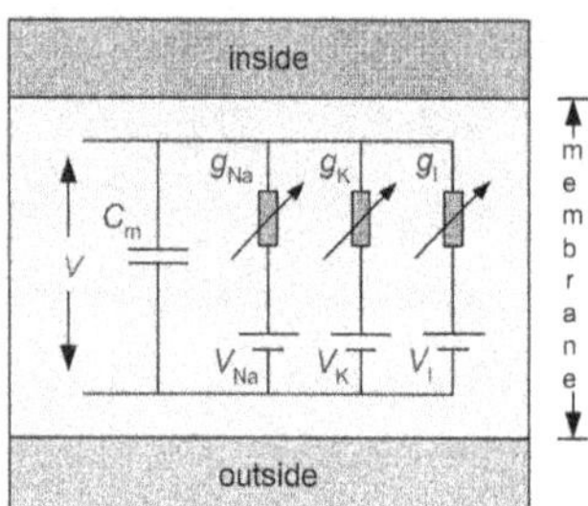

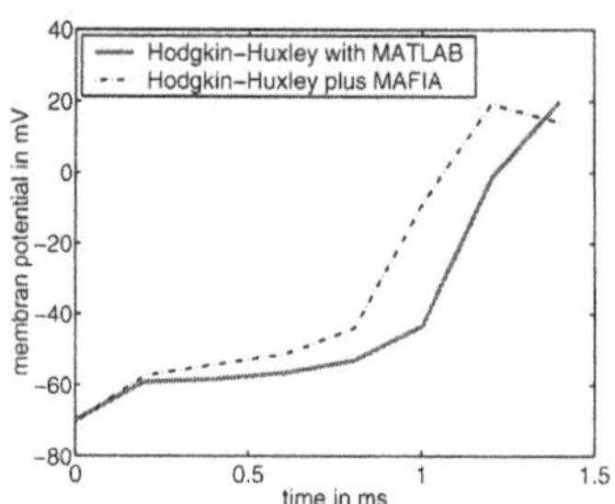

FIGURE 6. Left: Equivalent circuit for the membrane with voltage V, capacity C_m, conductances g_{Na}, g_K, g_l and resting potentials V_{Na}, V_K, V_l. Right: Ascending branch of the action potential determined directly from the Hodgkin-Huxley (HH) equations and by the coupled HH-EQS model.

4. Numerical Investigations for an electro-quasistatic application

As the work on low-frequent electromagnetic fields in the human body and on neuronal microelectrode arrays is still in the stage of modeling, implementation and testing, we would like to present some results obtained recently in fundamental studies on aging processes of humid insulators under high-voltage. As sample an epoxy resin block of 100 mm x 100 mm x 20 mm is used in accordance to accompanying experiments [25]. The block holds horizontally embedded electrodes with a center distance of 35 mm and a radius of 7.5 mm. In one of the experiments two water droplets are put on top. The droplets' diameter is 6 mm (hemispheres), their center distance is 10 mm. This setup, displayed in Fig. 4, is discretized with 450,241 mesh points. Open boundary conditions are used.

As linear solvers the Krylov-subspace methods BiCGCR [26, 27], QMR and one denoted here as CSYM [28] are compared, each combined with Algebraic Multigrid (AMG) or Jacobi as preconditioner. Explicit descriptions of the algorithms, further references and a detailed investigation for complex symmetric systems can be found in [26, 27, 17]. Both preconditioners are implemented in the software package PEBBLES [17, 29]. The characteristic convergence behaviour for the methods is shown in Fig. 8. QMR and BiCGCR both perform very similar

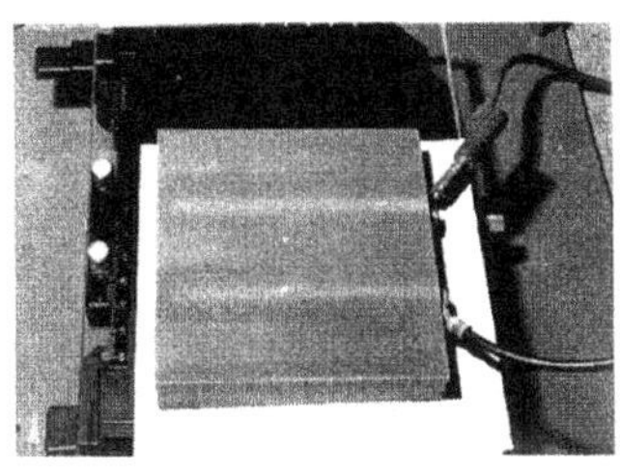

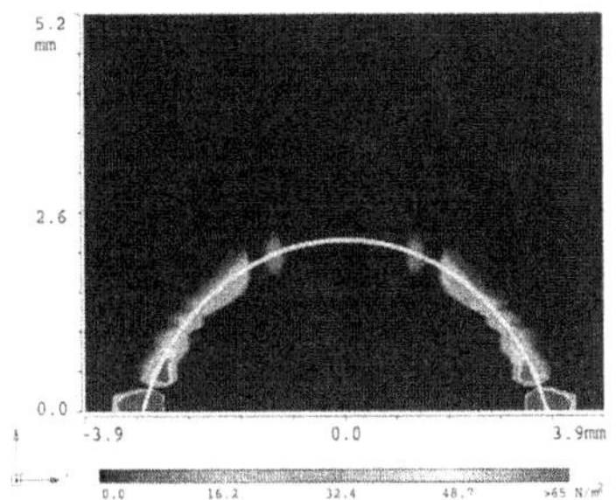

FIGURE 7. Left: High-voltage test setup with two water droplets on the epoxy resin surface. Right: Electric force density in a cutting plane through one of the droplets and the epoxy resin block.

	ϵ	setup (sec)	solver (sec)	total time (sec)	# iter.
AMG–BiCGCR	10^{-10}	187.77	703.44	891.21	15
AMG–QMR	10^{-10}	186.83	493.21	680.04	15
AMG–CSYM	10^{-10}	187.14	>2,000	>2,000	>65
Jacobi–BiCGCR	10^{-2}	10.45	856.15	866.60	41
Jacobi–QMR	10^{-2}	10.43	810.17	820.60	40
Jacobi–CSYM	10^{-2}	10.61	1,390.01	1,400.62	90

TABLE 1. Number of mesh points $N_h = 450,241$, 6 levels, $\alpha = 0.01$, mesh size ratio $\rho_h = 1$; SUN UltraSPARC–1 with 296 MHz.

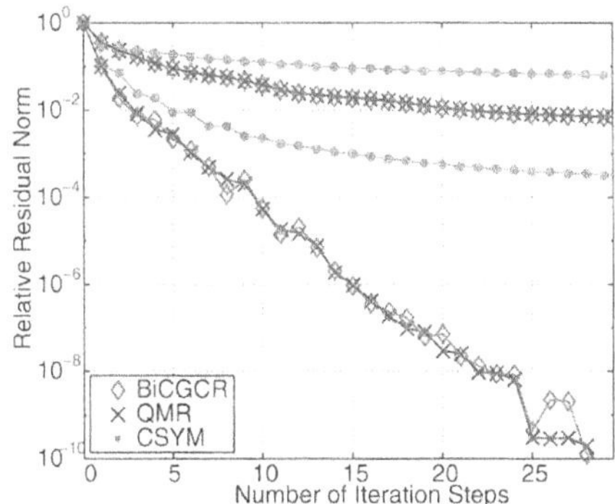

FIGURE 8. Comparison of different Krylov-subspace methods – once preconditioned with Jacobi (upper curves) and once with the algebraic multigrid algorithm PEBBLES (lower curves).

with respect to the number of iterations with the AMG-preconditioner PEBBLES. Table 1 specifies CPU-times. Overall, the AMG-preconditioner obviously accelerates the iteration process in spite of the relatively large setup times compared to classical iterative solvers.

5. Summary and outlook

The mathematical models for electro-quasistatic fields and for the transmembrane potential of neuronal cells have been described. Based on that and on the Finite Integration Technique discrete equations for time-harmonic and transient simulation have been set up. A possible layout for the coupled simulation of the Hodgkin-Huxley equation and the transient electro-quasistatic equations has been introduced and first preliminary results shown for the computation of the neuron's action potential. Some results for a high-voltage application underlined the speed up of linear solvers by multigrid preconditioning.
Modelling and simulation of the electro-quasistatic applications treated here as well as in the context of electromagnetic environmental compatibility will be continued. The next steps will cover among others preparation for parallel computation and for optimization of the neuronal microelectrode arrays.

References

[1] H.A. Haus, J.R. Melcher, *Electromagnetic Fields and Energy*, Prentice-Hall. Inc., 1989. See also: http://web.mit.edu/6.013_book/www/

[2] H.K. Dirks, *Quasi-Stationary Fields for Microelectronic Applications*, Electrical Engineering, 79 (1996): 145–155.

[3] L. Borcea, *Nonlinear Multigrid Algorithm for Imaging Electrical Conductivity and Permittivity at Low Frequency*, Inverse Problems, 17 (2001): 329–359.

[4] T. Weiland, *Eine Methode zur Lösung der Maxwellschen Gleichungen für sechskomponentige Felder auf diskreter Basis*, AEÜ, 31 (1977): 116–120.

[5] T. Weiland, *Time Domain Electromagnetic Field Computation with Finite Difference Methods*, JNM (Wiley), 9 (1996): 295–319.

[6] K.S. Yee, *Numerical Solution of Initial Boundary Value Problems Involving Maxwell's Equations in Isotropic Media*, IEEE-AP, 14 (1966): 302–307.

[7] U. van Rienen, T. Weiland, *Triangular Discretization Method for the Evaluation of RF-Fields in Cylindrically Symmetric Cavities*, IEEE-T.Mag., 21 (1985): 2317–2320.

[8] U. van Rienen, *Frequency Domain Analysis of Waveguides and Resonators with FIT on Non-Orthogonal Triangular Grids*, PIER, 32 (2001), Special Vol. *Geometrical Methods for Computational Electromagnetics*: 357–381.

[9] U. van Rienen, *Numerical Methods in Computational Electrodynamics – Linear Systems in Practical Applications*, Springer-LNCSE, Vol. 12, 2000.

[10] B. Krietenstein, P.Thoma, R. Schuhmann, T. Weiland, *Facing the Big Challenge of High Precision Field Computation*, Proc. 19th LINAC Conf., Chicago, 1998: 860–862.

[11] M. Clemens and T. Weiland, *Magnetic Field Simulation Using Conformal FIT Formulations*, IEEE-T.Mag., 38 (2002): 389–392.

[12] R. Schuhmann, P. Schmidt, T. Weiland, *A New Whitney-Based Material Operator for the Finite-Integration Technique on Triangular Grids*, IEEE-T.Mag., 38 (2002): 409–412.

[13] F. Sachse et al., *Comparison of solutions to the forward problem in electrophysiology with homogeneous, heterogeneous and anisotropic impedance model*, Biomedizinische Technik, 42 (1997): 277–280.

[14] U. van Rienen, M. Clemens, T. Weiland, *Simulation of Low-Frequency Fields on High-Voltage Insulators with Light Contaminations*, IEEE-T.Mag., 32 (1996): 816–819.

[15] U. Schreiber, J. Flehr, V. Motrescu, U. van Rienen, *The Electro-Quasistatic Model in Different Applications*, to appear in Proc. 4th Int. Conf. on Scientific Computing in Electrical Engineering, Springer-LNCSE.

[16] H. Krüger, *Zur numerischen Berechnung transienter elektromagnetischer Felder in gyrotropen Materialien*, PhD thesis, TU Darmstadt 2000.

[17] S. Reitzinger, U. Schreiber, U. van Rienen, *Algebraic Multigrid Methods for Complex Symmetric Matrices and Applications*, Journal for Computational and Applied Mathematics, to appear.

[18] D.A. Stenger, T.M. McKenna (Eds.), *Enabling Technologies for Cultured Neural Networks*, Academic Press, 1994.

[19] G.W. Gross, *Internal Dynamics of Randomized Mammalian Neuronal Networks in Culture*, 277–317 in [18].

[20] J. Keener, J. Sneyd, *Mathematical Physiology*, Springer, 2nd ed., 2001.

[21] http://www.chscene.ch/nervenzelle/data/index.htm (media gallery).

[22] A.L. Hodgkin, A.F. Huxley, *A quantitative description of membrane current and its application to conduction and excitation in nerve*, J. Physiology, 117 (1952): 500–544.

[23] MATLAB®, The MathWorks, Inc., 3 Apple Hill Drive Natick, MA 01760-2098, USA.

[24] MAFIA 4, CST – Computer Simulation Technology, Bad Nauheimer Str. 19, D-64289 Darmstadt, Germany.

[25] M. Kneuer, Diploma Thesis, TU Darmstadt, 2000.

[26] M. Clemens, R. Schuhmann, U. van Rienen, T. Weiland, *Modern Krylov Subspace Methods in Electromagnetic Field Computation Using the Finite Integration Theorie*, ACES J., 11 (1996): 70–84.

[27] M. Clemens, T. Weiland, U. van Rienen, *Comparison of Krylov-Type Methods for Complex Linear Systems Applied to High-Voltage Problems*, IEEE-T.Mag., 34 (1998): 3335–3338.

[28] A. Bunse-Gerstner, R. Stöver, *On a Conjugate Gradient-Type Method for Solving Complex Symmetric Linear Systems*, Lin.Alg.Appl., 287 (199): 105–123.

[29] S. Reitzinger, *PEBBLES – User's Guide*, 1999, http://www.sfb013.uni-linz.ac.at.

Institut für Allgemeine Elektrotechnik
Universität Rostock
Albert-Einstein-Str. 2
D-18051 Rostock
Germany
email: `ursula.van-rienen@etechnik.uni-rostock.de`

Modeling, Simulation and Optimization of Integrated Circuits
(Eds.) K. Antreich, R. Bulirsch, A. Gilg and P. Rentrop

International Series of Numerical Mathematics Vol. 146, 33–47

Radio Frequency Integrated Circuit System Design for UMTS Terminal Receivers

Robert Weigel, Andreas Springer, Linus Maurer

Abstract. The standardization phase for third generation wideband CDMA systems like UMTS which will add broadband data to support video, Internet access, and other high speed services for untethered devices is running towards its finalization. As is typical for mobile communication systems standardization, sufficient RF performance has been assumed and most efforts have been put to digital baseband issues. This is especially true for the mobile phone transceivers the RF part of which is, although its baseband part is much more complex in terms of number of devices, still the bottleneck of the entire system. In the RF concept engineering of today's commercial products an accurate prediction of the needed RF performance by using RF system simulation is meanwhile indispensable. This is in particular the case with third generation wireless systems which, from the RF design point of view, are quite different from second generation TDMA/FDMA systems due to the fact that the user signals are now separated in the code domain rather than in the time and/or frequency domain. The paper gives an insight of how to derive receiver requirements for 3G mobiles in terms recognizable by microwave designers.

1. Introduction

The radio frequency integrated circuit (RFIC) market has expanded greatly during the last few years. Especially wireless devices such as cellular and cordless phones, pagers, global positioning system (GPS) devices, and RF identification tags are rapidly penetrating all aspects of our daily lives. Two sources for this remarkable establishment of the RFIC market can be identified. First of all, the development of low-cost production facilities for the mass-production of highly integrated silicon-based circuits with bipolar transistors capable of operating at GHz frequencies, which appeared about 10 years ago. A second reason for the establishment of the RFIC market came up about the same time with the introduction

---

Key words and phrases. CDMA, RFIC, Receiver Architectures, Spread Sprectrum Communication, UMTS System Design.
The authors would like to thank all their colleagues of the University of Linz, Austria, the University of Erlangen-Nuremberg, Germany, DICE Danube Integrated Circuit Engineering, Linz, Austria, and Infineon Technologies, Munich, Germany who contributed to this paper.

of the GSM (Global System for Mobile Communications) system in Europe [1]. Since the telecommunication authorities of the European countries agreed to implement this standard, a big market was created which required high-quality and low-cost chip-sets. In the following years, both market and technology developed rapidly because of this fruitful technology push/market pull situation. It is expected that the advent of third generation (3G) mobile communication systems, summarized as IMT-2000 (International Mobile Telecommunication) systems, as well as other wireless applications like, e.g., Bluetooth [2], Wireless Local Area Networks (WLAN's) [3], Wireless Local Loop (WLL) [4], etc., will further support the development of the RFIC market.

Besides supplying the required functionality, any successful RFIC solution has to be cheap, small, and must run at low power. The main key to attain these features is the choice of a suitable system architecture since it determines to a high degree the level of integration and the power consumption. Because of the ever increasing integration level of RFIC's, the design process has to improve, too. Traditionally, for each building block (e.g., LNA, mixer, filter, etc.) certain target specifications have been derived from the communication system demands and subsequently each RF building block has been independently optimized to meet these system specifications with sufficient reserve. This modular design flow has started to change into a more integral design approach. Only by considering the entire communications system including both RF and baseband functionalities, an optimum design solution is possible. This requires the involvement and understanding of baseband signal processing techniques (e.g., coding, RAKE receiver, etc.), IC technologies, mixed signal design issues, layout techniques, analog (RF and baseband) and digital circuit design, etc., and, therefore, RFIC design has become a highly multidisciplinary task. In particular the interdependency between RF and baseband signal processing will become more critical due to the steadily increasing computational power of the digital circuitry.

2. Spread spectrum fundamentals

The basics of spread spectrum (SS) technology can be derived from Shannon's well-known channel capacity formula [5]

$$C = W \log_2 \left(1 + \frac{S}{N}\right), \tag{2.1}$$

where C is the capacity of an additive white Gaussian noise (AWGN) channel in bits/Hz, W the bandwidth, N the noise power, and S the signal power. As can be seen from Eq. (2.1), a low S/N can be compensated for by means of an increased transmission bandwidth. Two criteria define a spread spectrum system [6]:

- The transmitted bandwidth is much higher than the bandwidth of the information signal being sent.
- The spreading signal is independent of the information bearing signal.

An important reason for using SS is the linear dependency of the channel capacity C on the bandwidth W (see Eq. (2.1)), whereas C increases only with the logarithm of the signal to noise ratio (SNR) S/N. SS techniques have an inherent resistance against interference and jamming which is illustrated in Fig. 1 for a direct sequence spread spectrum (DS-SS) system. Suppose that a narrowband

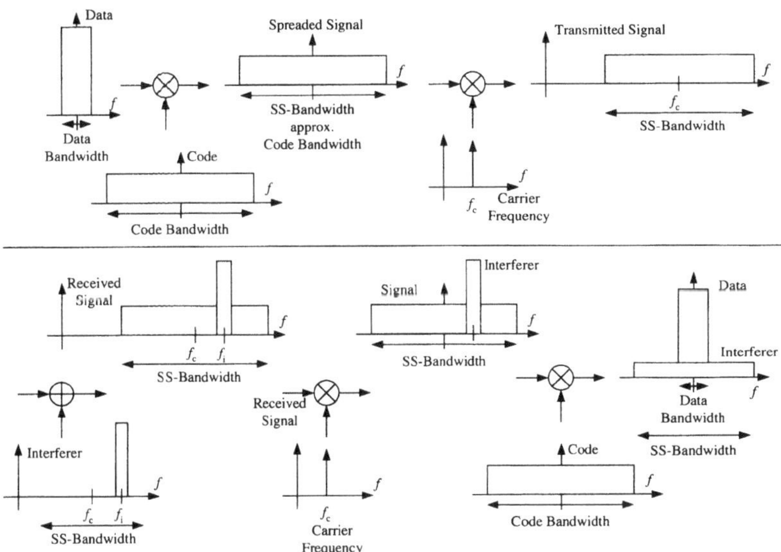

FIGURE 1. Suppression of a narrowband interferer in a direct sequence spread spectrum system.

interferer is present in the received signal. The despreading in the receiver recovers the original spectrum of the data signal which has a high spectral power density and a small bandwidth. At the same time the interference signal is spread over a much larger bandwidth and, therefore, the interference power within the receiver bandwidth decreases. To effectively obtain this increase of the wanted signal level, the locally generated code in the receiver, which is the same as is used in the transmitter, has to be exactly synchronized to the incoming wanted signal.

The best-known advantages of DS-SS systems for cellular system design include the possibility of selective addressing (Code Division Multiple Access (CDMA) [7]) and the ability to eliminate the effect of multipath propagation by using RAKE receiver techniques [8] in the mobile station. Drawbacks incorporate the relatively complex structure of the RAKE receiver, a high synchronization effort, and the need of an accurate output power control in order to deal with the near-far problem [9]. The wideband nature of the signal also leads to the necessity of wideband modems and wideband baseband amplifier stages. Furthermore, fast and accurate automatic gain control (AGC) circuitry is a prerequisite for an efficient handling of the multipath phenomena [6].

Neglecting imperfections and implementation details, the basic functionality and performance of SS systems can be described by a few equations. Important terms in the context of spread spectrum are the so-called spreading factor SF and the spreading gain SG. SF describes the ratio of the information data rate (represented by the bit duration T_{bit}) to the rate of the spreading code (represented by the chip duration T_{chip}). This ratio ranges from 4 to 512 in 3GPP systems.

$$SF = \frac{T_{\text{bit}}}{T_{\text{chip}}} \tag{2.2}$$

$$SG[\text{dB}] = 10 \log SF. \tag{2.3}$$

Let us denote the chip energy to interference power spectral density (PSD) ratio E_{c}/I and the bit energy to interference PSD ratio E_{b}/I. E_{c}/I and E_{b}/I appear before and after despreading in the receiver, respectively. Then E_{c}/I, E_{b}/I, and SG are related by

$$\frac{E_{\text{b}}}{I} = \frac{E_{\text{c}}}{I} + SG + OF, \tag{2.4}$$

where the orthogonality factor OF describes the degree of orthogonality between the wanted user signal and the interference signal. For example in the case of Gaussian noise, OF equals 0 dB. Therefore, in a Gaussian noise environment the wanted user signal level is increased by an amount of SG dB. If only interference from other users is considered, OF approaches infinity for perfectly orthogonal signals. Thus, the choice of codes employed for the spreading of the user signals greatly influences the overall performance of a CDMA system.

These considerations are only valid for perfect synchronization of the received signal and the locally generated code used for despreading. It can be shown that a timing error of, e.g., one half of the chip duration T_{chip} results in an SNR loss of 6 dB. Therefore, obtaining initial synchronization and keeping the code synchronized by a code tracking loop can be considered as key problems in SS system design [6].

3. Receiver requirements and 3GPP front-end test cases

In what follows, we will briefly discuss four 3GPP front-end test cases. By way of these examples it is demonstrated how RF key parameters can be derived from the 3GPP specifications (3GPP: Third Generation Partnership Project). The complete set of the RF specific test cases for the 3GPP FDD mode can be found in [10], and are further discussed in [11]. In Table 1 common terms from the 3GPP specifications used in the following section are described. Unless otherwise stated all parameters are specified at the antenna connector of the UE. They are defined using the 12.2 kbps down link (DL) reference measurement channel [10]. The total received power spectral density $\hat{I}_{\text{or}}$ for all test cases described in the following

TABLE 1. Common 3GPP parameters

DPCH_E_C	Average energy per chip of a dedicated physical channel (DPCH)
$\hat{I}_{or}$	Received (DL) power spectral density measured at the UE antenna connector
I_{or}	Total DL transmit power spectral density at the base station antenna connector
I_{oac}	Power spectral density of the adjacent channel measured at the UE antenna connector.
I_{ouw}	Unwanted signal power level.
OCNS	Orthogonal Channel Noise Simulator, a mechanism used to simulate users or control signals on the other orthogonal channels of a DL.

is composed of the actual data carrying signal (DPCH) to be detected and so-called common downlink channels (pilot channel, synchronization channel, etc.), necessary for establishing and maintaining a link between base station and UE. Therefore, DPCH_E_C is usually several dB below $\hat{I}_{or}$.

3.1. Reference sensitivity level test case

In 3GPP, the reference sensitivity is the minimum receiver input power measured at the antenna port at which the bit error rate (BER) does not exceed a value of 10^{-3}. This test case determines the tolerable noise figure (NF) of the receiver front-end. $\hat{I}_{or}$ and DPCH_E_C are -106.7 dBm/3.84 MHz and -117 dBm, respectively. The 12.2 kbps reference measurement channel used for this test case has a symbol rate of 30 ksps and an SF of 128, i.e., an SG of approximately 21 dB. Let us assume that the required bit energy to interference PSD ratio $E_{b,req}/I$ is 5 dB [12], that the insertion loss (IL) for the baseband implementation is 2 dB, and that the coding gain (CG) is 4 dB (coding gain estimation is difficult for the used convolutional coder; 4 dB seems to be rather conservative). Then the acceptable interference signal level after despreading (P_I) results in

$$P_I = \mathrm{DPCH\_E_C} + SG + CG - \frac{E_{b,req}}{I} - IL. \tag{3.1}$$

Inserting the above given values we have $P_I = -99$ dBm. This leaves a margin for the front-end noise figure (NF) of

$$NF = P_I - 10\log(kTB) = -99\ \mathrm{dBm} + 108\ \mathrm{dBm} = 9\ \mathrm{dB}, \tag{3.2}$$

with the Boltzmann constant k, the ambient temperature $T = 300$ K, and the bandwidth $B = 3.84$ MHz.

3.2. Adjacent channel selectivity test case

Adjacent channel selectivity (ACS) is a measure of a receiver's ability to receive a W-CDMA signal at its assigned channel frequency in the presence of an adjacent channel signal at a given frequency offset from the center frequency of the assigned channel. ACS is the ratio of the receive filter attenuation at the assigned channel frequency to the receive filter attenuation at the adjacent channel frequencies. The ACS has to be better than 33 dB. Simultaneously, the bit error rate shall not exceed 10^{-3} for the following test parameters (see also Fig. 2): $\hat{I}_{\mathrm{or}}$ and DPCH_-$\mathrm{E_C}$ are -92.7 dBm/3.84 MHz and -103 dBm/3.84 MHz, respectively. The same reference measurement channel with a symbol rate of 30 ksps and an SG of 21 dB is used as was the case with the reference sensitivity level test case. The power spectral density of the adjacent channel signal ± 5 MHz away from the wanted channel is $I_{\mathrm{oac}} = -52$ dBm/3.84 MHz. The signal levels for the ACS test case are summarized in Fig. 2

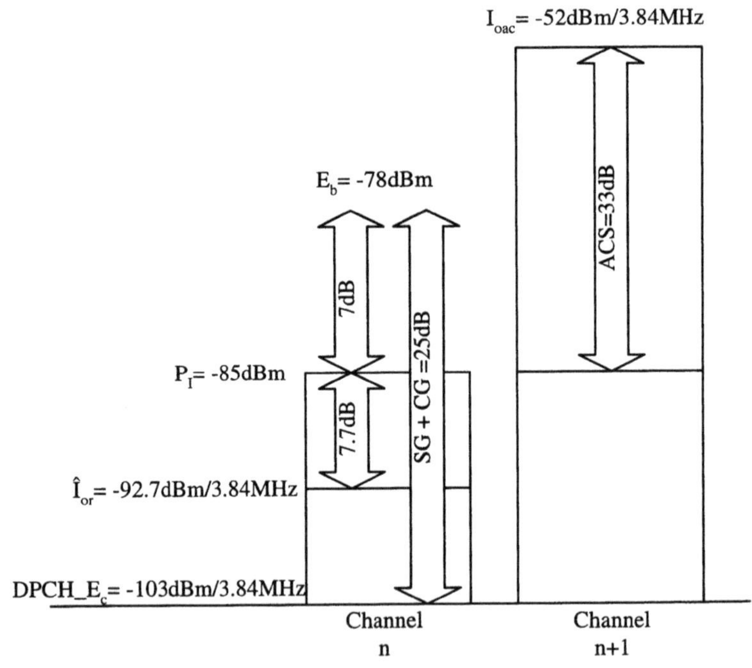

FIGURE 2. Signal levels for the Adjacent Channel Selectivity test case.

We assume again $E_{\mathrm{b,req}}/I$ to be 5 dB, IL for the baseband implementation to be 2 dB, and CG to be 4 dB which leads to an acceptable interference level P_{I} of

$$P_{\mathrm{I}} = \mathrm{DPCH\_E_C} + SG + CG - \frac{E_{\mathrm{b,req}}}{I} - IL = -85 \text{ dBm}. \tag{3.3}$$

If the adjacent channel interference signal is treated as Gaussian noise-like interference, the required ACS can be derived to be

$$ACS = I_{\text{oac}} - P_{\text{I}} = -52 \text{ dBm} + 85 \text{ dBm} = 33 \text{ dB}. \tag{3.4}$$

3.3. Intermodulation test case

For the intermodulation test case two types of interferers are specified: a CW interferer (I_{ouw1}) and a W-CDMA interference signal (I_{ouw2}). Both interferers have a power of -46 dBm, with the CW signal spaced 10 MHz away from the wanted signal and the modulated interferer having a spacing of 20 MHz. The power of the wanted channel is DPCH_E$_\text{C}$ = -114 dBm/3.84 MHz and $\hat{I}_{\text{or}} = -103.7$ dBm/3.84 MHz. The modulated interference signal consists of the necessary common channel for any connection and 16 dedicated data channels with uncorrelated user data and the channelization codes for data channels are chosen to optimally reduce the peak-to-average ratio. Fig. 3 illustrates these test case conditions. The sum of both interfering signals is transferred by means of

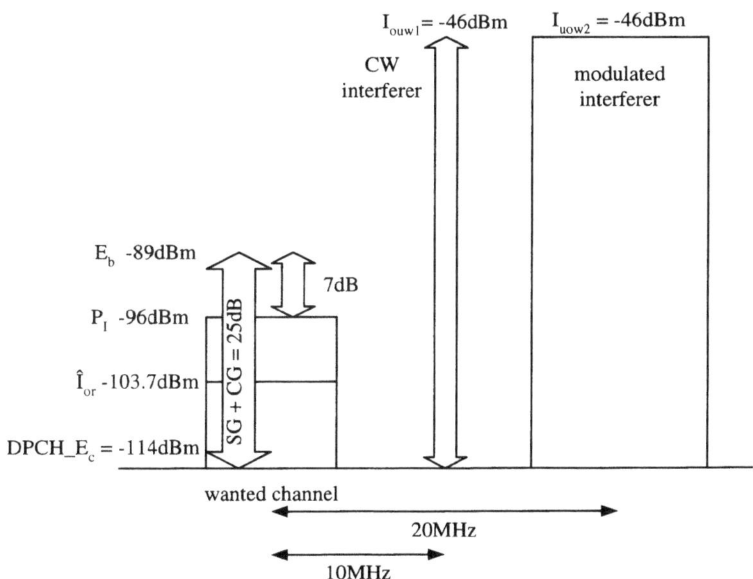

FIGURE 3. Intermodulation test case.

a third-order nonlinearity into the desired channel. Therefore, this test case defines the required input intercept point of third order (IIP3) of the receiver. The acceptable noise-plus-interference level $P_{\text{N,I}}$ in the desired channel must not exceed -96 dBm/3.84 MHz if we assume a combined spreading and coding gain of 25 dB. Before determining the required IIP3 we have to assign the total noise-plus-interference power $P_{\text{N,I}}$ to their sources. According to [13], we assume the following: 50% noise power (−3 dB), 15% intermodulation power (−8 dB), 15% blocking from CW interferer (−8 dB), 15% blocking from modulated interferer (−8 dB), and 5% power from oscillator noise (−13 dB). Furthermore, we neglect second order products. From these assumptions we derive a tolerable level for the third order intermodulation power of $P_{\mathrm{I},3} = -104.2$ dBm/3.84 MHz. The required IIP3 can be derived as

$$\mathrm{IIP3} = \frac{2I_{\mathrm{ouw1}} + I_{\mathrm{ouw2}} - P_{\mathrm{I},3}}{2} = -16.9\,\mathrm{dBm}\,. \tag{3.5}$$

The equations used in Sections IV.A to IV.C exemplify how the signal levels are influenced by the despreading operation and by interference sources. Further estimations like the above-mentioned ones can be found in [13]. However, one should keep in mind that these results can only serve as coarse estimates. What has been neglected in, e.g., the IIP3 calculation is the fact that Eq. (3.5) is based on pure sinusoidal signals. But the IIP3 of a nonlinear building block with respect to a W-CDMA signal is different from the IIP3 for sinusoidal signals [14]. A second point neglected in the above estimation is the fact that the modulated interference signal is a W-CDMA signal spread with other OVSF (orthogonal variable spreading factor) codes than the wanted signal. Depending on how good the orthogonality between signals which are spread with different codes is preserved (modeled by *OF*, see Eq. 2.4) this type of interference can behave strongly different to the Gaussian noise model which is often used for CDMA signals. These simplifications make the above computation of the required IIP3 only an estimation. Furthermore, the actual coding gain has to be simulated to achieve realistic values. Altogether, this leads for certain test cases to the necessity of accurate computer-aided system design using an appropriate combination of commercial baseband and microwave simulation tools like COSSAP© and ADS© which, in some cases, have to be backed up with self-written user-defined codes. The maximum input level test case is such an example.

3.4. Maximum input level test case

The maximum input level test case defines the maximum input power at the antenna port of the mobile station at which a coded bit error rate (BER) of at least 10^{-3} must be achieved. According to the 3GPP specifications [10] the majority of the interference consists of a signal derived from a so-called Orthogonal Channel Noise Simulator (OCNS). This signal is used to simulate other user and control signals on the orthogonal channels in the downlink. $\hat{I}_{\mathrm{or}}$ is specified to be −25 dBm/3.84 MHz with the wanted user signal level being 19 dB below. If we assume the interference coming only from the common channels and the OCNS and perform the same simplified estimations as in the above-described test cases, we would require a combined spreading and coding gain of 26 dB to achieve the necessary E_{b}/I of 7 dB for a BER of 10^{-3}. However, the system simulation results displayed in Fig. 4 show different results [15]. Even an E_{b}/I of below 1.5 dB is sufficient to achieve a BER of 10^{-3}. It can be seen, that the difference between

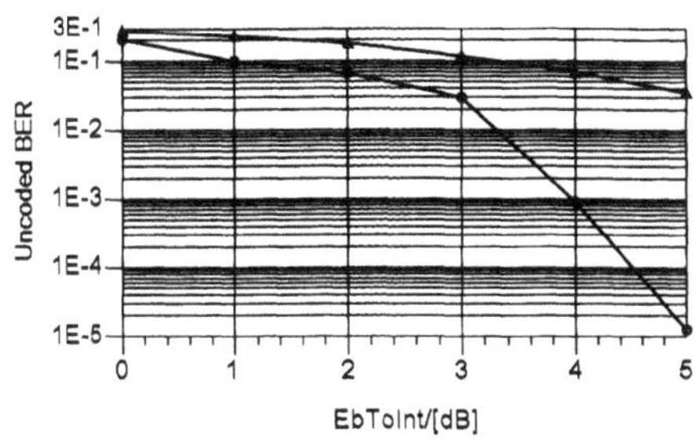

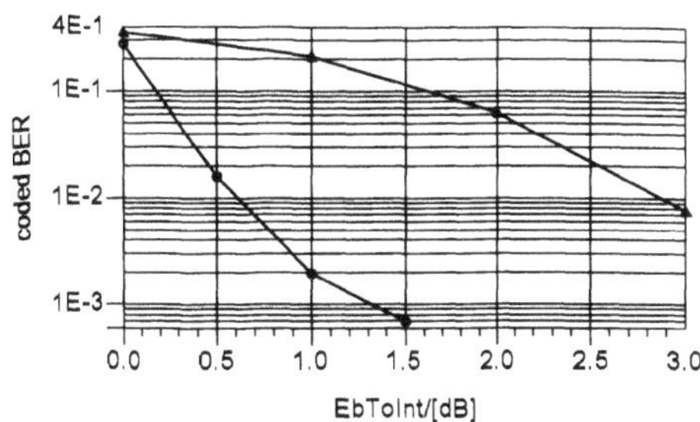

FIGURE 4. Uncoded and coded bit error rate (BER) for interference modeled as Gaussian noise (△) and Orthogonal Channel Noise Simulator (OCNS) signal (∘).

the coded and the uncoded BER strongly depends on the value of E_b/I. The distinct BER difference between a Gaussian noise model and the OCNS interference illustrates the effect of spreading the different user signals with the theoretically orthogonal OVSF codes. It becomes clear that using filtered Gaussian noise to model the interference from other users is not valid in any case and must be considered carefully.

From the ACS test case the required 1 dB compression point P_{1dB} of the low-noise amplifier can be determined as is shown in Fig. 5

4. Receiver architectures

The selection of a receiver architecture is determined by several criteria like number of external components, cost, power dissipation, complexity, and of course the expertise and intellectual property rights of each manufacturer's RF design team. In the following, we will review two important receiver architectures with respect to their applicability to UMTS.

4.1. Heterodyne receiver

Fig. 6 shows the heterodyne receiver structure. This architecture first translates the received signal band down to some intermediate frequency (IF) which is usually much lower than the initially received frequency band. Channel selection filtering is performed at this IF frequency which relaxes the requirements for the channel selection filter. The choice of the IF frequency incorporates a principal design trade-off in heterodyne receiver design (see Fig. 7).

A major advantage of the heterodyne receiver structure is its adaptability to many different receiver requirements. This yields superior performance with respect to selectivity and sensitivity. Therefore, it has been the dominant choice in RF systems for many decades. However, the complexity of the structure and the need for a large number of external components (e.g., the IF filters which in today's cellular phone systems are usually surface acoustic wave (SAW) filters [16])

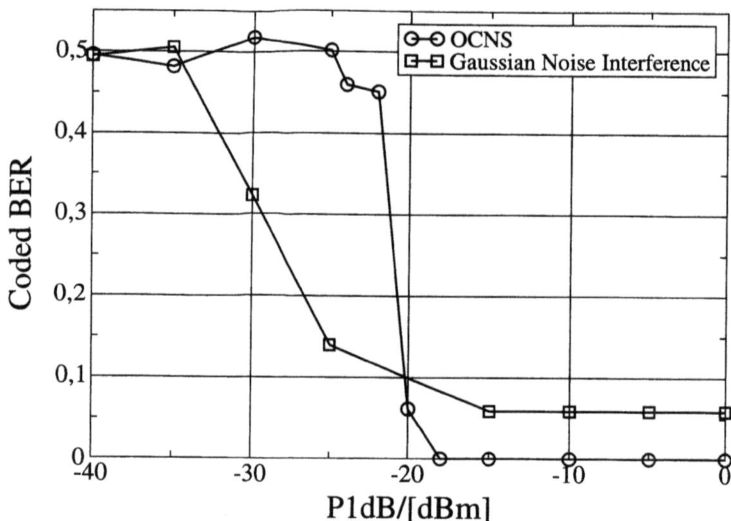

FIGURE 5. Coded bit error rate (BER) as function of the 1 dB compression point for interference modeled as Orthogonal Channel Noise Simulator (OCNS) signal (○) and Gaussian noise (□).

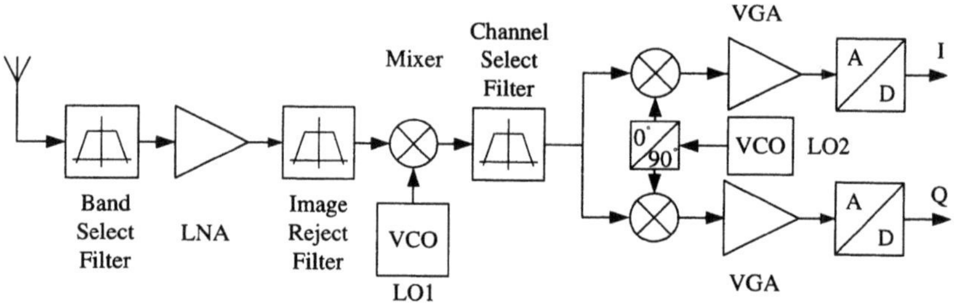

FIGURE 6. Heterodyne receiver.

make problems if a high level of integration is necessary. This is also the major drawback from the costs point of view. Furthermore, amplification at a high IF frequency can cause high power consumption.

The most important disadvantage of the heterodyne receiver architecture is the missing adaptability of a single design to different wireless standards and modes. Since the external IF filter is optimized for a certain mode of operation resulting in a fixed bandwidth and center frequency, it cannot be re-used for, e.g., a different mobile communication standard. Mobile terminals for UMTS will most likely be dual-mode devices supporting also a 2G standard. Such a device

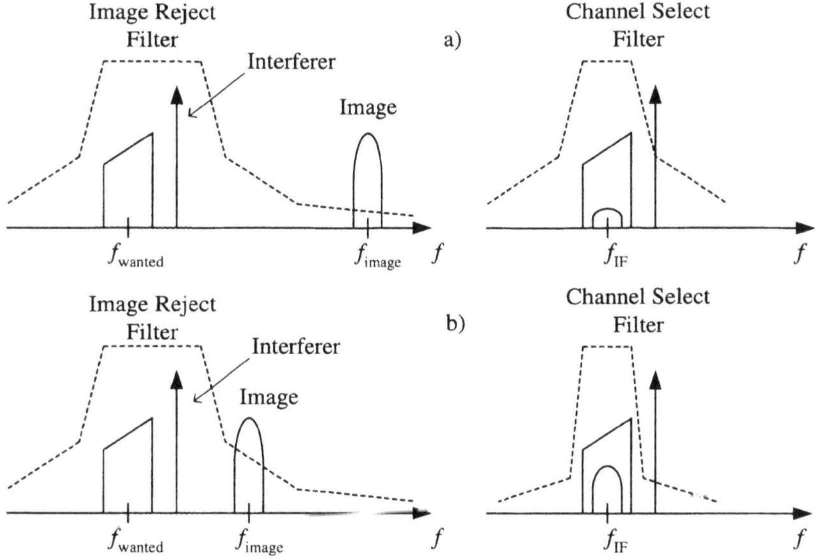

FIGURE 7. Compromise selectivity for sensitivity in a heterodyne receiver. a) high IF b) low IF.

would need two different receive paths each with its own external IF filter, if a heterodyne structure is used. Designing heterodyne receiver architectures for multi-mode operation would probably result in too costly and complex solutions.

4.2. Homodyne receiver

The homodyne receiver architecture (also called zero-IF or direct conversion architecture) avoids the disadvantages of the heterodyne concept by reducing the IF to zero. This saves the (first) mixer, the (first) LO, and the IF channel selection filter, as can be seen from Fig. 8, and moreover, also the image problem vanishes if a quadrature down-converter is used [17]. Thus, the simplicity of this structure offers two important advantages over its heterodyne counterpart. First the problem of image is circumvented because $f_{\text{IF}} = 0$. As a result, no image filter is required. This may also simplify the LNA (low noise amplifier) design, because there is no need for the LNA to drive a 50 Ω load which is often necessary when dealing with image rejection filters. Second, the IF SAW filter and subsequent down-conversion stages can be replaced by low-pass filters and baseband amplifiers that can easily be integrated. The possibility of changing the bandwidth of the integrated low-pass filters (and thus changing the receiver bandwidth) is a major advantage if multimode and multiband applications are of concern.

On the other hand the zero-IF receiver topology incorporates a number of issues that do not play a role or are at least not as serious in a heterodyne receiver.

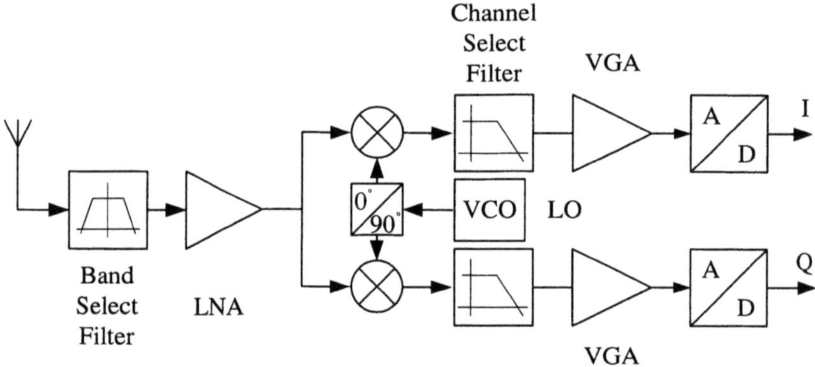

FIGURE 8. Homodyne receiver.

Since in a homodyne topology the down-converted band extends to zero frequency, offset voltages can corrupt the signal and, more importantly, saturate the following stages. There are three main ways how DC offsets can be generated.

- Mixing of the LO leakage signal with the LO signal. If the LO signal leaks to the antenna, is radiated, and subsequently reflected from a moving target back to the receiver, a time varying part of the DC offset can be generated.
- Self-mixing of a strong interferer due to leakage from the LNA or mixer input to the LO port.
- Demodulation of a large amplitude modulated (AM) signal via second order nonlinearity of the mixer which generates a time varying DC offset.

The three described sources can easily generate a DC offset voltage of some mV. Since a UMTS receiver requires about 80 dB gain most of which is assigned to the baseband amplifiers, the analog-to-digital converter (ADC) will be driven into saturation even by such a small DC offset. In time-division multiple access (TDMA) systems idle time intervals can be used to carry out offset cancellation. This would be a practical solution for the 3GPP-TDD mode but cannot be used for offset cancellation in the FDD mode because of the continuous signal reception. Here, the natural solution for DC offset cancellation is high-pass filtering. Since the signal band extends from DC to approximately 2 MHz, a high-pass filter with a cut-off frequency of several kHz results in an acceptable degradation of the system performance [18]. This approach is only possible because of the wideband nature of the signal. A more general DC offset compensation approach which is also applicable to narrow-band systems like the Japanese PHS (Personal Handy-phone System) is experimentally demonstrated in [19].

I/Q mismatch is another critical issue for the zero-IF receiver topology. Any mismatch distorts the constellation diagram of the baseband signal, resulting in an enhanced BER. Fortunately, pilot symbol assisted channel estimation is used in W-CDMA systems. Irrespective of the pilot symbols used (either the time multiplexed pilot symbols or the common pilot signal), this estimation can be used to compensate for I/Q phase and amplitude mismatches.

A further severe problem associated with the homodyne receiver is flicker noise. In a wideband system like UMTS the impairments are lower than in narrowband systems, however, especially for CMOS implementations a careful design of the mixer and the following baseband circuitry is necessary for a successful implementation. Finally, LO leakage produces not only DC offsets but also spurious emissions. In case of UMTS these leakage must be less than −60 dBm/3.84 MHz for UE receivers.

Due to the high integration level and the possibility to electronically adapt the receiver to different bands and modes of operation, the homodyne receiver is the most promising candidate for future 3G terminals in a long term perspective. The described problems can clearly be solved by means of improved semiconductor technology and a system level based optimization of the circuits and their design.

5. Conclusion

The paper gives a review of cellular phone receiver concepts for IMT 2000 system use. It is shown how basic equations lead to an estimation of the transceiver system requirements. Examples exhibit the influence of 3GPP test cases on key parameters for the receiver design. Furthermore, two popular receiver architectures have been investigated. Their suitability for W-CDMA systems has been evaluated and possible problems have been addressed. Further receiver architectures as well as recent work on receiver RFIC's for use in W-CDMA mobiles are presented in [20].

References

[1] A. Mehrotra, *GSM System Engineering*, Artech House, Norwood, 1997

[2] J.C. Haartsen, "The Bluetooth Radio System", *IEEE Personal Communications*, vol. 7, no. 1, Feb. 2000, pp. 28–36

[3] R. van Nee, G. Awater, M. Morikura, H. Takanashi, M. Webster, K.W. Halford, "New High-Rate Wireless LAN Standards", *IEEE Communications Magazine*, vol. 37, no. 12, Dec. 1999

[4] H. Sari, "A State-of-the-Art-Review of Broadband Wireless Access", *Proc. 2000 Europ. Conf. on Wireless Technology*, Oct. 2000, Paris, France, pp. 310–315

[5] C.E. Shannon, "A Mathematical Theory of Communication", *Bell Syst. Tech. J.*, vol. 27, July 1948, pp. 379–423 and 623–656

[6] R.C. Dixon, *Spread Spectrum Systems with Commercial Applications*, John Wiley & Sons, New York, 3rd Edition, 1994

[7] S. Glisic, B. Vucetic, *Spread Spectrum CDMA Systems for Wireless Communications*, Artech House, Norwood, 1997

[8] G.L. Stüber, *Principles of Mobile Communication*, Kluwer Academic Publishers, Boston, 1994

[9] M.K. Simon, J.K. Omura, R.A. Scholtz, B.K. Levitt, *Spread Spectrum Communications Handbook*, McGraw-Hill, New York, Revised Edition, 1994

[10] 3GPP, Technical Specification Group (TSG) RAN WG4, "UE Radio Transmission and Reception (FDD)", *TS 25.101 v4.0.0*, March 2001

[11] 3GPP, Technical Specification Group (TSG) Terminal, "Terminal Conformance Specification; Radio Transmission and Reception (FDD)", *TS 34.121 v3.5.0*, June 2001

[12] TSG-RAN Working Group 4, Nokia mobile Phones, "MS Receiver Sensitivity in UTRA FDD Mode", *Document TSGW4 #1(99)005*, Jan. 1999

[13] O.K. Jensen, T.E. Kolding, C.R. Iversen, S. Laursen, R.V. Reynisson, J.H. Mikkelsen, E. Pedersen, M.B. Jenner, T. Larsen, "RF Receiver Requirements for 3G W-CDMA Mobile Equipment", *Microwave Journal*, vol. 43, no. 2, Feb. 2000, pp. 22–46

[14] H. Pretl, L. Maurer, W. Schelmbauer, R. Weigel, B. Adler, J. Fenk "Linearity Considerations of W-CDMA Frontends for UMTS", *Proc. of the 2000 IEEE MTT-S Microw. Symp.*, Boston, USA, July 2000, pp. 433–436

[15] L. Maurer, W. Schelmbauer, H. Pretl, A. Springer, B. Adler, Z. Boos, R. Weigel, "Influence of Receiver Frontend Nonlinearities on W-CDMA Signals", *Proc. 2000 Asia-Pacific Microwave Conference*, Dec. 2000, Sidney, Australia, pp. 249–252

[16] L. Maurer, W. Schelmbauer, H. Pretl, Z. Boos, R. Weigel, A. Springer, "Impact of IF-SAW Filtering on the Performance of a W-CDMA Receiver", *Proc. 2000 IEEE International Ultrasonics Symposium*, Puerto Rico, Oct. 2000, pp. 375–378

[17] B. Razavi, *RF Microelectronics*, Prentice Hall, Upper Saddle River, 1999

[18] J.H. Mikkelsen, T.E. Kolding, T. Larsen, T. Klingenbrunn, K.I. Pedersen, P. Mogensen, "Feasibility Study of DC Offset Filtering for UTRA-FDD/WCDMA Direct-Conversion Receiver", *Proc. 17th IEEE NORCHIP Conf.*, Oslo, Norway, Nov. 99, pp. 34–39

[19] H. Tsurumi, M. Soeya, H. Yoshida, T. Yamaji, H. Tanimoto, Y. Suzuki, "System-Level Compensation Approach to Overcome Signal Saturation, DC Offset, and 2nd-Order Nonlinear Distortion in Linear Direct Conversion Receiver", *IEICE Trans. Electron.*, vol. E82-C, no. 5, May 1999, pp. 708–716

[20] A. Springer, L. Maurer, R. Weigel, "RF System Concepts for Highly Integrated RFICs for W-CDMA Mobile Radio Terminals", *IEEE Transactions Microwave Theory Tech.*, vol. MTT-50, no. 1, pp. 254–267

Robert Weigel
University of Erlangen-Nuremberg
Institute of Electronics Engineering
Cauerstrasse 9
D-91058 Erlangen, Germany and
DICE Danube Integrated Circuit Engineering
Freistädterstrasse 401
A-4040 Linz, Austria
email: `r.weigel@ieee.org`

Andreas Springer
University of Linz
Institute for Communications and Information Engineering
Altenbergerstrasse 69
A-4040 Linz, Austria,
email: `a.springer@icie.jku.at`

Linus Maurer
DICE Danube Integrated Circuit Engineering
Freistädterstrasse 401
A-4040 Linz, Austria
email: `linus.maurer@infineon.com`

Modeling, Simulation and Optimization of Integrated Circuits
(Eds.) K. Antreich, R. Bulirsch, A. Gilg and P. Rentrop

International Series of Numerical Mathematics Vol. 146, 49–64

Finite-element Discretizations of Semiconductor Energy-transport Equations

Stefan Holst, Ansgar Jüngel and Paola Pietra

Abstract. Energy-transport models describe the flow of electrons in a semiconductor crystal. Several formulations of these models, in the primal or dual entropy variables or in the drift-diffusion-type variables, are reviewed. A numerical discretization of the steady-state drift-diffusion-type formulation using mixed-hybrid finite elements introduced by Marini and Pietra is presented. The scheme is first applied to the simulation of a one-dimensional ballistic diode with non-parabolic band diagrams. Then a two-dimensional deep submicron MOSFET device with parabolic bands is simulated, using an adaptively refined mesh.

1. Introduction

Energy-transport models are used in the modeling of submicron semiconductor devices to describe the charge transport of electrons in the semiconductor crystal. The models consist of the continuity equations for the mass and energy of the electrons and are coupled to the Poisson equation for the electrostatic potential. The energy-transport equations are more complex than the drift-diffusion equations since they include an energy equation and are thus capable of modeling hot electron effects in submicron devices. On the other hand, they are less complex than the hydrodynamic models which contain hyperbolic modes. This intermediate complexity makes the energy-transport system appropriate for fast and accurate semiconductor applications.

In this paper, we review several formulations of energy-transport models. Furthermore, we describe an adaptively refined mixed finite-element scheme of the drift-diffusion formulation of the energy-transport model and simulate two submicron semiconductor devices.

2000 *Mathematics Subject Classification.* Primary 65N30; Secondary 82D37, 78A35.
Key words and phrases. Mixed finite elements, exponential fitting, adaptive scheme, two-dimensional, non-parabolic bands, ballistic diode, submicron MOSFET, semiconductors.
The authors acknowledge partial support from the IHP Project HPRN-CT-2002-00282 of the European Union and from the German-Italian DAAD-Vigoni Project. The first and second author are partly supported by the Gerhard-Hess Program of the Deutsche Forschungsgemeinschaft, grant JU395/3.

In the physical literature, energy-transport equations, first presented by Stratton in 1962 [37], have been derived from hydrodynamic models usually by neglecting certain convection terms (see, e.g., [33] and references therein). Another approach is to derive energy-transport models from the Boltzmann equation by means of the Hilbert expansion method [2]. Denoting the chemical potential, the electron temperature and the electrostatic potential by μ, T and V, respectively, the scaled stationary equations in the so-called *primal entropy variables* μ/T and $-1/T$ read as follows:

$$
\begin{aligned}
-\mathrm{div} J_1 &= 0, && (1.1)\\
-\mathrm{div} J_2 &= -J_1 \cdot \nabla V + W(\mu, T), && (1.2)\\
J_1 &= L_{11}\Big(\nabla \frac{\mu}{T} - \frac{\nabla V}{T}\Big) + L_{12} \nabla \Big(\frac{-1}{T}\Big), && (1.3)\\
J_2 &= L_{21}\Big(\nabla \frac{\mu}{T} - \frac{\nabla V}{T}\Big) + L_{22} \nabla \Big(\frac{-1}{T}\Big), && (1.4)\\
\lambda^2 \Delta V &= n - C(x). && (1.5)
\end{aligned}
$$

Here, J_1 and J_2 are the particle and energy current densities, respectively, $n = n(\mu, T)$ is the electron density, and $W(\mu, T)$ is the energy relaxation term. The diffusion coefficients $L_{ij} = L_{ij}(\mu, T)$ form a symmetric, positive definite (2×2) matrix. The precise dependence of these dimensionless functions on μ and T will be given in Section 2.1. The space-dependent function $C = C(x)$ is the doping profile modeling fixed charged background ions in the crystal, and the positive constant λ is the scaled Debye length.

The above equations have to be solved in a bounded domain $\Omega \subset \mathbb{R}^d$ $(d \leq 3)$. We complement the equations with physically motivated mixed Dirichlet-Neumann boundary conditions:

$$
\begin{aligned}
n = n_D, \quad T = T_D, \quad V = V_D \qquad & \text{on } \Gamma_D,\\
J_1 \cdot \nu = J_2 \cdot \nu = \nabla V \cdot \nu = 0 \qquad & \text{on } \Gamma_N,
\end{aligned}
$$

modeling the contacts Γ_D and the insulating boundary parts Γ_N. We have assumed that $\partial\Omega = \Gamma_D \cup \Gamma_N$ and $\Gamma_D \cap \Gamma_N = \emptyset$. The exterior normal unit vector on $\partial\Omega$ is denoted by ν. In the mathematical literature, there are only a few results concerning the existence and uniqueness of solutions to (1.1)–(1.5) (see, e.g., [11, 12, 14, 17]), proving existence of solutions for uniformly bounded diffusion coefficients L_{ij} or in near-equilibrium situations.

The numerical discretization of energy-transport models has been investigated in the physical literature for quite some years [1, 9, 15, 36, 38]. Mathematicians started to pay attention to these models in the 1990s, using different numerical methods. Jerome and Shu employed ENO (essentially non-oscillatory) numerical schemes [22]. Finite-difference discretizations have been used by Ringhofer [32], based on an entropy-decaying property, and by Fournié [16], with compact schemes of fourth order. Bosisio et al. used mixed finite-volume techniques [3]. Mixed finite-element methods applied to different formulations of the energy-transport models have been employed in [13, 18, 19, 25, 29, 30] (see [5] for an overview).

In the first part of the paper we make precise the definitions of L_{ij}, n and W as functions of μ and T for general and parabolic band structures (Section 2.1) and give three different formulations of the energy-transport model (i) in the primal entropy variables (see above), (ii) in the dual entropy variables, and (iii) in the drift-diffusion-type variables. We explain the advantages and disadvantages regarding the numerical approximation of the equations for each of these formulations (Section 2.2).

The second part of the paper is devoted to the numerical approximation of the equations in the drift-diffusion formulation using mixed-hybrid finite elements introduced by Marini and Pietra [26, 27] (Section 3). The resulting algebraic system is solved employing a variant of the Gummel method. In the two-dimensional situation we adaptively refine the mesh using an error estimator motivated by the results of [20] and [39]. Finally, we apply the resulting scheme to the numerical simulation of a 1-D ballistic diode with non-parabolic bands (Section 4.1) and to a 2-D MOSFET device with parabolic bands (Section 4.2).

2. Formulation of the model

2.1. Physical assumptions and non-parabolic band approximations

In order to give precise definitions for the coefficients L_{ij} in terms of μ and T, we impose the following physical assumptions:

- The energy band diagram ε of the semiconductor crystal is spherically symmetric and a strictly monotone function of the modulus $k = |\vec{k}|$ of the wave vector $\vec{k}$.
- The momentum relaxation time is defined by

$$\tau(\varepsilon) = \left(\phi_0 \varepsilon^\beta N(\varepsilon)\right)^{-1}, \quad \beta > -2,$$

  where $\phi_0 > 0$ is some constant and $N(\varepsilon) = 4\pi k^2/|\varepsilon'(k)|$ is the density of states of energy $\varepsilon = \varepsilon(k)$ [2].
- The electron density n is given by non-degenerate Boltzmann statistics.

For the physical background, we refer to the textbooks [4, 35], for instance.

The energy-transport model (1.1)–(1.5) is completed by giving the definitions of the functions $L_{ij}(\mu, T)$, $n(\mu, T)$, and $W(\mu, T)$. In [2] the following expressions have been derived:

$$L_{ij}(\mu, T) = e^{\mu/T} \int_0^\infty d(\varepsilon)\varepsilon^{i+j-2} e^{-\varepsilon/T} d\varepsilon, \quad i, j = 1, 2, \tag{2.1}$$

$$n(\mu, T) = e^{\mu/T} \int_0^\infty e^{-\varepsilon/T} N(\varepsilon) d\varepsilon, \tag{2.2}$$

where

$$d(\varepsilon) = \frac{4\pi}{3}\tau(\varepsilon)|\varepsilon'(k)|k^2 \quad \text{and} \quad \varepsilon = \varepsilon(k).$$

The energy relaxation term in the Fokker-Planck approximation of the phonon collision operator is given by [13]

$$W(\mu, T) = -\phi_1 e^{\mu/T} \int_0^\infty \varepsilon^\beta N(\varepsilon)^2 e^{-\varepsilon/T} d\varepsilon \left(1 - \frac{T_0}{T}\right),$$

where $T_0 > 0$ is the (constant) lattice temperature and $\phi_1 > 0$ is some constant.

The advantages of the model are that the assumptions on the band diagram are rather general (the above hypotheses can be even weakened, see [2]) and that *analytic* expressions can be obtained under the assumption of non-parabolic bands in the sense of Kane [24], where $\varepsilon(k)$ is defined by

$$\varepsilon(1 + \alpha\varepsilon) = k^2/2,$$

and $\alpha > 0$ is the non-parabolicity parameter. We call the band structure *parabolic* if $\alpha = 0$; in this situation $\varepsilon(k) = k^2/2$.

For non-parabolic bands in the sense of Kane, we can rewrite the expressions for L_{ij}, n and W in terms of the following integrals:

$$L_{ij}(\mu, T) = \frac{2}{3\phi_0} T^{i+j-\beta} e^{\mu/T} \int_0^\infty \frac{1 + \alpha T u}{(1 + 2\alpha T u)^2} u^{i+j-\beta-1} e^{-u} du, \tag{2.3}$$

$$n(\mu, T) = 2\pi(2T)^{3/2} e^{\mu/T} \int_0^\infty (1 + \alpha T u)^{1/2}(1 + 2\alpha T u) u^{1/2} e^{-u} du, \tag{2.4}$$

$$W(\mu, T) = 8(2\pi)^2 \phi_1 e^{\mu/T} T^{1+\beta}(T_0 - T) \times \int_0^\infty (1 + \alpha T u)(1 + 2\alpha T u)^2 u^{1+\beta} e^{-u} du. \tag{2.5}$$

We obtain *explicit* expressions for parabolic band diagrams. In fact, we have just to set $\alpha = 0$ in the above equations which yields

$$\begin{aligned} L_{ij}(\mu, T) &= \frac{2}{3\phi_0}\Gamma(i + 1 - \beta) T^{i+j-\beta} e^{\mu/T}, \\ n(\mu, T) &= (2\pi)^{3/2} T^{3/2} e^{\mu/T}, \\ W(\mu, T) &= 8(2\pi)^2 \phi_1 \Gamma(2 + \beta) T^{1+\beta} e^{\mu/T}(T_0 - T), \end{aligned} \tag{2.6}$$

where Γ denotes the Gamma function. Chen et al. [9] used $\beta = 1/2$ in the simulation of MOS devices. Then it follows

$$(L_{ij}) = \mu_0 n \begin{pmatrix} 1 & 3T/2 \\ 3T/2 & 15T^2/4 \end{pmatrix} \tag{2.7}$$

and

$$W(\mu, T) = \frac{3}{2} \frac{n(T_0 - T)}{\tau_0}, \tag{2.8}$$

where $\mu_0 = 1/(6\sqrt{2\pi}\phi_0)$ and $\tau = 1/(4\sqrt{2\pi}\phi_1)$. Generally, we call an energy-transport model, in which the expressions (2.3)–(2.5) and $\beta = 1/2$ are used, the *Chen model.*

2.2. Formulations of the energy-transport models

We describe three different formulations of the energy-transport equations:

- formulation in the *primal entropy variables* μ/T, $-1/T$ (see (1.1)–(1.5)),
- formulation in the *dual entropy variables* $(\mu - V)/T$, $-1/T$, and
- *drift-diffusion* formulation in the variables L_{11}, L_{12}.

The first formulation has been already presented in Section 1. The second formulation of the energy-transport equations is obtained by introducing the *dual entropy variables*

$$w_1 = \frac{\mu - V}{T}, \quad w_2 = -\frac{1}{T},$$

yielding the system

$$-\mathrm{div} I_1 = 0, \qquad I_1 = D_{11}\nabla w_1 + D_{12}\nabla w_2, \tag{2.9}$$

$$-\mathrm{div} I_2 = W(\mu, T), \qquad I_2 = D_{21}\nabla w_1 + D_{22}\nabla w_2, \tag{2.10}$$

where the new diffusion coefficients are given by

$$D_{11} = L_{11}, \quad D_{12} = D_{21} = L_{12} - VL_{11}, \quad D_{22} = L_{22} - 2VL_{12} + V^2 L_{11}.$$

The diffusion matrix consisting of the coefficients D_{ij}, $i, j = 1, 2$, is again symmetric and positive definite. In this formulation, the Joule heating term $J_1 \cdot \nabla V$ is eliminated and the problem is symmetrized. The advantage of this formulation is twofold. First, the theory of systems of quasilinear elliptic equations can be directly applied. This has been used in [11, 12] in order to prove the existence of solutions to the transient and stationary equations (however, under the assumption of *uniformly bounded* coefficients $D_{ij}(w)$). Second, the system can be discretized by a mixed finite-element method. However, as the diffusion matrix is not diagonal, the resulting nonlinear equations have to be solved generally using the Newton method which makes the final code unflexible to changes of the parameters or the semiconductor band structure.

This disadvantage can be avoided in the third formulation which is given in the *drift-diffusion variables* g_1 and g_2:

$$-\mathrm{div} J_1 = 0, \tag{2.11}$$

$$-\mathrm{div} J_2 = -J_1 \cdot \nabla V + W(\mu, T), \tag{2.12}$$

$$J_i = \nabla g_1 - g_i \frac{\nabla V}{T}, \quad i = 1, 2, \tag{2.13}$$

where $g_1 = L_{11}$ and $g_2 = L_{12}$. Indeed, from (1.3)–(1.4) and (2.1) follows

$$J_i = \int_0^\infty d(\varepsilon) \left(\nabla e^{(\mu-\varepsilon)/T} + \nabla V \frac{\partial}{\partial \varepsilon} e^{(\mu-\varepsilon)/T} \right) \varepsilon^{i-1} d\varepsilon$$

$$= \nabla \int_0^\infty d(\varepsilon) e^{(\mu-\varepsilon)/T} \varepsilon^{i-1} d\varepsilon - \frac{\nabla V}{T} \int_0^\infty d(\varepsilon) e^{(\mu-\varepsilon)/T} \varepsilon^{i-1} d\varepsilon, \quad i = 1, 2,$$

which equals (2.13), setting

$$g_1 = \int_0^\infty d(\varepsilon) e^{(\mu-\varepsilon)/T} d\varepsilon, \quad g_2 = \int_0^\infty d(\varepsilon) e^{(\mu-\varepsilon)/T} \varepsilon d\varepsilon,$$

and from (2.1) follows the assertion.

We have to show that n and T can be uniquely defined as functions of g_1 and g_2. We define the function $f(T) = L_{12}(\mu, T)/L_{11}(\mu, T) = g_2/g_1$. In [13] it is shown that the function f can be always inverted since $f'(T) = \det(L_{ij})_{ij}/(Tg_1)^2 > 0$. Thus, the function $T = T(g_1, g_2) = f^{-1}(g_2/g_1)$ can be uniquely defined. Now, the density n is defined through (2.2), i.e.,

$$n(g_1, g_2) = g_1 \int_0^\infty N(\varepsilon) e^{-\varepsilon/T(g_1,g_2)} d\varepsilon \Big/ \int_0^\infty d(\varepsilon) e^{-\varepsilon/T(g_1,g_2)} d\varepsilon.$$

In the parabolic Chen model, the temperature can be explicitely computed from g_1 and g_2. Indeed, since $g_1 = L_{11} = \mu_0 n$ and $g_2 = L_{12} = (3/2)\mu_0 nT$, it follows $T = 2g_2/3g_1$.

The relaxation term can be written as follows [13]:

$$W = \gamma_1(T) g_1 - \gamma_2(T) g_2,$$

where

$$\frac{1}{\gamma_1(T)} = \frac{T}{\phi_1 T_0} \int_0^\infty d(\varepsilon) N(\varepsilon) d\varepsilon, \quad \frac{1}{\gamma_2(T)} = \frac{1}{\phi_1} \int_0^\infty d(\varepsilon) N(\varepsilon) \varepsilon d\varepsilon.$$

Notice that the quotients $e^{\mu/T}/L_{1i}(\mu, T)$ do not depend on μ (see (2.1)).

The formulation (2.11)–(2.13) has the advantages that first, the equations can be easily discretized using mixed finite elements as already employed for the drift-diffusion model (see, e.g., [6, 7]) and second, a Gummel iteration technique can be applied to the final nonlinear system (see Section 3).

3. Numerical discretization

We discretize the energy-transport model in the drift-diffusion formulation (2.11)–(2.13). For this, we recast the equations in the following form

$$-\text{div} J_i + \sigma_i g_i = f_i, \quad J_i = \nabla g_i - \beta g_i \quad \text{in } \Omega, \; i = 1, 2, \tag{3.1}$$

$$g_i = g(n_D, T_D) \quad \text{on } \Gamma_D, \quad J_i \cdot \nu = 0 \quad \text{on } \Gamma_N, \tag{3.2}$$

where

$$\sigma_1 = 0, \quad \sigma_2 = \gamma_2, \quad f_1 = 0, \quad f_2 = -J_1 \cdot \nabla V + \gamma_1 g_1, \quad \beta = \nabla V / T(g_1, g_2).$$

By freezing the coefficients (obtained from the previous iteration in the global iteration procedure), the energy-transport model can be written (in each iteration step) as a system of convection-diffusion equations. It is well known in semiconductor modeling that this system is usually convection dominated due to high electric fields. Therefore, special care has to be taken in the numerical approximation. We discretize the equations using an exponentially fitted mixed finite-element method.

We describe now the discretization of the energy conservation equation (3.1) with $i = 2$ in detail. The discretization of the mass conservation equation (3.1) with $i = 1$ is similar but simpler since the zeroth-order term and the right-hand side vanish.

Let $\mathcal{T}_h$ be a regular family of decompositions of Ω into triangles K such that there is no element across the interface between Γ_D and Γ_N. Furthermore, let $\mathcal{E}_h$ be the set of edges e of $\mathcal{T}_h$. We discretize the Poisson equation using nonconforming Crouzeix-Raviart elements yielding V^h [10]. Furthermore, we assume that the temperature is a piecewise constant function $\overline{T}$ obtained in the iteration procedure. As ∇V^h is piecewise constant, also $\beta^h = \nabla V^h/\overline{T}$ is constant in each triangle. Therefore, we can suppose that β in (3.1) is piecewise constant. Consequently, we can symmetrize the problem in each triangle K by means of the local Slotboom variable $y = e^{-\beta\cdot\mathbf{x}}u$. The symmetric form of the current density equation reads

$$e^{-\beta\cdot\mathbf{x}}J \quad \nabla y = 0 \quad \text{in } K. \tag{3.3}$$

The vector-valued approximation space for the mixed scheme developed in [27] is built on the following set of polynomial vectors:

$$\Sigma(K) = \text{span}(\tau_1, \tau_2, \tau_3),$$

with $\tau_1 = (1, 0)$, $\tau_2 = (0, 1)$, and $\tau_3 = (\omega_1, \omega_2)$. To define the properties that determine the polynomials ω_i we need the following notations. The edges of each triangle K are numbered counter-clockwise starting with e_1 that connects the vertices with the smallest values of $\beta \cdot \mathbf{x}$ in K. We denote by ν_i the exterior unit normal vector of K associated to the edge e_i $(i = 1, 2, 3)$. Then we choose $\tau_3 = (\omega_1, \omega_2)$ with ω_1, $\omega_2 \in P_2(K)$ fulfilling the following conditions:

$$\tau_3 \cdot \nu_2|_{e_2} = \tau_3 \cdot \nu_3|_{e_3} = 0, \quad \tau_3 \cdot \nu_1|_{e_1} = 1, \tag{3.4}$$

$$\int_K \tau_3 \cdot \text{curl}(\lambda_1\, \lambda_2\, \lambda_3)dxdy = 0, \tag{3.5}$$

$$\int_K \omega_1 dxdy = \int_K \omega_2 dxdy = 0, \tag{3.6}$$

where λ_i denotes the i-th barycentric coordinate. It can be easily checked that $\dim(\text{div}\, \Sigma(K)) = 1$ (see (3.4)).

We introduce the following finite-dimensional spaces:

$$\Sigma_h = \{\tau \in [L^2(\Omega)]^2 : \tau|_K \in \Sigma(K)\ \forall K \in \mathcal{T}_h\},$$
$$W_h = \{\phi \in L^2(\Omega) : \phi|_K \in P_0(K)\ \forall K \in \mathcal{T}_h\},$$
$$\Lambda_{h,\xi} = \{\mu \in L^2(\mathcal{E}_h) : \mu|_K \in P_0(e)\ \forall e \in \mathcal{E}_h;\ \textstyle\int_e (\mu - \xi)ds = 0\ \forall e \in \Gamma_D\},$$

where ξ is any function in $L^2(\Gamma_D)$. The spaces $P_i(K)$ and $P_i(e)$ denote the sets of polynomials of degree i in K and on e, respectively. The mixed-hybrid formulation of (3.1)–(3.2) then reads as follows:

$$\begin{cases} \text{Find } J_2^h \in \Sigma_h,\ \bar{g}_2^h \in W_h \text{ and } g_2^h \in \Lambda_{h,u_D} \text{ such that:} \\ \displaystyle\sum_{K\in\mathcal{T}_h}\Big(\int_K A_K J_2^h\cdot\tau dx + \int_K B_K \bar{g}_2^h \mathrm{div}\tau dx - \sum_{e_i(K)}\int_{e_i} B_{e_i} g_2^h \tau\cdot\nu ds\Big) = 0, \\ \displaystyle\sum_{K\in\mathcal{T}_h}\Big(-\int_K \mathrm{div} J_2^h \phi dx + \int_K \sigma_2 \bar{g}_2^h \phi dx\Big) = \int_K f_2 \phi dx, \\ \displaystyle\sum_{K\in\mathcal{T}_h}\int_{\partial K} \mu J_2^h\cdot\nu ds = 0 \end{cases} \tag{3.7}$$

for all $\tau\in\Sigma_h$, $\phi\in W_h$, and $\mu\in\Lambda_{h,0}$. The functions A and B are piecewise constant on a triangle K or on one of its edges $e_i(K)$, respectively, and are defined by the mean values

$$A_K = \frac{1}{|K|}\int_K e^{-\beta\cdot\mathbf{x}}dx, \tag{3.8}$$

$$B_{e_i(K)} = \frac{1}{|e_i|}\int_{e_i} e^{-\beta\cdot\mathbf{x}}ds, \qquad B_K = \max_i B_{e_i(K)} = B_{e_1(K)}. \tag{3.9}$$

These terms account for the convective part of the flux. One important property is that strong convective effects are reproduced by the scheme with the correct order of magnitude [27].

The first equation in (3.7) is a weak discrete version of (3.3), the second equation corresponds to the discrete version of the first equation of (3.1), and the third equation imposes the continuity requirement of the normal component of J_2^h at the interelement boundaries.

Due to the piecewise regularity of the approximation spaces, a static condensation procedure applied to the linear system associated with (3.7) allows to eliminate the unknowns for J_2^h and $\bar{g}_2^h$ from the algebraic system leading to a final matrix acting only on the Lagrange multipliers g_2^h. This matrix is an M-matrix if the triangulation is weakly acute. The nonlinear algebraic system then is solved using a variant of the Gummel method. We recall that the Gummel iteration procedure can be regarded as an approximate Newton method, where the information about the strong coupling of the unknowns is incorporated into the Poisson equation, and the Jacobian of the system is replaced by a diagonal matrix [21].

In order to refine the mesh adaptively, we have to define a suitable error estimator. We use an error estimator consisting of two parts: an indicator for the primal variables of the mixed-hybrid formulation in the L^2 norm, motivated by the error estimators used for Raviart-Thomas elements [20], and an indicator for the Poisson equation, which is solved using Crouzeix-Raviart elements, taken from [39]. The error estimator for the Crouzeix-Raviart elements is a residual based error indicator. The estimator derived for the Raviart-Thomas elements can be interpreted as a residual based error indicator. It is motivated by a saturation assumption involving the difference $g_2^h - \bar{g}_2^h$. In [19] the saturation assumption has been verified numerically for the above finite elements introduced by Marini and Pietra, applied to some convection-dominated test problems.

After convergence of the Gummel method, the triangles belonging to the set $\{K \in \mathcal{T}_h : \eta_{K,ET} \leq 0.7 \max_K \eta_{K,ET}\}$ are regularly refined, removing any hanging nodes by using the algorithm of Bank and regularizing the mesh. Then new vectors of nodal values are generated by interpolating the solution from the previous mesh and the Gummel algorithmus is again applied until the error estimator is smaller than a prescribed value or until the maximal number of elements is reached (see [19] for details).

4. Numerical examples

In this section we present numerical simulations of a silicon ballistic diode and a silicon MOSFET device. The ballistic diode is simulated using the energy-transport model (2.11)–(2.13) including the physical parameters (see, e.g., [23]) with non-parabolic band diagrams in the sense of Kane (see Section 2.1) and the coefficients (2.3)–(2.5) with $\beta = 1/2$. In this case, the temperature T is a nonlinear function of g_1 and g_2. The MOSFET device is solved using the parabolic Chen model of Section 2.1 (i.e., using (2.6)–(2.8)) including electron-hole recombination-generation effects. More precisely, the equations (2.11) and (2.12) now read

$$\begin{aligned} -\mathrm{div} J_1 &= -R(n,p), \\ -\mathrm{div} J_2 &= -J_1 \cdot \nabla V + W(n,T) - \frac{3}{2} T R(n,p), \end{aligned}$$

where p denotes the hole particle density, and $R(n,p)$ is the Shockley-Read-Hall recombination-generation term given by

$$R(n,p) = \frac{np - n_i^2}{\tau_p(n+n_i) + \tau_n(p+n_i)},$$

with the (scaled) intrinsic density n_i and the electron and hole life times τ_n, τ_p, respectively. The hole density is computed using the drift-diffusion model

$$-\mathrm{div} J_p = R(n,p), \quad J_p = -\nabla p - p \nabla V.$$

The relevant numerical values of the physical parameters used for both devices are collected in Table 1.

4.1. A 1-D ballistic diode with non-parabolic bands

First, we present numerical simulations of a one-dimensional n^+nn^+ ballistic diode which is a simple model for the channel of a MOS transistor. The semiconductor domain is given by the interval $(0, \ell)$. The maximal doping concentration in the n^+-regions equals $5 \cdot 10^{17}\mathrm{cm}^{-3}$; in the n-channel the minimal doping profile is $2 \cdot 10^{15}\mathrm{cm}^{-3}$. The length of the n^+-regions is $0.1\mu\mathrm{m}$, whereas the length of the channel region equals $0.4\mu\mathrm{m}$. On the boundary points we assume that the total space charge $C - n$ vanishes and that the temperature equals the ambient temperature, i.e., (in unscaled variables) $n = C$ and $T = T_0$. Denoting by $U > 0$ the applied potential

Parameter	Physical meaning	Numerical value
$\tau_p \,/\, \tau_n$	carrier life times	$10^{-5} \,/\, 10^{-6}$s
$\mu_p \,/\, \mu_n$	low-field carrier mobilities	$450 \,/\, 1500$ cm^2/Vs
n_i	intrinsic density	$1.4 \cdot 10^{10}$ cm^{-3}
τ_0	energy relaxation time	0.4 ps
$\varepsilon_r(Si)$	relative permittivity of silicon	11.7
$\varepsilon_r(SiO_2)$	relative permittivity of the oxide	3.8
T_0	ambient temperature	300K

TABLE 1. Material and model parameters.

we set $V(0) = 0$ and $V(\ell) = -U$. The non-parabolicity parameter is taken to be $\alpha = 0.5/\text{eV}$.

The numerical experiments are performed with a uniform mesh of 100 grid points. In Figure 1 (left) we present the electron temperature for vanishing and non-vanishing non-parabolicity parameter. As expected the temperature in the n-channel is large, corresponding to "hot" electrons. The temperature is reduced due to the non-parabolicity effects. Another effect of the non-parabolicity is the reduction of the spurious velocity overshoot peak in Figure 1 (right).

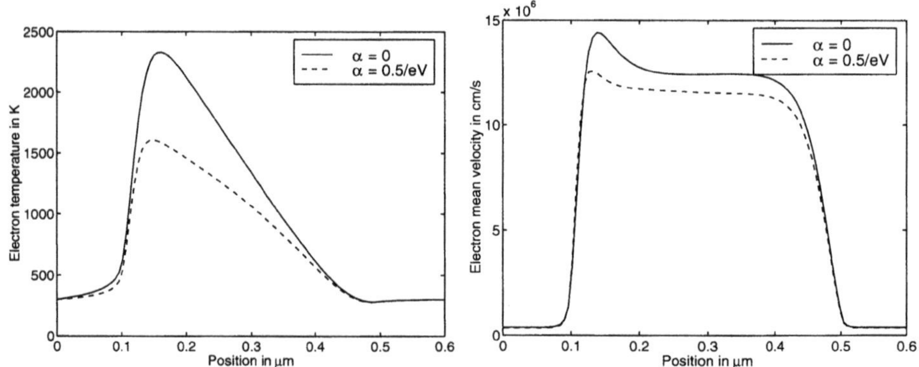

FIGURE 1. Electron temperature T (left) and mean velocity J/n versus position in a ballistic diode.

4.2. A 2-D MOSFET device with parabolic bands

A MOSFET (metal-oxide semiconductor field-effect transistor) device can be used as a voltage-driven switch and is the most used device in computer technology. We simulate a transistor of size 420nm×210nm with an oxide thickness of 1.5nm. The length of the source and drain contacts is 30nm and the length of the channel varies between 70nm and 110nm (see Figure 2). The doping profile is given by a step function with values 10^{19}cm^{-3} in the n^+-region and -10^{17}cm^{-3} in the p bulk region. The voltage between source and drain is $V_{SD} = 1\text{V}$. The geometry and data of this device are adapted from the work of Cassan et al. [8] where the use of the energy-transport model for such device geometries has been justified.

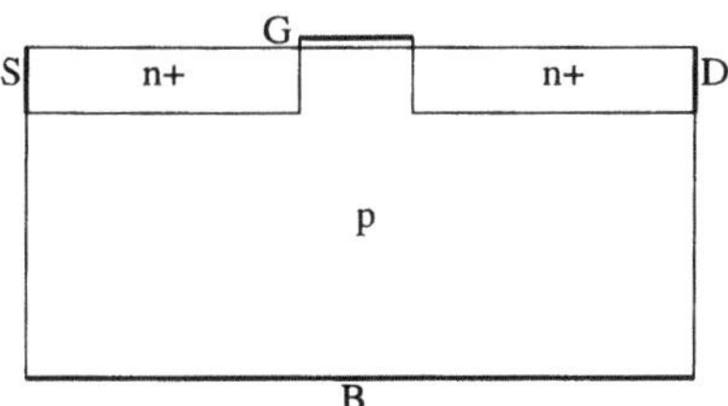

FIGURE 2. Geometry of the MOSFET with source S, drain D, gate G and bulk B contacts.

We model the influence of the oxide by assuming that the particles do not penetrate the oxide region Ω_O. Thus, the particle and energy densities are only computed in the semiconductor domain Ω_S. We choose Dirichlet boundary conditions at the source, drain and bulk contacts, i.e., the densities are set to their equilibrium values (see [28]). On the remaining boundary parts (including the silicon/silicon oxide interface), homogeneous Neumann boundary conditions $J_i \cdot \nu = 0$, $i = 1, 2, p$, are imposed. The Poisson equation is solved in the domain $\Omega_S \cup \Omega_O$ with a space-dependent permittivity which is constant in each Ω_i, $i = S, O$, and we specify homogeneous Neumann boundary conditions on the interface.

Before we present the numerical results we show the meshes generated by our refinement strategy in Figure 3. The initial mesh is constructed without any knowledge of the location of the junctions. After the adaptive procedure, the final grid is refined near the junctions and near the gate oxide. The importance of the adaptive scheme becomes apparent when comparing the current-voltage curves for different meshes (Figure 4 (left)). The characteristics seem to stabilize for meshes with about 4000 triangles.

Figure 4 (right) shows the dependence of the drain current on the effective channel length. For smaller channel lengths, not only the values of the drain current are larger but also the slope in the nearly linear region (for $V > 0.3\text{V}$) increases, due to higher electric fields in the channel. The electron density in a MOSFET with 70nm and 110nm channel lengths is presented in Figure 5. The voltage between gate and bulk is chosen to be $V_{GB} = 1\text{V}$. Our numerical scheme is able to resolve the large gradients near the junctions. Due to the high temperature region at the drain junction, the effective electron mobility $\mu(T) = \mu_n/T$ decreases and a region with a higher electron concentration is formed near the gate. The temperature T in a 110nm MOSFET is depicted in Figure 6. The values for T in the channel are larger in the closed state than in the open state. However, notice that in the closed state, the channel consists only of a few electrons.

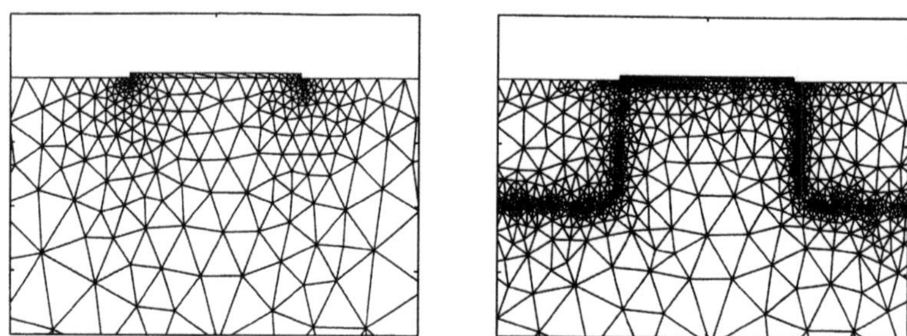

FIGURE 3. Adaptively refined triangulations for the MOSFET (zoom). Left: initial mesh (about 1000 triangles); right: final mesh (about 4000 triangles).

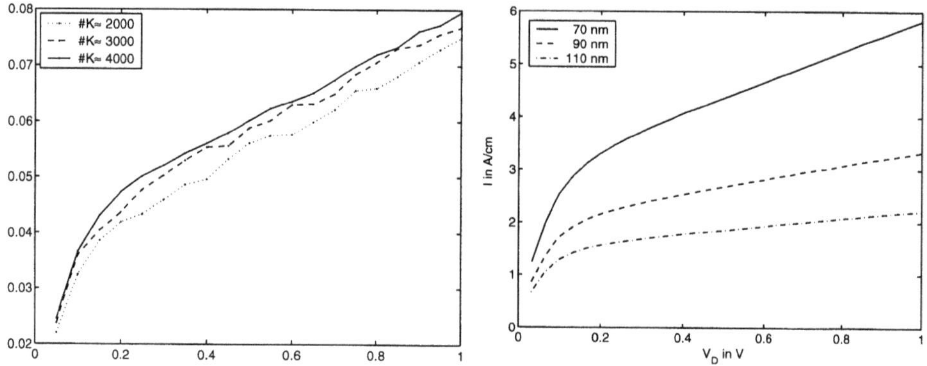

FIGURE 4. Drain current depending on the drain voltage for different meshes with $V_{GB} = 1\text{V}$ (left) and for different channel lengths (right).

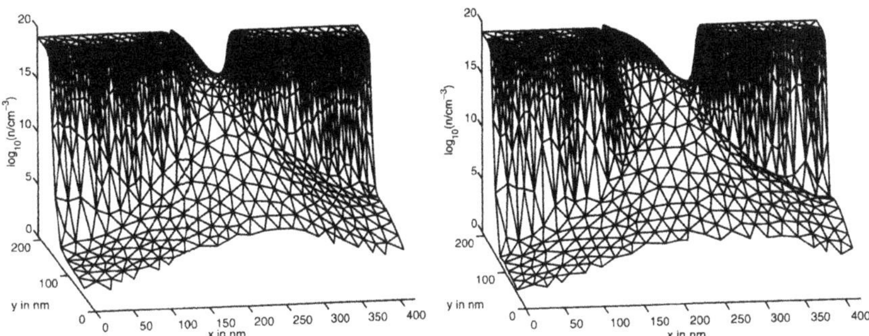

FIGURE 5. Electron density in a MOSFET with 70nm channel length (left) and 110nm channel length (right). Notice that a part of the bulk region is not shown.

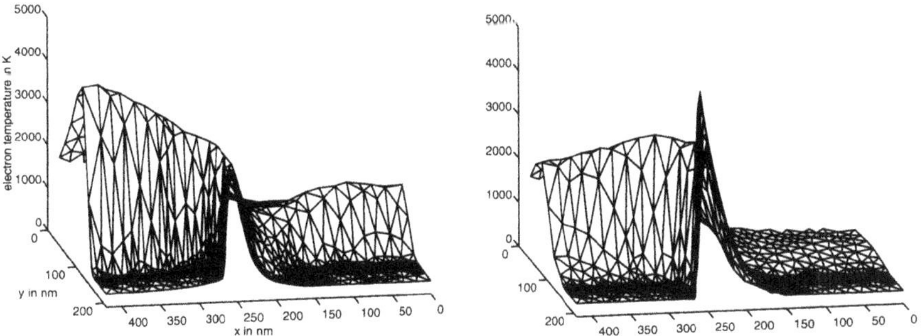

FIGURE 6. Electron temperature in a MOSFET with 110nm channel length in an open state ($V_{GB} = 1$V, left) and in a closed state ($V_{GB} = 0$V, right).

References

[1] Y. Apanovich, P. Blakey, R. Cottle, E. Lyumkis, B. Polsky, A. Shur, and A. Tcherniaev. Numerical simulations of submicrometer devices including coupled nonlocal transport and nonisothermal effects. *IEEE Trans. Electr. Dev.*, 42:890–897, 1995.

[2] N. Ben Abdallah and P. Degond. On a hierarchy of macroscopic models for semiconductors. *J. Math. Phys.*, 37:3308–3333, 1996.

[3] F. Bosisio, R. Sacco, F. Saleri, and E. Gatti. Exponentially fitted mixed finite volumes for energy balance models in semiconductor device simulation. In H. Bock et al., editors, *Proceedings of ENUMATH 97*, pages 188–197, Singapore, 1998. World Scientific.

[4] K. Brennan. *The Physics of Semiconductors.* Cambridge University Press, Cambridge, 1999.

[5] F. Brezzi, L. Marini, S. Micheletti, P. Pietra, R. Sacco, and S. Wang. Discretization of semiconductor device problems. In W. Schilders and E. ter Maten, editors, *Handbook of Numerical Analysis. Numerical Methods for Electrodynamical Problems*, to appear, 2002. North-Holland-Elsevier Science Publishers.

[6] F. Brezzi, L. Marini, and P. Pietra. Numerical simulation of semiconductor devices. *Comp. Meth. Appl. Mech. Engrg.*, 75:493–514, 1989.

[7] F. Brezzi, L. Marini, and P. Pietra. Two-dimensional exponential fitting and applications to drift-diffusion models. *SIAM J. Num. Anal.*, 26:1342–1355, 1989.

[8] E. Cassan, S. Galdin, P. Dollfus, and P. Hesto. Comparison between device simulators for gate current calculation in ultra-thin gate oxide n-MOSFETs. *IEEE Trans. Elect. Dev.*, 83:1194–1202, 2000.

[9] D. Chen, E. Kan, U. Ravaioli, C. Shu, and R. Dutton. An improved energy transport model including nonparabolicity and non-Maxwellian distribution effects. *IEEE Electr. Dev. Letters*, 13:26–28, 1992.

[10] M. Crouzeix and P.A. Raviart. Conforming and nonconforming finite element methods for solving the stationary Stokes equation. *RAIRO*, 7:33–76, 1973.

[11] P. Degond, S. Génieys, and A. Jüngel. A system of parabolic equations in nonequilibrium thermodynamics including thermal and electrical effects. *J. Math. Pures Appl.*, 76:991–1015, 1997.

[12] P. Degond, S. Génieys, and A. Jüngel. A steady-state system in nonequilibrium thermodynamics including thermal and electrical effects. *Math. Meth. Appl. Sci.*, 21:1399–1413, 1998.

[13] P. Degond, A. Jüngel, and P. Pietra. Numerical discretization of energy-transport model for semiconductors with non-parabolic band structure. *SIAM J. Sci. Comp.*, 22:986–1007, 2000.

[14] W. Fang and K. Ito. Existence of stationary solutions to an energy drift-diffusion model for semiconductor devices. *Math. Models Meth. Appl. Sci.*, 11:827–840, 2001.

[15] A. Forghieri, R. Guerrieri, P. Ciampolini, A. Gnudi, M. Rudan, and G. Baccarani. A new discretization strategy of the semiconductor equations comprising momentum and energy balance. *IEEE Trans. Comp. Aided Design Integr. Circuits Sys.*, 7:231–242, 1988.

[16] M. Fournié. Numerical discretization of energy-transport model for semiconductors using high-order compact schemes. *Appl. Math. Lett.*, 15:727–734, 2002.

[17] J. Griepentrog. An application of the implicit function theorem to an energy model of the semiconductor theory. *Z. Angew. Math. Mech.*, 79:43–51, 1999.

[18] S. Holst, A. Jüngel, and P. Pietra. A mixed finite-element discretization of the energy-transport model for semiconductors. To appear in *SIAM J. Sci. Comp.*, 2002.

[19] S. Holst, A. Jüngel, and P. Pietra. An adaptive mixed scheme for energy-transport simulations of field-effect transistors. Submitted for publication, 2002.

[20] R. Hoppe and B. Wohlmuth. A comparison of a posteriori error estimators for mixed finite element discretization by Raviart-Thomas elements. *Math. Comp.*, 68:1347–1378.

[21] J. Jerome. *Analysis of charge transport. A mathematical study of semiconductor devices.* Springer, Berlin, 1996.

[22] J. Jerome and C.-W. Shu. Energy models for one-carrier transport in semiconductor devices. In W. Coughran et al., editors, *Semiconductors, Part II*, volume 59 of *IMA Volumes in Math. and Its Appl.*, pages 185–207, New York, 1994. Springer.

[23] A. Jüngel. *Quasi-hydrodynamic Semiconductor Equations.* Birkhäuser, Basel, 2001.

[24] E. Kane. Band structure of indium-antimonide. *J. Phys. Chem. Solids*, 1:249–261, 1957.

[25] C. Lab and P. Caussignac. An energy-transport model for semiconductor heterostructure devices: application to AlGaAs/GaAs MODFETs. *COMPEL*, 18:61–76, 1999.

[26] L.D. Marini and P. Pietra. An abstract theory for mixed approximations of second order elliptic equations. *Mat. Aplic. Comp.*, 8:219–239, 1989.

[27] L.D. Marini and P. Pietra. New mixed finite element schemes for current continuity equations. *COMPEL*, 9:257–268, 1990.

[28] P.A. Markowich, C.A. Ringhofer, and C. Schmeiser. *Semiconductor Equations.* Springer, Vienna, 1990.

[29] A. Marrocco and P. Montarnal. Simulation des modèles energy-transport à l'aide des éléments finis mixtes. *C. R. Acad. Sci. Paris*, 323:535–541, 1996.

[30] A. Marrocco, P. Montarnal, and B. Perthame. Simulation of the energy-transport and simplified hydrodynamic models for semiconductor devices using mixed finite elements. In *Proceedings ECCOMAS 96*, London, 1996. John Wiley.

[31] P. Raviart and J. Thomas. A mixed finite element method for second order elliptic equations. In *Mathematical Aspects of the Finite Element Method*, volume 606 of *Lecture Notes in Math.*, pages 292–315. Springer, 1977.

[32] C. Ringhofer. An entropy-based finite difference method for the energy transport system. *Math. Models Meth. Appl. Sci.*, 11:769–796, 2001.

[33] M. Rudan, A. Gnudi, and W. Quade. A generalized approach to the hydrodynamic model of semiconductor equations. In G. Baccarani, editor, *Process and Device Modeling for Microelectronics*, Amsterdam, 1993. Elsevier.

[34] C. Schmeiser and A. Zwirchmayr. Elastic and drift-diffusion limits of electron-phonon interaction in semiconductors. *Math. Models Meth. Appl. Sci.*, 8:37–53, 1998.

[35] K. Seeger. *Semiconductor Physics. An Introduction.* 6th edition, Springer, Berlin, 1997.

[36] K. Souissi, F. Odeh, H. Tang, and A. Gnudi. Comparative studies of hydrodynamic and energy transport models. *COMPEL*, 13:439–453, 1994.

[37] R. Stratton. Diffusion of hot and cold electrons in semiconductor barriers. *Phys. Rev.*, 126:2002–2014, 1962.

[38] P. Visocky. A method for transient semiconductor device simulation using hot-electron transport equations. In J. Miller, editor, *Proc. of the Nasecode X Conf.*, Dublin, 1994. Boole Press.

[39] B. Wohlmuth. A residual based error estimator for mortar finite element discretizations. *Num. Math.*, 84:143–171, 1999.

S. Holst and A. Jüngel:
Fachbereich Mathematik und Informatik
Universität Mainz
Staudingerweg 9
55099 Mainz, Germany
email: {holst,juengel}@mathematik.uni-mainz.de

P. Pietra:
Istituto di Matematica
Applicata e Tecnologie Informatiche, C.N.R.
Via Ferrata 1
27100 Pavia, Italy
email: pietra@dragon.imati.cnr.it

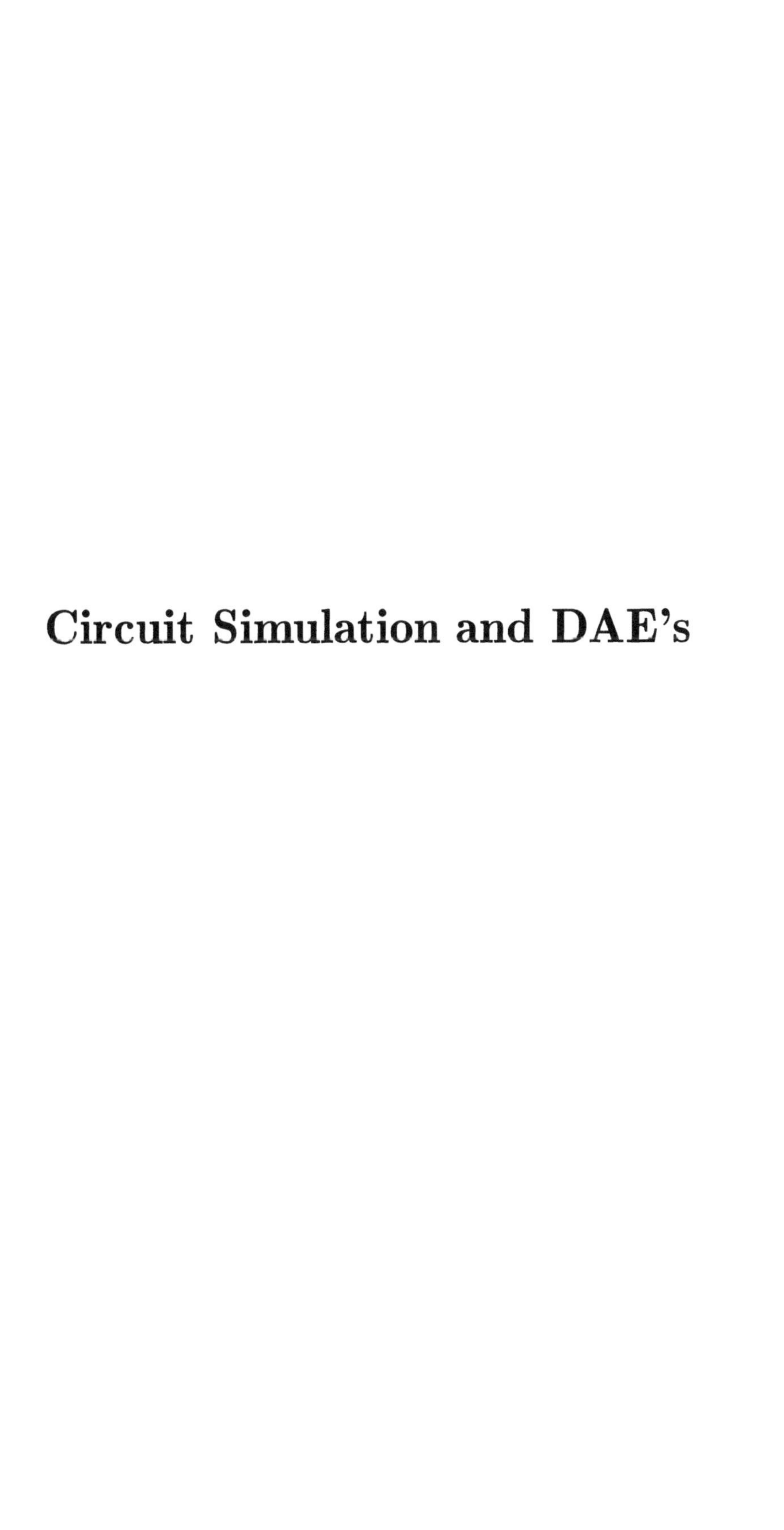

Circuit Simulation and DAE's

Modeling, Simulation and Optimization of Integrated Circuits
(Eds.) K. Antreich, R. Bulirsch, A. Gilg and P. Rentrop

International Series of Numerical Mathematics Vol. 146, 67–81

An Embedding Approach for the Simulation of Electronic Circuits with Widely Seperated Time Scales

Hans Georg Brachtendorf, Angelika Bunse-Gerstner, Barbara Lang, Rainer Laur

Abstract. Widely separated time scales in electronic circuits occur in many cases, especially in radio frequency circuits, making analysis with standard numerical methods very difficult and costly. Low and high frequency parts of the solution are often superimposed enforcing very tiny time steps over a long period in the computation of the numerical solution. In this paper we present a general method of embedding the differential-algebraic equations (DAEs) which describe the electronic circuit in a system of partial differential equations (PDEs), such that a restriction of the solution of the PDEs onto a suitable path yields the desired solution of the DAEs. This allows to treat such different frequency-parts seperately in different dimensions. In many cases the physical background provides knowledge about the frequencies and the solution of the PDEs is periodic in all or some dimensions with often known periodicity lengths. The solution of the PDEs then needs only to be computed on its basic periodicity domain, where it is often a very slowly varying function. Therefore computing the solution of the PDEs can be much more efficient. Here theoretical results are presented as well as new approaches to numerical methods, based on the embedding method.

1. Introduction

An electronic circuit is simulated by solving a system of differential algebraic equations, which is derived via the modified nodal analysis from Kirchhoff's laws. The system is of the form

$$\frac{d}{dt}q(x(t)) + f(x(t)) + b(t) \quad = \quad 0, \tag{1.1}$$

with node voltages and branch currents of impedant elements $x : \mathbb{R} \to \mathbb{R}^n$, charge and flux terms of impedant elements $q : \mathbb{R}^n \to \mathbb{R}^n$, static currents $f : \mathbb{R}^n \to \mathbb{R}^n$

2000 *Mathematics Subject Classification.* Primary 65L80; Secondary 94C99.
Key words and phrases. circuit simulation, widely separated time scales, numerical methods, embedding method.

and input signals $b : \mathbb{R} \to \mathbb{R}^n$, where the stimulus $b(t)$ is constant for autonomous circuits and time-dependent otherwise. Because the derivative of q is in general not invertible, there are not only dynamical but also algebraic conditions and the system is a proper system of DAEs.

Electronic circuits often have solutions consisting of oscillations with widely separated time scales. Mixers for example have solutions which are superpositions of fast and slow oscillations which require very tiny stepsizes over a long period of time when calculating the solution numerically. To give an simple highly oscillating example we take the ring modulator described in [9] and choose artificially extreme input signals $U_{\text{in}1}(t) = 0.5\sin(2000\pi t)$ and $U_{\text{in}2}(t) = 2\sin(200000\pi t)$. The frequencies differ by a factor of 100 and the solutions are superimposed oscillations, an example for one node voltage being given in Figure 1. Assuming that the solution is of the form

$$x(t) = \sum_{z \in \mathbb{Z}^2} X_z \; e^{i(z_1\omega_1 t + z_2\omega_2 t)},$$

with $X_z \in \mathbb{R}^n$ and $\omega_1 = 2000\pi, \omega_2 = 200000\pi$, we can associate an n-dimensional function in two variables of the form

$$\hat{x}(\tau_1, \tau_2) = \sum_{z \in \mathbb{Z}^2} X_z \; e^{i(z_1\tau_2 + z_2\tau_2)}.$$

This function is $2\pi-$periodic in both dimensions. If we know $\hat{x}(\tau_1, \tau_2)$ on $[0, 2\pi] \times [0, 2\pi]$ we know it on all of $\mathbb{R}^2$. Given the path $\gamma(t) := (\omega_1 t, \omega_2 t)$ in $\mathbb{R}^2$ we see that

$$\hat{x}(\gamma(t)) = x(t).$$

Note that $\gamma(t)$ has a very steep slope and is close to the τ_2-axis. If we had $\hat{x}(\tau_1, \tau_2)$ on $[0, 2\pi] \times [0, 2\pi]$ we could easily retrieve $x(t)$. In this situation $\hat{x}(\tau_1, \tau_2)$ is a slowly varying function on $[0, 2\pi] \times [0, 2\pi]$ and can be approximated much easier than $x(t)$ on a sufficiently large time intervall.

Switching to a multivariate function in this way we seperate the two time scales. In Section 3 we will see in a much more general situation that a special system of partial differential equations can be associated to the DAEs such that the PDE solution is $\hat{x}(\tau_1, \tau_2)$. Figure 2 shows the part of the PDE solution which corresponds in this way to the node voltage in Figure 1.

With the principle of seperating time scales as in our example solutions with several superimposed oscillations or oscillating solutions modulated by an envelope can be considered as well. Behavior of this kind can be described by somewhat more general functions than considered above. A function showing a qualitative behavior like some transient responses of oscillators is given in Figure 3. We need at least 750 sampling points to represent the function's behavior. The associated two-dimensional function is given in Figure 4. The latter is periodic in only one direction, τ_2. It is very slowly varying on $\mathbb{R} \times [0, 2\pi]$ and even though it is a two-dimensional function its behavior can be represented by much less sampling points than the very heavily oscillating original one-dimensional function.

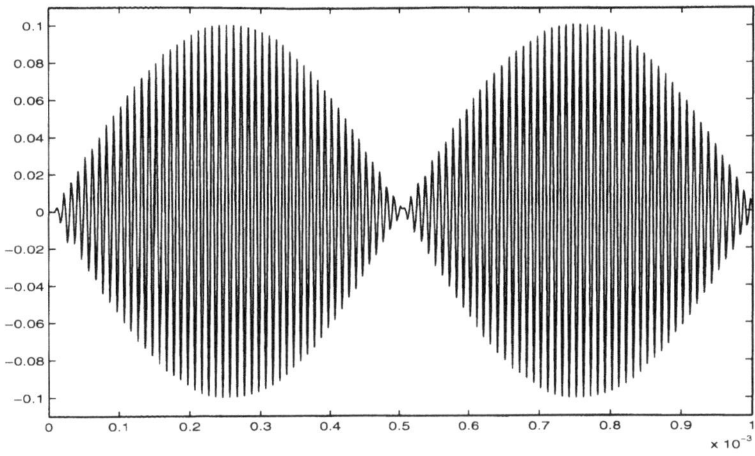

FIGURE 1. Reference solution of one node voltage, ring modulator

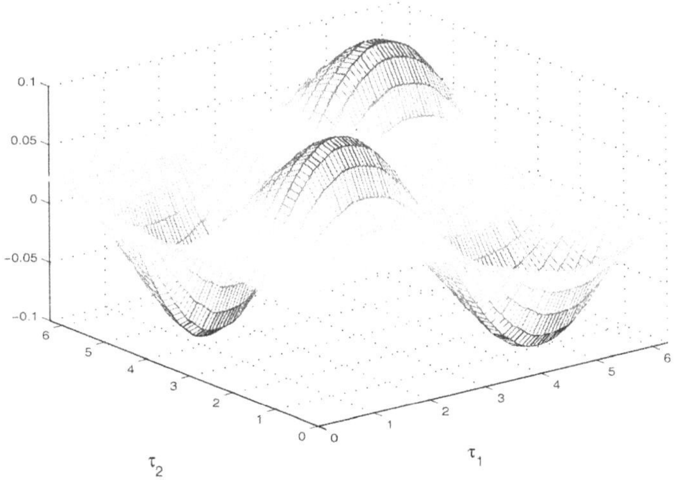

FIGURE 2. Solution as a multivariable function

The system of DAEs can be considered as being embedded in the associated system of PDEs. This embedding idea first occured in [8], where it was used to give an meaningful interpretation for a method in which a two-dimensional Fourier transformation was used in a harmonic balance method for multi-tone systems. Since then various embedding methods have been presented in the literature, considering these embeddings as a basis for new numerical methods for the simulation of the circuits (see [2], [7], [15], [4], [12], [5], [6], [14] [16], [3], [13], [10]).

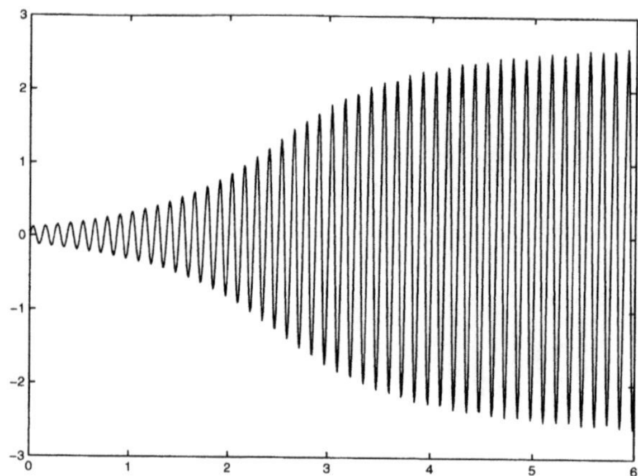

FIGURE 3. $x(t) = (1.3 + \arctan(t - 2.5))\sin((50 + \frac{1}{t+1})t)$ on $[0, 6]$. Plot with 750 sampling points

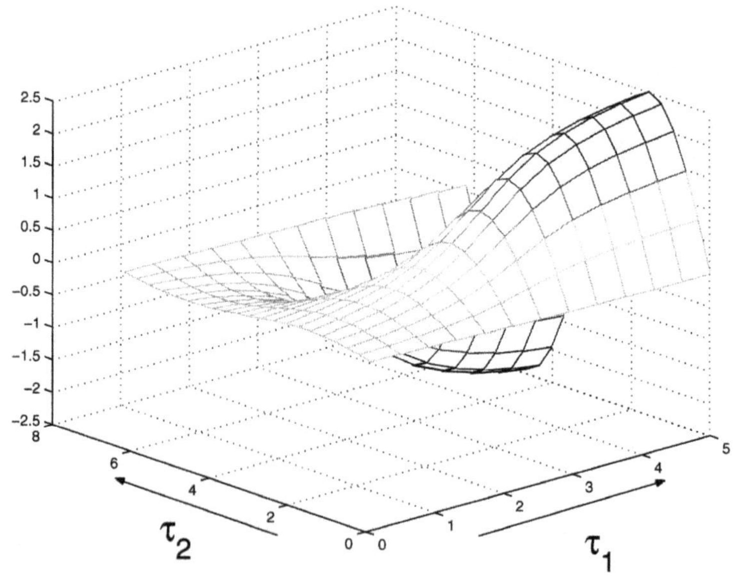

FIGURE 4. $\hat{x}(\tau_1, \tau_2) = (1.3 + \arctan(\tau_1 - 2.5)) \cdot \sin(\tau_2)$ on $[0, 6] \times [0, 2\pi]$. Plot with 150 sampling points

Essentially these numerical methods are “multi-tone harmonic balance” methods and standard methods to solve the PDEs directly.

These embedding methods differ in the kind of partial differential equation, the path of evaluation to retrieve the DAE solution and the considered domain.

In this paper we will derive a general embedding method under which all these approaches can be unified. The relation between the DAEs and the PDEs can be used to develop new numerical methods and to give new view on existing methods.

In the following generalized quasiperiodic functions are introduced, the embedding method is described and theoretical results are given. Applications of this embedding method are presented and numerical algorithms are considered. A detailed derivation can be found in [10].

2. Generalized quasiperiodic functions

Consider functions $x : \mathbb{R} \to \mathbb{R}^n$ of the form

$$x(t) = \sum_{z \in \mathbb{Z}^m} X_z(t) \, e^{i(\sum_{j=1}^{m} z_j \omega_j(t))t} \tag{2.1}$$

where $X_z(t) : \mathbb{R} \to \mathbb{R}^n$ and $\omega = (\omega_1, \omega_2, \ldots, \omega_m)$ with $\omega : \mathbb{R} \to \mathbb{R}^m$.

We call them **generalized quasiperiodic functions**. If $X_z(t)$ and $\omega_j(t)$ do not depend on t, we get the well-known quasiperiodic functions. Throughout this paper these sums are assumed to be finite.

For each generalized quasiperiodic function x from (2.1) and each given interval $I \subseteq \mathbb{R}$ we can define a path γ in $\mathbb{R}^m$:

$$\gamma(t) := (\omega_1(t)t, \omega_2(t)t, \ldots, \omega_m(t)t) : I \to \mathbb{R}^m,$$

where $\omega = (\omega_1, \omega_2, \ldots, \omega_m)$ is from $x's$ representation (2.1).

To each generalized quasiperiodic function x we associate a function $\hat{x}$ with multi-dimensional domain of the following form:

$$\hat{x}(\tau) = \sum_{z \in \mathbb{Z}^m} \hat{X}_z(\tau) \, e^{i(\sum_{j=1}^{m} z_j \tau_j)},$$

where $\tau = (\tau_1, \tau_2, \ldots, \tau_m) \in \mathbb{R}^m$ and the $\hat{X}_z(\tau)$ satisfy

$$\hat{X}_z(\gamma(t)) = X_z(t) \quad \text{for all} \quad z \in \mathbb{Z}^m. \tag{2.2}$$

Then $\hat{x}(\gamma(t)) = x(t)$. We call such a function **multivariate function associated with the generalized quasiperiodic function x(t).** Because different functions $\hat{X}_z$ could satisfy property (2.2) above, an associated multivariate function is in general not unique. It depends on the choice of $\hat{X}_z$.

Note, that if the $\hat{X}_z$ and ω_j are constant the associated multivariate function is 2π-periodic in each coordinate by construction.

3. Embedding method and theoretical investigations

For the circuit system (1.1) we define a corresponding embedding system, a system of partial differential equations:

$$\sum_{j=1}^{m} \alpha_j(\tau) \frac{\partial}{\partial \tau_j} q(\hat{x}(\tau)) + f(\hat{x}(\tau)) + \hat{b}(\tau) = 0, \tag{3.1}$$

where $\hat{x} : \Omega \to \mathbb{R}^n$ ($\Omega \subseteq \mathbb{R}^m$, Ω open and connected), $\tau = (\tau_1, \dots, \tau_m) \in \Omega$, $\alpha_j : \Omega \to \mathbb{R}, \alpha_j(\tau) > 0$, $m \in \mathbb{N}$. In practice we have in most cases $m \leq 3$. q and f are directly from (1.1), but $\hat{b}$ must be chosen such that if evaluated along a certain path

$$\gamma : I \to \bar{\Omega} : t \mapsto \gamma(t) = (\gamma_1(t), \gamma_2(t), \dots, \gamma_m(t)) \tag{3.2}$$

it coincides with b ($I \subset \mathbb{R}$ interval with $0 \in I$), i.e., $\hat{b}(\gamma(t)) = b(t)$. The path γ is the solution of the initial value problem

$$\dot{\gamma}_j(t) = \alpha_j(\gamma(t)), \quad \text{for all} \quad j \in \{1, \dots, m\}, \tag{3.3}$$

$$\gamma(0) = \gamma_0 \in \bar{\Omega}. \tag{3.4}$$

Equation (3.1) is a system of quasilinear partial differential equations of first order.

In this situation, the following can be shown:

Theorem 3.1. *Consider the differential equation system* (1.1). *Let $\hat{x}$ be a solution of the corresponding embedding system* (3.1) *constructed above and let γ be a path which satisfies conditions* (3.3) *and* (3.4).

Then $x := \hat{x} \circ \gamma$ is a solution of system (1.1).

Proof. Simple insertion provides the result. □

In special situations the reverse can be shown using the theory of characteristics. The principle idea is to decouple system (1.1). This is possible if Dq, the Jacobian of q, is invertible. This approach was first presented for a special system of partial differential equations in [14]. It is very likely that this decoupling idea can be modified to handle problems of index 1 as well, but this has not been proven yet.

We get the following theorem. The mathematically strict formulation of the theorem and its proof demand for a lot of notation. We therefore present here only the main ideas. For details see [10].

Theorem 3.2. *Consider the system of partial differential equations* (3.1) *and a non-characteristic initial surface Σ and let $\underline{\Sigma}$ be it's projection on the τ-plane with $\underline{\Sigma} \subset \Omega$.*

Then there exists a special domain where the solution can be assembled from solutions of a suitable family of systems of ordinary differential equations of the form (1.1). *The initial conditions for these systems of DAEs are given by Σ.*

Proof. For each $\sigma \in \underline{\Sigma}$ we construct characteristics pathes $\gamma_\sigma \in \Omega$ such that

$$\gamma_\sigma(0) = \sigma \quad \text{and} \tag{3.5}$$

$$\dot{\gamma}_\sigma(t) = \alpha(\gamma_\sigma(t)). \tag{3.6}$$

For each σ we consider a systems of DAEs of the form (1.1). f and q are given by (3.1) but b depends on σ :

$$b_\sigma = \hat{b} \circ \gamma_\sigma.$$

Then using the theory of characteristics, the solution of the system of partial differential equations can be assembled from the solutions of the family of solutions of the systems of ordinary differential equations on a (possibly very small) domain. □

We point out that in [3] a similar approach is aimed for in a special case using quasiperodic functions.

Up to now the solution of the PDE can still break down by blow ups of the solutions of the DAEs or by crossing of the characteristics. But the next theorem shows, that the latter can be excluded.

Theorem 3.3. *The characteristics do not intersect.*

Proof. Let $\hat{x}$ be a solution of the system of partial differential equations (3.1), Σ a non-characteristic surface and $\underline{\Sigma}$ its projection on the τ-plane. Let γ_a and γ_b be characteristics, with $\gamma_a(0), \gamma_b(0) \in \underline{\Sigma}$ and $\gamma_a(0) \neq \gamma_b(0)$. Suppose they intersect and let t_a and t_b be the smallest positiv real numbers with $\gamma_a(t_a) = \gamma_b(t_b)$. The situation is described in Figure 5.

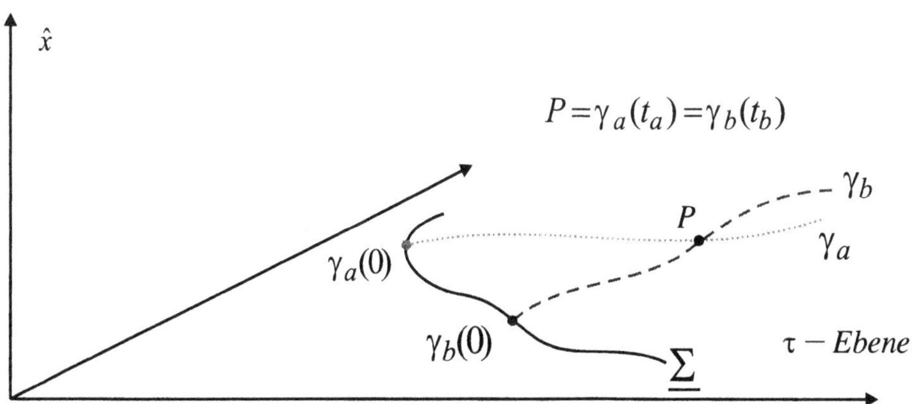

FIGURE 5. Crossing characteristics

Now let

$$\delta(s) := s - t_a + t_b$$

and

$$\tilde{\gamma} \quad := \gamma_b \circ \delta.$$

Then $\tilde{\gamma}$ is a solution of the differential equation (3.2) and (3.3), because

$$\begin{aligned} \frac{d}{ds}\tilde{\gamma}(s) &= \delta'(s)\gamma_b(\delta(s)) \\ &= \dot{\gamma}(\delta(s)) \\ &= (\alpha_1(\delta(s)), \alpha_2(\delta(s)), \dots, \alpha_m(\delta(s)), 0). \end{aligned}$$

Furthermore

$$\tilde{\gamma}(t_a) = \gamma_b(t_a - t_a + t_b) = \gamma_b(t_b) = \gamma_a(t_a).$$

Consider the system of differential equations (3.2) and (3.3) with initial value

$$\gamma(t_a) = \gamma_a(t_a)$$

for $t \leq t_a$. Because $\tilde{\gamma}$ and γ_a are solutions, there exists according to the Picard-Lindelöff Theorem an interval

$$[t_a - \varepsilon, t_a] \subseteq \mathbb{R},$$

where these solutions coincide. Without loss of generality let ε be such that

$$t_a - \varepsilon \quad > \quad 0, \tag{3.7}$$

$$\varepsilon \quad < t_b. \tag{3.8}$$

Let $\tilde{t}_a \in (t_a - \varepsilon, t_a)$, then

$$\tilde{\gamma}(\tilde{t}_a) \quad = \quad \gamma_a(\tilde{t}_a),$$

but also

$$\tilde{\gamma}(\tilde{t}_a) \quad = \quad \gamma_b(\delta(\tilde{t}_a)).$$

Defining $\tilde{t}_b := \delta(\tilde{t}_a)$, we get with the inequalities (3.7) and (3.8):

$$0 < t_b - \varepsilon = t_a - \varepsilon - t_a + t_b < \underbrace{\tilde{t}_a - t_a + t_b}_{=\tilde{t}_b} < t_b.$$

Because of

$$\gamma_b(\tilde{t}_b) = \tilde{\gamma}(\tilde{t}_a) = \gamma_a(\tilde{t}_a),$$

$\tilde{t}_a$ and $\tilde{t}_b$ are also a pair of values where γ_a and γ_b coincide. Because of

$$0 \leq \tilde{t}_a < t_a$$
$$0 \leq \tilde{t}_b < t_b$$

this contradicts our choice of the pair t_a und t_b. □

4. Applications

Various embedding methods have been presented in the literature (see [2], [7], [15], [4], [12], [5], [6], [14], [16], [3], [13], [10]). They differ in the partial differential equation, the path of evaluation and the considered domain. The embedding method presented above allows a general derivation and provides a unified theoretical framework for these approaches.

If we specify the coefficient functions α_j, $j = 1, \ldots, m$, the path of evaluation is determined. Together with the specification of the domain of the PDEs various circuit simulation problems can be treated. These are, e.g., steady states of non-autonomous and mixed autonomous/non-autonomous circuits with quasi periodic stimulus (see [1],[2],[3],[8],[12],[16]). Also transient behavior (see [11],[15],[16]) as well as steady state analysis for autonomous circuits with quasi periodic steady states can be treated. Furthermore one can choose an embedding system to analyze the transient behavior of high Q oscillators (see [2],[4]). In [3] also an embedding system for the studying "injection locking" is presented.

Several numerical methods are considered in the literature as well. Essentially these are "multi-tone harmonic balance" and standard methods to solve the PDE directly.

5. Numerical methods

The embedding can be used to derive a new approach to solve the DAE numerically which can roughly be described as follows. The embedding PDE is only solved implicitly by a modified method of characteristics, working directly on the DAE and computing approximations of only every k-th oscillation. This general idea can be used for various problem classes.

The following algorithm, which is a modification of the method of characteristics presented in [12], calculates the steady state solution of electric circuits with two superimposed oscillations. It is based on the embedding method and works on the circuit equation directly. The system of partial differential equations provides periodicity conditions on time intervals where the circuit equation is evaluated. The frequencies are first assumed to be known. Then an extension to the case of unknown frequencies is discussed.

We consider steady state solutions of the form

$$x(t) = \sum_{z \in \mathbb{Z}^2} X_z \, e^{i(z_1\omega_1 + z_2\omega_2)t}, \tag{5.1}$$

with $\omega_2 \gg \omega_1$ both known. The solution is embedded in the 2π-biperiodic function

$$\hat{x}(\tau) = \sum_{z \in \mathbb{Z}^2} \hat{X}_z \, e^{i(z_1\tau_1 + z_2\tau_2)}, \tag{5.2}$$

which returns x when evaluated on the path $\gamma(t) = (\omega_1 t, \omega_2 t)$. The circuit is described by the differential-algebraic system (1.1) which corresponds to the system of partial differential equations

$$\omega_1 \frac{\partial}{\partial \tau_1} q(\hat{x}(\tau)) + \omega_2 \frac{\partial}{\partial \tau_2} q(\hat{x}(\tau)) + f(\hat{x}(\tau)) + \hat{b}(\tau) = 0. \tag{5.3}$$

The PDE is considered on the domain $[0, 2\pi) \times [0, 2\pi)$ with periodic boundary conditions. The evaluation path γ is projected on this square, see Figure 6. For the algorithm we select every k-th of these projected lines (Figure 7). The PDE-solution restricted to the selected lines is exactly the DAE-solution on corresponding time intervals (Figure 8). These parts of the DAE-solution are here computed by a multiple shooting method (any other numerical method may be used as well). This, together with the PDE's periodicity conditions, yields the nonlinear system of equations to be solved.

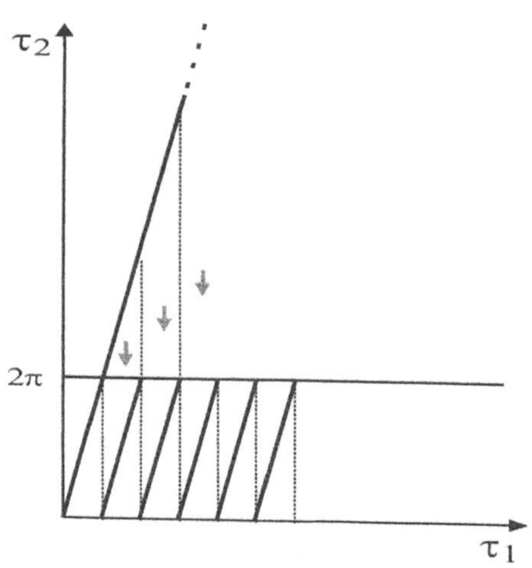

FIGURE 6. Projection of the characteristic γ

As an example we tested the algorithm on the ring modulator example mentioned in the introduction with frequencies $\omega_1 = 2000\pi$ and $\omega_2 = 200000\pi$. Figure 9 shows the result of the algorithm after one Newton-step. As initial values we chose a 10% perturbation of the steady state solution in 40 points. Again, the calculated oscillations are marked. The oscillations in between are obtained by linear interpolation and therefore the oscillation peaks are piecewise linear which could be improved without significant increase of calculations, e.g., by quadratic interpolation. Figure 10 illustrates the computed result as the corresponding multivariate function.

The method described above can be expanded to the case of unknown frequencies. This problem occurs for autonomous circuits. The frequencies are now treated as additional unknowns. Because the characteristics depend on the frequencies, it is necessary to adjust them at certain states and the algorithm must be embedded in an outer iteration loop. As the frequencies add to the number

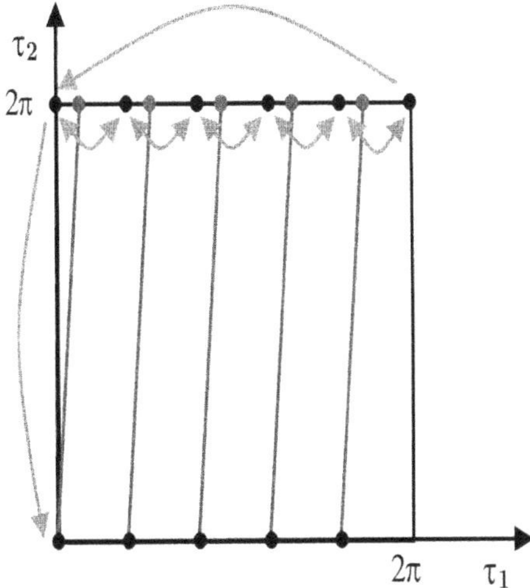

FIGURE 7. Characteristics and periodic conditions

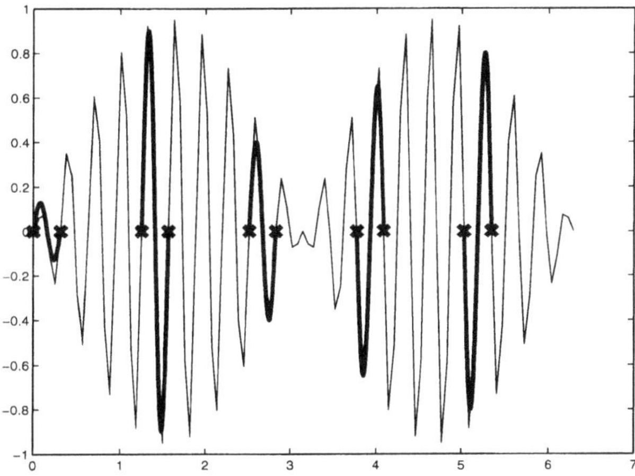

FIGURE 8. Solution: Marked oscillations correspond to the characteristics which are used for calculation

of unknowns, two more equations are needed. The value of one component of the solution can be fixed at a certain time and/or the derivative of some components can be set to zero. The advantage of the first choice is, that the zero-solution is excluded (see [13]). However it cannot be guaranteed that the fixed value can be reached by the solution. Another possibility is to choose a calibration condition of

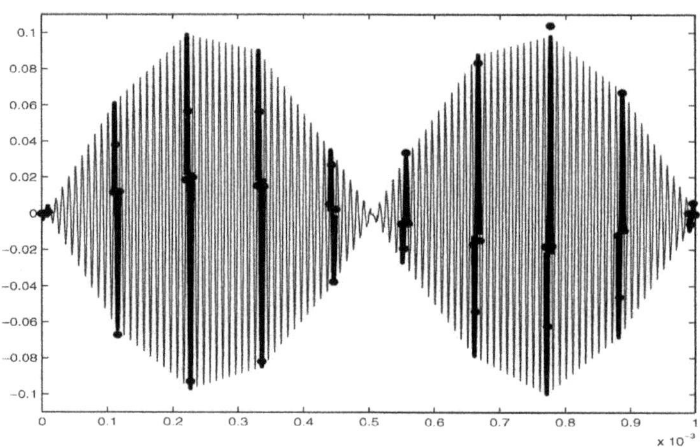

FIGURE 9. Solution by method of characteristics

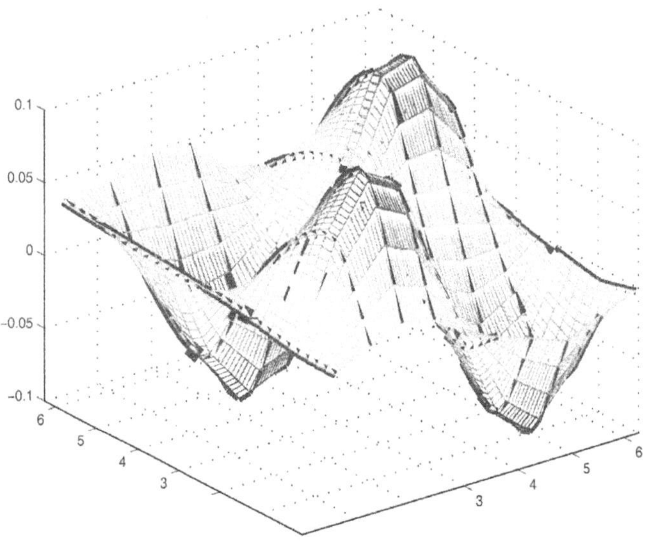

FIGURE 10. Solution as a multivariate function

the form

$$\frac{1}{T}\int_0^T x(t)\cos(\frac{2\pi}{T}t)dt = 0 \tag{5.4}$$

for each parameter, which has been proposed in [3] in a slightly different situation.

After each Newton-step it has to be checked that the current frequencies are still admissible. Otherwise a next step of the outer loop must be started.

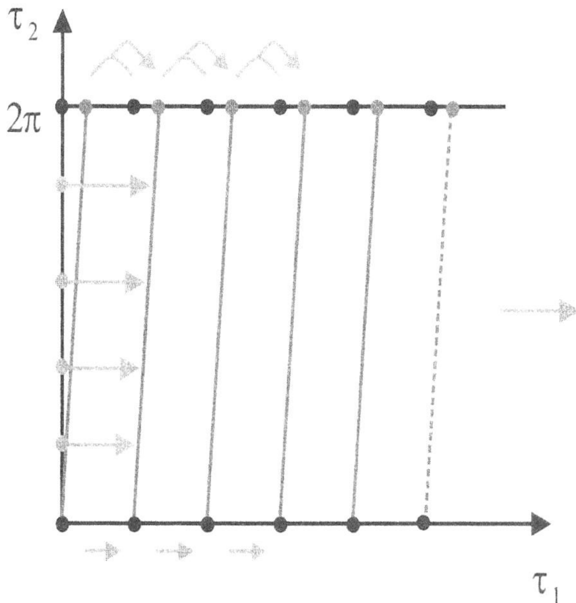

FIGURE 11. Characteristics and periodicity in τ_2-direction

Analogously an algorithm for the transient analysis of oscillators can be developed. Here the principal idea is presented and difficulties are discussed.

The start-up behavior of oscillators is described by the DAE (1.1) together with the initial condition $x(0) = x_0$. It is assumed that the solution is a fast oscillation which is modulated by a slowly varying function. We use the ansatz

$$x(t) = \sum_{z \in \mathbb{Z}^m} X_z(t)\, e^{i(\sum_{j=1}^{m} z_j \omega_j)t}, \tag{5.5}$$

where $m = 2$, $z = (0, z_2)$ and ω_2 is known. The embedding system is

$$\omega_1 \frac{\partial}{\partial \tau_1} q(\hat{x}) + \omega_2 \frac{\partial}{\partial \tau_2} q(\hat{x}) + f(\hat{x}) + \hat{b} = 0 \quad \text{on} \quad [0, 2\pi) \times \mathbb{R}_+ . \tag{5.6}$$

The solution is assumed to be 2π-periodic in τ_2. This yields the boundary condition $\hat{x}(\tau_1, 0) = \hat{x}(\tau_1, 2\pi)$. We have the initial condition $\hat{x}(0,0) = \hat{x}_0(0) = x_0$ from the DAE. But in addition we need initial values on $\hat{x}(0, \tau_2) = \hat{x}_0(\tau_2)$ on $[0, 2\pi)$ to solve the PDE numerically.

With numerical algorithms proceeding in τ_1-direction approximate values on the k-th characteristic can be obtained. (Multiple) shooting methods can correct these extrapolated values. This is then continued in the τ_1-direction. This approach is illustrated in Figure 11 and Figure 12.

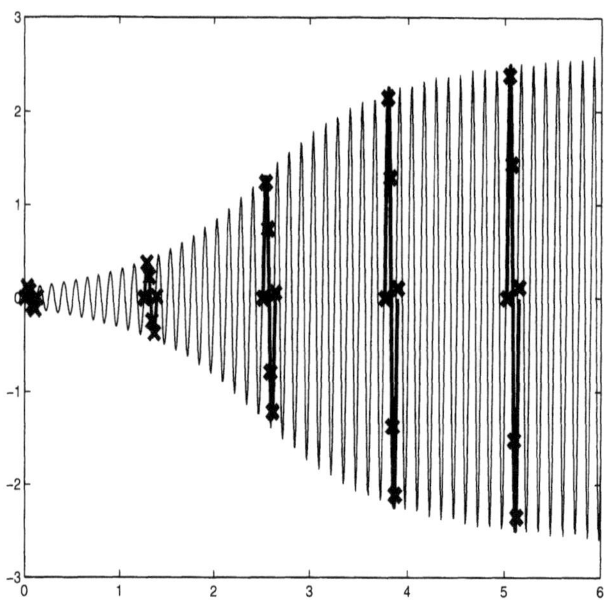

FIGURE 12. Solution: Marked oscillations correspond to the characteristics which are used for calculation

This method has still some problems. First of all, different choices of initial values on the τ_2-axis lead to the same solution of the DAE. It is important for numerical stability and efficiency to choose initial conditions such that the solution is smooth. Extrapolation methods to proceed in τ_1-direction are not efficient. BDF methods for the PDE are highly damping which is unsuitable for solutions with increasing oscillations (see [3]). When correcting with the multiple shooting method, different solutions are possible. Again further calibration conditions have to be used.

References

[1] H.G. Brachtendorf. *Simulation des eingeschwungenen Verhaltens elektronischer Schaltungen.* Shaker, Aachen, 1994.

[2] H.G. Brachtendorf. *On the relation of certain classes of ordinary differential algebraic equations with partial differential algebraic equations.* Technical Report 1131G0-791114-19TM, Bell-Laboratories, 1997.

[3] H.G. Brachtendorf. *Theorie und Analyse von autonomen und quasiperiodisch angeregten elektrischen Netzwerken.* Habilitationsschrift, Universität Bremen, 2001.

[4] H.G. Brachtendorf and R. Laur. *Transient simulation of oscillators.* Technical Report 1131GO-980410-09TM, Bell-Laboratories, 1998.

[5] H.G. Brachtendorf and R. Laur. Multi-rate PDE methods for high Q oscillators. In N. Mastorakis, editor, *Problems in Modern Applied Mathematics, World Scientific (2000)*, 391–398, Athen, July 2000. CSCC 2000, MCP 2000, MCME 2000 Multiconference.

[6] H.G. Brachtendorf and R. Laur. *On Consistent Initial Conditions for Circuit's DAEs with Higher Index.* Proc. IEEE Int. Symp. on Circuits and Systems, Genf, May 2000

[7] H.G. Brachtendorf, G. Welsch, and R. Laur, *A novel time-frequency method for the simulation of the steady state circuits driven by multitone signals.* ProcIEEESCS, June 1997.

[8] H.G. Brachtendorf, G. Welsch, R. Laur, and A. Bunse-Gerstner. *Numerical steady state analysis of electronic circuits driven by multi-tone signals.* Electronic Engineering, **79**(2):103–112, April 1996.

[9] W. Kampowski, P. Rentrop, and W. Schmidt. *Classification and numerical simulation of electric circuit.* Surveys on Mathematics for Industry, **2**(1):23–65, 1992.

[10] B. Lang. *Einbettungsverfahren für Netzwerkgleichungen.* Shaker-Verlag Aachen, 2002.

[11] E. Ngoya and R. Larcheváque. *Envelope transient analysis: A new method for the transient and steady state analysis of microwave communications circuit and systems.* In Proc. IEEE MTT-S Int. Microwave Symp., pages 1365–1368, San Francisco, 1996.

[12] R. Pulch. *Numerische Simulation von PDE-Modellen in der Analyse von RF-Schaltungen.* 1999.

[13] R. Pulch. *PDE techniques for finding quasi-periodic solutions of oscillators.* Preprint 09, IWRMM, Universität Karlsruhe, 2001.

[14] R. Pulch and M. Günther. *A method of characteristics for solving multirate partial differential equations in radio frequency application.* Preprint 07, IWRMM, Universität Karlsruhe, 2000.

[15] J. Roychowdhury. *Efficient methods for simulating highly nonlinear multi-rate circuits.* In Proc. IEEE Design Automation Conf., pages 269–274, 1997.

[16] J. Roychowdhury. *Analyzing circuits with widely separated time scales using numerical pde methods.* IEEE Transactions on Circuits and Systems I – Fundamental Theory and Applications, **48**:578–594, May 2001.

A. Bunse-Gerstner
Zentrum für Technomathematik
FB3 (Mathematik)
Universität Bremen
PF 33 04 40
28334 Bremen, Deutschland
email: `Bunse-Gerstner@math.uni-bremen.de`

Modeling, Simulation and Optimization of Integrated Circuits
(Eds.) K. Antreich, R. Bulirsch, A. Gilg and P. Rentrop

International Series of Numerical Mathematics Vol. 146, 83–99

Actual Problems of Circuit Simulation in Industry

Diana Estévez Schwarz and Uwe Feldmann

Abstract. In the first part of this article actual trends in microelectronics are highlighted. These give rise to identify some topics of mathematical research, which are desirable to be solved in order to cope with future circuit simulation problems.

Parts of this research – which are fairly beyond the scope of industrial activities – are carried out by university partners, and are reported in other contributions to this book. Other parts, however, can only be pursued in close contacts to applications, which makes them better suited for industrial research. As an example, we will focus in the second part of this paper on actual own work on diagnosis. We give an overview of attempts to provide user oriented diagnostics in case of faulty circuit models, problems with coupling structural and functional descriptions, and failure of numerical integration due to high DAE index.

1. Introduction

Classical circuit simulation plays a key role of still growing importance in the design process of electronic circuits. Nevertheless we observe that it is no longer in the focus of attention, neither at large design automation conferences for microelectronics, nor in the mind of CAD (computer aided design) managers at industry. The reason is that circuit simulation tools have obtained a high degree of maturity, and are more and more embedded and hidden in a complex design flow, where they are merely used as an engine to calculate circuit behaviour for a large variety of bias conditions and of environment, technological and design parameters.

Until recently, the key factors for the success of circuit simulation have been

- the powerful background of network theory;
- well-established modelling and parameter extraction methodologies;
- perfect tuning of a robust but very conservative mathematical kernel;
- a comfortable user interface and optimal embedding into design flows.

2000 *Mathematics Subject Classification.* Primary 94-02; Secondary 65L80, 94C15, 34A12.
Key words and phrases. Circuit simulation, differential-algebraic equations, diagnostics, bipartite graphs.

However, in face of visible trends in semiconductor technology and design requirements, the actually used mathematical and modelling methods seem to approach their limits. Even the classical network theoretical background will no longer be sufficient, since coupling issues with other domains become critical. Hence considerable amount of research is necessary, since otherwise circuit simulation may no longer be able to cope with upcoming requirements.

It is one purpose of this contribution to highlight these trends in microelectronics, as far as they have an impact on circuit simulation. Therefrom directions of mathematical research are derived, which could help to solve future challenges in this field. We will see that parts of this research are fairly beyond the scope of industrial activities, as well as of commercial vendors offering CAD tools. Our hope is that mathematical research groups in academia will find interest to solve some of the problems presented. Based on our very positive experience in the past, we are optimistic that their contributions can well help to bring innovative mathematics into practical use.

Other directions of research, however, are strongly driven by applications and/or require deep knowledge about thinking of circuit and system designers. Hence, they are better pursued directly in industrial research activities. An actual example is presented in the second part of this paper, where some diagnosis methods are briefly described. Their objective is to provide user oriented diagnostics in case of numerical difficulties during simulation, which identify critical circuit parts and suggest measures to regularize them. The underlying assumption – which is supported by practical experience – is that numerical algorithms in general are sufficiently robust to cope with any circuit, provided it is reasonably designed and modelled, and driven at reasonable bias conditions.

The field of diagnostics has numerous, very different aspects, and a general framework is not established yet. We will restrict here on the class of methods which are based on topological and structural considerations. These methods were implemented in Infineon's circuit simulator TITAN; they turned out to be very fast and were quite helpful in a variety of real life applications, especially in connection with the use of circuit models written in a high level hardware description language like VHDL-AMS [2].

2. Actual trends in microelectronics

The purpose of this section is to motivate and to stimulate mathematical research into directions, which are clearly visible from ongoing trends in microelectronic technologies and design requirements for the next two to five years. We worry that without such help from mathematics – and of course other disciplines as well – circuit simulation tools may become limiting factors in the design process, when designers try to exploit technological possibilities for developing advanced microelectronic circuits and systems on chip.

From our origin we can only restrict to systems where microelectronics are dominating. So domains like power electronics and mechatronics are not considered here. We think, however, that the latter disciplines are doing much pioneering work: They deal just now with problems which are likely to be – at a much larger and more complex scale – problems of microelectronics in the next future.

2.1. Highlighting the Trends

Shrinking goes on

Actually, the channel length of transistors in production lines is $\approx 0.14\mu m$, and in three years it is expected to be $\approx 0.07\mu m$, such that Moore's law about scaling down of device sizes over the years will most probably be still valid. Unfortunately, accurate compact modelling of such small sized electric devices for circuit simulation becomes more and more difficult:

- The different mechanisms governing current and capacitance characteristics of the device can no longer be separated and described independently from each other;
- the device characteristics are more and more dominated by parasitic effects.

This is true for both active and passive devices and even for interconnect; actually, MOS transistor modelling is most critical:

- The basic model core is 30 years old;
- each new effect is modelled by a new layer around this model core, or – even worse – hooked as a backpack onto one of these layers;
- 150 model parameters for the core – and more than 600 model parameters overall – require huge efforts for parameter extraction from measurements;
- depending on parameter configuration and bias conditions, even fundamental requirements like passivity are violated.

Hence, a fundamental back orientation to the physical roots is needed. As a consequence, it may turn out that numerical solution of the semiconductor equations becomes inevitable for each transistor in the circuit netlist. Fortunately, there are first promising approaches into this direction which indicate that this does not necessarily cause prohibitive computational efforts, if the semiconductor equations are sufficiently simplified (cf. [12, 9]). This may open a solution for handling the majority of transistors with implicit models even in large circuit simulation problems.

A difficulty however arises, if such large circuits contain transistors which are driven at their physical limits, e.g., their maximal switching frequency. Certain few transistors of the HF[1] driver and receiver part in data communication chips are typical representatives for this kind of problem. In future, full confidence into correct functionality may only be obtainable by full coupling of device simulation into circuit simulation. This has successfully been done already in the past (cf. [3]); the challenge here is to review the coupling mechanisms and discretization under

[1] HF means the high frequency range, which is actually beyond several GHz.

numerical and DAE[2] aspects, and to upgrade the circuit simulator part for a circuit size of 10^5 conventionally handled transistors.

Increasing power density

Since transistors of shrinking size deliver almost the same current as larger ones, but can be put much closer together, power density increases roughly quadratically with the inverse of the shrinking factor. Actually, it has obtained a value of $\approx$ 80 Watt/100 mm^2, and we can see from any high performance PC that this causes significant thermal problems. Although several measures have been undertaken to bring heat off chip, thermal issues are estimated to be extremely difficult to solve. Especially, it is important to avoid so-called hot spots, i.e., regions where temperature is significantly higher than elsewhere on the chip.

Since most power dissipation takes place in the active regions, i.e., in the channel region of transistors, it is important that their bulk is well connected to material of low thermal resistance. This can be easily done in case of conventional transistors with fixed bulk potential. However, technology trend is towards transistors with floating bulk, so-called SOI (silicon on insulator) technologies. These are characterized by a thin electrically insulating layer between the bulk body and substrate. Unfortunately, thermal resistance of this insulating layer is quite large either, such that direct heat flow into the substrate is no longer possible, and the danger of getting unwanted hot spots is much larger than with fixed bulk technologies.

In order to control heat flow in SOI technologies, it is mandatory to couple circuit and thermal simulation. The problems with this coupling are twofold:

- Heat flow is in principle a 3D-problem;
- the time constants of the electrical and the thermal network differ by $3 \ldots 6$ orders of magnitude.

In power electronics, such simulation problems have already been solved. The new challenge here is, that we have to deal with a circuit size up to 10^5 transistors.

Decreasing voltage swing

In order to reduce power density and to minimize the risk of breakthrough effects in short channel transistors, operating voltages are continuously lowered. While five years ago power supply was somewhat higher than 3 Volt, it is actually $\approx$ 1.3 Volt, and a further decrease down to $\approx$ 1 Volt has to be expected. As a consequence, we observe a trend of decreasing noise margins. In analog circuits the signal/noise ratio gets even worse due to more ambitious requirements with respect to resolution, linearity, jitter, phase noise, etc.

This has two impacts on circuit simulation:

- Accuracy has to be improved; in particular, parasitics have to be carefully modelled and included into simulation. This is mainly a question of extraction

[2]Differential Algebraic Equations (DAEs) are implicit ordinary differential equations (ODEs) of the form $F(x', x, t) = 0$, where $F'_{x'}$ is singular.

from layout, but for circuit simulation the netlists get larger, and sparseness decreases.
- Device noise becomes more important, and noise analysis in the time domain may become an issue. This includes to add stochastic terms to the network equations, and requires significant modelling and algorithmic efforts. Fortunately, the circuits tend not to be quite large for this kind of applications.

Increasing frequencies
Due to decreasing device size at almost constant currents, the circuits get faster. In a standard CMOS technology the rise and fall time of a gate is actually $\approx 50\,\text{psec}$, and the gate delay is similar; the maximal frequency of a CMOS based VCO (voltage controlled oscillator) is beyond 40 GHz.

At these speeds the wave character of signal propagation and field effects may become important. For transmission lines there exist special models in circuit simulation, and also special solvers of the telegraph equation are implemented. The problem with field effects is that they couple "signal noise" of high frequency into slower parts of the chip. These effects are difficult to simulate in general, unless a coupling of field solver and circuit simulation is being used.

Increasing variety of designs
For serving the needs of the market, an increasing variety of designs has to be developed as spin offs from a lead product in a rather short time frame. Infineon's 256 MBit DRAM (dynamic random access memory) chips, e.g., are sold in more than 10 variants, which differ with respect to access time, power consumption, memory addressing, package, etc.

It is natural that existing designs are reused as far as possible for getting the spin off products. So the modifications to be done are moderate, in general, but extensive simulation has to verify that these modifications fit into the rest of design and provide the desired functionality.

From this methodology it is clear that "full chip simulation" at utmost reliability and accuracy is of great interest in the designers community. This requires for circuit simulation to cope in a reasonable time with a circuit size of 500k–20M transistors. While the lower end of this range is actually possible with some efforts using parallel simulation, we think that the upper end will only be obtained with advanced algorithms which exploit the fact that not all parts of the circuit are active at the same time.

Increasing system functionality on chip
There is an ongoing trend to integrate analog functions like sensors, actuators, converters onto complex digital chips, in order to provide an interface to the mostly analog environment. Since the analog part is often controlled in a complex manner by the digital part, mixed signal simulation is useful for checking the interface between both parts. Due to proper decoupling of both parts, this kind of simulation usually does not impose major algorithmic problems. In order to speed up analog simulation – which otherwise would be 2–3 orders of magnitude slower than digital

simulation – a high level behavioural model description of the analog circuit, e.g., in VHDL-AMS may be used.

Verification of the analog circuit itself is more difficult due to the trend to realize more and more analog functions as controlled systems with feedback loops. These are driven by a high speed clock, while the analog functionality settles at moderate speed. Moreover, often parts of these feedback systems are purely digital, but interact closely with the really analog circuitry. Typical examples are analog/digital converters (ADCs) and phase locked loops (PLLs).

The key for tightly coupled mixed signal simulation is the handling of events: A change of a digital signal may cause a discontinuity of the analog solution[3]; and if an analog signal crosses a given threshold, we have an analog event[4], which causes to switch a digital signal. Only if both analog and digital part are stable, the simulator can restart from that actual switching point. It should be noted, that extensions of existing solutions to circuit equations of higher index[5] – which we often have in practice – are not simply possible.

2.2. Issues of mathematical research

From the previously discussed trends we derive some directions, where mathematics could contribute to solve industrial problems of the near future in the area of circuit simulation. Fortunately, parts of this research are being well addressed in actual academic and/or industrial activities.

Multiscale nature of problems

From most of the trends discussed before follows a clear trend towards a multiscale nature of mathematical problems: Time constants may differ by six and more orders of magnitude, and due to nonlinearity, even the rapidly changing signals have to be sampled with sufficiently small timesteps, while global behaviour of the circuit is dominated by the slowly varying states.

So it is a fundamental requirement for the progress of circuit simulation to improve adaptivity. This can be best done by decoupling problems and/or by applying hierarchical concepts. Examples are:

- Multirate integration, exploitation of latency;
- envelope methods;
- separation of timescales by formulating the problem as set of partial differential equations[6].

Reuse: Exploitation of previously computed results

In practice, the same circuit parts – and even complete circuits – are analyzed again and again, eventually in slightly different embeddings or with slightly different

[3]This requires to compute a new consistent solution, which satisfies some conservation requirements for charges and fluxes.

[4]Its computation requires a root calculation, eventually with backtracking analog simulation.

[5]The index of a DAE, in all its variants, measures in some sense the deviation of a DAE from a regular ODE and characterizes the numerical problems one has to face when solving the equations.

[6]See, e.g., the contributions of H.G. Brachtendorf and A. Bunse-Gerstner and of R. Pulch in this issue.

environment parameters. However, numerics each time start more or less from scratch, without any information about the results of previous calculations. In order to avoid this waste of efforts, the development of reuse techniques would be highly desirable.

Model reduction and model verification
A particular form of reuse is to apply model reduction techniques, and to employ the reduced models in subsequent circuit simulations rather than the expensive original circuit blocks. This approach is well suited for a bottom-up design methodology, and will become a key issue for future circuit simulation[7].

An alternative – which requires more modelling efforts, but then can be used for top-down design methodologies as well – is to generate a parameterized high level model of functional blocks, e.g., with the hardware description language VHDL-AMS[8]. The problem with this approach is to guarantee that the model matches the real circuit behaviour for a sufficient large domain of parameters and bias conditions. This is the task of analog verification, which is just in the state of very beginning (cf. [8]).

Coupling models of different domains
The coupling of models of different domains in practice implies the coupling of a variety of numerical methods, which operate in a nonuniform solution space. As a consequence, the numerical theory of the coupled systems needs to be further developed. In circuit simulation the following kinds of couplings are of major interest:

- **DAEs and PDEs**: Circuit simulation has to be coupled with device, field and thermal simulation; even coupling with mechanics may be needed. Mathematically this means coupling of DAEs with partial differential equations, PDEs, which gives rise to PDAEs, partial differential algebraic equations.

  Actually there is a lot of activity in this field, driven by industrial needs. In most practical implementations, simulation packages from the different domains are coupled together. These have obtained a remarkably powerful state meanwhile, but an essential mathematical question is not yet answered in most cases: How to set up and then discretize the equations, such that the coupled problem overall has favourable mathematical properties, e.g., with respect to DAE index, stability and uniqueness of solution, and can be solved efficiently[9].
- **Digital and analog, mixed signal simulation**: The main problems in this field are efficient calculation of the switching points (analog events), the computation of consistent initial values, and efficient restart from events. Most of the actually available solutions are restricted to DAEs of index 0 or 1, while solutions for DAEs of index 2 are needed. Furthermore, for efficient restarts,

[7]See, e.g., the contribution of M. Kahlert in this issue.
[8]An example is given in the contribution of G. Stehr and H. Gräb in this issue.
[9]See, e.g., the contributions of A. Bartel, M. Günther and C. Tischendorf in this issue.

higher order one-step methods may become more important than they were up to now in circuit simulation.

Stochastic DAEs
Once issues of modelling noise in semiconductor circuits have been solved[10] and an integration method for circuits including noise sources has been developed[11], it is possible to perform noise analysis inclusive all nonlinear effects by computation of pathwise solutions in the time domain. However, for getting reliable information about noise densities or jitter etc., a large number of solution pathes has to be calculated. This would require prohibitive computational efforts, unless these calculations can be combined efficiently, such that the computational expense scales underlinearly with the number of traces.

Monitoring and diagnostics
For improving numerical robustness, the mathematical kernel should become more flexible. Especially monitoring the behaviour of the integrator and the development of diagnostic tools will be useful for the case of numerical anomalies: On the one hand, to identify and issue an erroneous part of the circuit model – if there is any at all; on the other hand, to switch back to more conservative numerical parameters/methods, if the circuit model appears to be reasonable.

We think that diagnostics in this context further contributes to use circuit simulators as a tool in a black box like mode, and substitutes painful try and error methods, when designers look for the source of problems in their circuit. The area is not really explored yet; a first glance at it is given in the following section.

3. Diagnosis

In this section, we will give a brief overview of recent work related to some diagnosis methods in circuit simulation. If we denote the circuit equations by $F(x', x, t) = 0$, then the diagnosis we will consider in this article is concerned with checking if the following criteria are given:

1. TR-solvability[12]: Is the TR-iteration matrix $\lambda F'_{x'} + F'_x$ regular for almost all step-sizes?
2. DC-solvability[13]: Is the DC-iteration matrix F'_x regular?
3. Low DAE-index: Is the index of F not greater than 2?

Properties 1 and 3 are crucial for the theoretical and numerical solvability. This is well known from a large body of literature. Property 2 is necessary for calculating the DC-operating point, i.e., x_0 fulfilling $F(0, x_0, t_0) = 0$. Although from the ODE/DAE point of view this property is not necessary, in circuit simulation

[10] See, e.g., the contributions of G. Denk et al., of W. Mathis and L. Weiss and of J. ter Maten in this issue.
[11] See, e.g., the contribution of R. Winkler and W. Römisch in this issue.
[12] TR stands for transient.
[13] DC stands for direct current.

this x_0 is usually taken as initial value. Hence, its computation is common practice for starting the simulation.

In case that these criteria are not given, the circuit simulation often fails. The goal of the diagnostic tools is to identify critical circuit parts or, even better, give hints for a regularization to overcome the numerical difficulties.

In this paper, we will only focus on topological and structural checks. In practice, these checks are of special interest due to their graph-theoretical nature. Consequently, in spite of the large dimensions, the above criteria can be checked very fast.

For a better understanding, we will give several rather trivial but illustrative examples.

3.1. Topological checks

For networks containing only conventional elements the above-mentioned properties can be checked with fast topological algorithms. These algorithms analyze the corresponding network graph with regard to the existence of loops and cutsets consisting of specific types of branches (capacitive branches, resistive branches, etc.).

For DC- and, therefore, TR-solvability these checks are well known (cf., e.g., [1]) and have been implemented in almost any circuit simulator since SPICE [13] and its predecessors. Regularization of inconsistencies is straightforward: Break a loop by inserting a small resistor, and bypass a cutset by a large resistor in parallel.

Example 3.1. *We consider the following circuit which contains a loop consisting of an inductance and a voltage source and the corresponding equations resulting by charge-oriented MNA*[14].

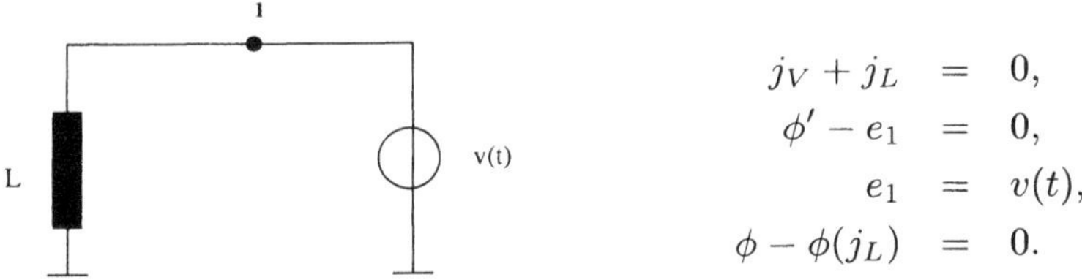

$$\begin{aligned} j_V + j_L &= 0, \\ \phi' - e_1 &= 0, \\ e_1 &= v(t), \\ \phi - \phi(j_L) &= 0. \end{aligned}$$

Obviously, the equations are not DC-solvable. The diagnosis tool implemented in TITAN gives the hint:

```
*** ERROR *** INDUCTOR/VOLTAGE-SOURCE LOOP FOUND, CONTAINING NODE 1
```

[14]**MNA** means Modified Nodal Analysis. e_1 is the node potential on node 1, j_L and j_V are the currents of the inductance and the voltage source, respectively, ϕ is the flux of the inductance, and v is the function for the voltage source.

Hence, the hint reveals where to introduce a low-ohmic resistor in order to regularize the equations. The regularized circuit and the corresponding equations are[15]*:*

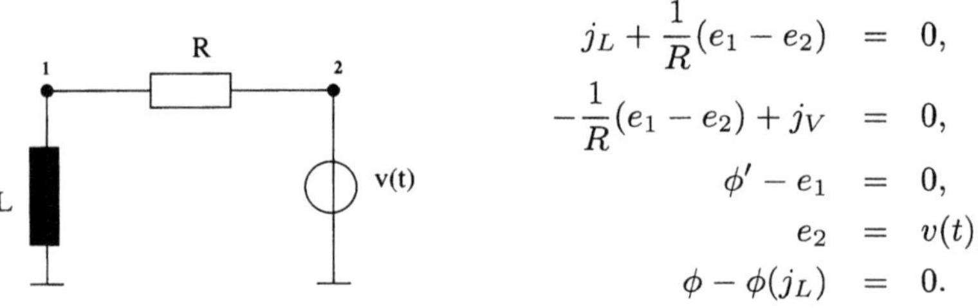

$$
\begin{aligned}
j_L + \frac{1}{R}(e_1 - e_2) &= 0,\\
-\frac{1}{R}(e_1 - e_2) + j_V &= 0,\\
\phi' - e_1 &= 0,\\
e_2 &= v(t),\\
\phi - \phi(j_L) &= 0.
\end{aligned}
$$

The equations corresponding to the regularized circuit are DC-solvable.

Recently, topological checks to determine the DAE-index and DAE-structure were established [5, 6]. Assuming the positive definiteness of the capacitance, inductance and conductance matrices[16], an easy analysis of the network graph permits to determine if the DAE index is 0, 1, or 2, or if the equations present a disadvantageous DAE-structure.

The disadvantageous DAE-structure arises only if the controlled sources do not fulfill certain topological restrictions regarding their position in the network and some other restrictions regarding the position of the controlling elements. In practice, if these conditions are not given, then a regularization becomes reasonable, since the DAE-index and DAE-structure are unknown[17]. Moreover, if critical sources are coupled to each other, then the index is probably greater than 2. Hence, special attention has been paid in implementing an algorithm that provides the user of the diagnosis tool concrete hints for a suitable regularization.

Example 3.2. *We consider the following example of a circuit that does not fulfill the conditions described in* [6] *for being of index* ≤ 2*. The circuit contains a CV loop*[18] *coupled with an LI cutset*[19] *by means of the current controlled current source (CCCS)* $F1$*. As a consequence, the integration fails after a breakpoint.*

---

[15] R is the resistance.

[16] Actually, a test for checking the positive definiteness of these matrices has also been implemented in TITAN. Due to nonlinear devices, this test is not an a priori check, but a local check that involves the numerical values related to the devices at each time-step. Consequently, it requires some extra computational efforts and hence in practice is only carried out for critical applications.

[17] Recall that the DAE-structure may also be disadvantageous although the index is 2. In [17] some simple examples are given that illustrate this for consistent initialization and for asymptotic stability.

[18] **CV loop** means a loop consisting of capacitances and voltage sources only.

[19] **LI cutset** means a cutset of inductances and/or current sources only.

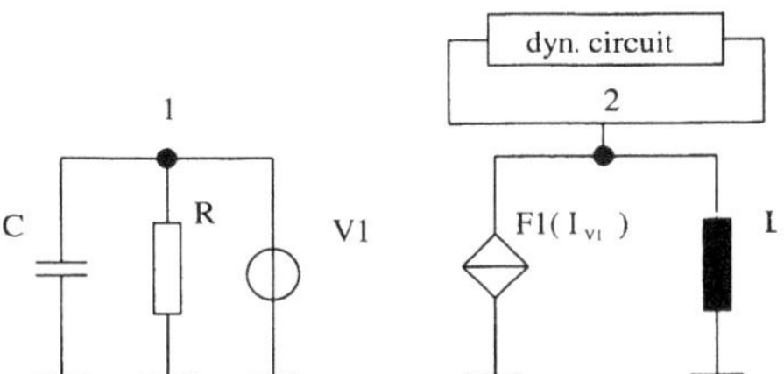

In fact, this circuit leads to an INDEX-3 DAE, which explains the failure of the integrator. To overcome this difficulty, the diagnosis tool implemented in TITAN provides the following information:

```
INDEXTEST:

WARNING: Reliable statements about index can not be made.
DISADVANTAGEOUS DAE-STRUCTURE due to CCCS F1.
  In order to REGULARIZE, introduce, e.g.,
        a low-ohmic resistor in series with: voltage source: V1

WARNING: Index of this system is perhaps greater than 2
DISADVANTAGEOUS DAE-STRUCTURE: CCCS F1 lies in an LI cutset.
  In order to REGULARIZE, introduce, e.g.,
        a high-ohmic resistor in parallel with: current source: F1
```

After introducing the proposed regularizing resistors, the DAE-index becomes 2 *and the integration does not fail anymore.*

3.2. Structural Checks

The fact that nowadays circuits may also contain elements described in a hardware description language like VHDL-AMS requires new diagnosis tools. Since in these description languages the user may describe arbitrary equations, a corresponding network graph does not necessarily exist and topological analysis is not practicable anymore. Hence, checking if the matrix is structurally singular by means of fast algorithms based on matchings for bipartite graphs became extremely helpful (cf. [15] and the references therein). The bipartite graph is set up as follows. Each node e of one partition corresponds to an equation and each node v of the other partition to a variable. There is an edge between node e and v if the concerned equation contains the corresponding variable. The system is considered to be structurally solvable if there exists a perfect matching for this bipartite graph. Roughly speaking, this means that we can assign an equation to each variable and that we can suppose, therefore, that this equation determines this variable.

In cases of singularity, the algorithm described in [15] provides information on the linear dependent rows and columns[20]. Thanks to this information, the user can

[20]If rows are structurally linear dependent then the corresponding equations do not contain enough variables to be solvable; if columns are structurally linear dependent then at least one of the corresponding variables is not determined.

often remove the inconsistencies. However, sometimes the obtained information on the linear dependent rows and columns is very extensive and confusing.

Three follow up issues are currently addressed in this context:

- **Local check of consistency.** Structural checks are also used in first attempts to analyze locally the VHDL-AMS elements with regard to their internal consistency. These checks are urgently needed since the IEEE Standard (1076.1) only prescribes to check equality of the number of variables and equations. Developing adequate checks for this purpose is not a trivial task due to almost arbitrary mutual interdependencies of the terminal quantities in VHDL-AMS models. Note, however, that although local checks are quite useful for assuring some minimal "quality" of each VHDL-AMS model they cannot completely substitute global tests, since inconsistencies may also arise from the coupling of the elements.

  The following two examples illustrate what we mean by internal inconsistency and by inconsistency arising from the coupling of the elements, respectively.

  Example 3.3. *Suppose that an VHDL-AMS model contains the equations*

$$\begin{aligned} u_1 &= 0, \\ u_1 &= 1, \end{aligned}$$

  and that a variable u_2 and the variable u_1 do not appear in other equations, nor their derivatives or integrals. By merely counting equations and variables, we would not detect the obvious inconsistency. In contrast, the matching algorithm identifies this dependence. In this case, the algorithm implemented in TITAN provides the user with the helpful information that the system is singular and that the variable u_2 is not determined.

  Example 3.4. *A voltage source can also be described as a VHDL-AMS element. Hence, although being consistent in itself, if it appears in parallel to an inductance, then we have the same adverse configuration as in Example* 3.1.

- **Consideration of numerical values.** If we also take into account numerical criteria for setting up the bipartite graphs, then considerably more valuable diagnosis information will be obtained.

  Example 3.5. *Suppose that an VHDL-AMS model contains the equations*

$$\begin{aligned} u_1 + \epsilon u_2 &= 0, \\ u_1 &= 1, \end{aligned}$$

  and that u_1 and u_2 do not appear in other equations, nor their derivatives or integrals. The numerical solvability of the corresponding part of the iteration matrix

$$F_x = \begin{pmatrix} 1 & \epsilon \\ 1 & 0 \end{pmatrix}$$

for $x = (u_1, u_2)$ depends, in practice, on the size of ϵ. Hence, in case of failure of the simulation, it would be helpful if in the bipartite graph used for the diagnosis the edge between the equation-node and the variable-node exists only if ϵ is somehow sufficiently large.

Of course, in practice the difficulties are due to the multi-scale nature of applications. In this context, some promising approaches are currently being developed.

It has to be mentioned that all structural checks are not only applicable to circuits containing VHDL-AMS models, but also to those containing only conventional elements. This becomes very interesting if numerical unsolvability is caused by an (almost) singular iteration matrix (e.g., resistance too large) or (almost) infinite entries in the matrix (e.g., resistance too small, capacitance too large).

- **Suitable formulation of the network equations.** Although MNA is adequate for circuit simulation, this scheme for setting up the equations turned out to be inconvenient for structural solvability checks. This becomes relevant for those circuits containing VHDL-AMS models and conventional elements, since global solvability checks consider the structure of the whole system of equations.

Example 3.6. *Let us consider the following circuit that contains a cutset consisting of two current sources only. The corresponding equations resulting by MNA are definitely unsolvable*[21]*:*

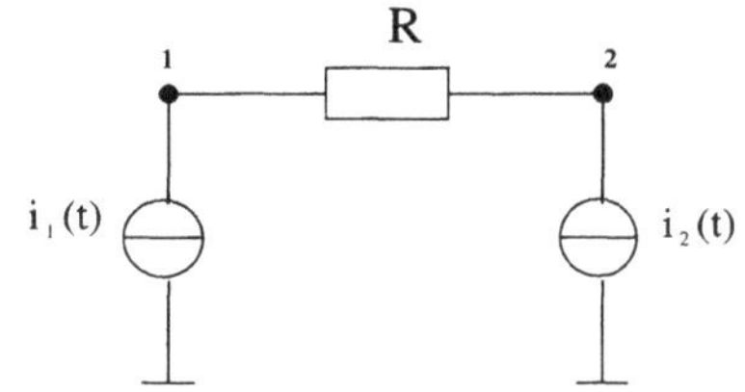

$$\begin{aligned} \frac{1}{R}(e_1 - e_2) + i_1(t) &= 0, \\ -\frac{1}{R}(e_1 - e_2) + i_2(t) &= 0. \end{aligned}$$

However, it is easy to see that these equations are structurally regular. In contrast, if we formulate the circuit equations with branch voltages instead of node potentials[22]*, we obtain the equations*

$$\begin{aligned} \frac{1}{R}u_1 + i_1(t) &= 0, \\ -\frac{1}{R}u_1 + i_2(t) &= 0, \\ u_1 - u_2 - u_3 &= 0, \end{aligned}$$

[21] e_1 and e_2 are the node potential on nodes 1 and 2, i_1, i_2 are the functions for the current sources, and R is the resistance.

[22] u_1, u_2, and u_3 are the branch voltages of the branches corresponding to the resistor and the current sources.

where the last equation is the mesh-equation. These equations are structurally singular.

In TITAN we have implemented a check using this alternative formulation. We emphasize that the alternative formulation is used for diagnosis only, while for simulation we use charge-oriented MNA as usual. Concretely, the structure of the system resulting from the circuit equations formulated with branch voltages instead of node potentials together with the mesh-equations is tested for singularity. On the one hand, this is more expensive, since we enlarge considerably the dimension of the analyzed system of equations. On the other hand, diagnosis is more precise. Nevertheless, it is still an open question which scheme for setting up the equations leads to the sharpest diagnosis.

With regard to structural methods for diagnosis of the DAE-index no satisfying algorithm is available so far. Pantelides [14] constructed an algorithm using graph theory methods to differentiate subsets of the system. By considering specific bipartite graphs, this algorithm determines the so-called structural index. However, already in [14] it was pointed out that the structural index may be smaller than the actual index. Recently, it was realized that the structural index may also exceed the actual index, even for DAEs with constant coefficients [16]. Consequently, the structural determination cannot be regarded to be reliable up to now. Moreover, since the index may also depend on parameter values or bias conditions [7], we conclude that purely structural information is not sufficient for index calculation in general.

New aspect for structural diagnosis: Regularization
For diagnosis, the inclusion of VHDL-AMS elements into circuit simulation does not only lead to difficulties, but also to completely new aspects concerning regularization.

The *break statement* of VHDL-AMS is a language construct that gives the user the possibility to prescribe values for certain variables in certain situations. Hence, by inserting and/or modifying this construct it often becomes possible to regularize singular matrices without modifying the model itself.

In fact, an algorithm has been developed that provides concrete hints for correcting the VHDL-AMS descriptions with break statements, in case that, for instance, DC-solvability is not given[23] (cf. [11, 10]). These proposals are determined by the analysis of a suitably constructed bipartite graph that reflects properly the different types of variables and equations as well as the corresponding hierarchical scopes. From this graph it is analyzed if possible break statements would lead to a perfect matching for the bipartite graph corresponding to F'_x. The resulting proposals can be introduced into the VHDL-AMS source code in interaction with the user, avoiding time-consuming manual error corrections.

[23] We already mentioned before, that from the ODE/DAE point of view DC-solvability is not necessary. Hence, after calculating an initial value with the help of break statements, the transient simulation itself may be possible also with the original model.

Example 3.7. *Suppose that an VHDL-AMS model contains the equations*

$$u_1' = 0, \tag{3.1}$$

$$u_2' = u_2 \tag{3.2}$$

and that u_1 *and* u_2 *do not appear in other equations, nor their derivatives or integrals. The corresponding part of the DC-iteration matrix*

$$F_x = \begin{pmatrix} 0 & 0 \\ 0 & -1 \end{pmatrix}$$

for $x = (u_1, u_2)$ *is obviously singular and the algorithm implemented in TITAN warns that the linear dependent equation is* (3.1) *and the linear dependent variable is* u_1. *If we insert the break statement proposed by the algorithm from* [11, 10], *that prescribes an initial value for* u_1, *then the corresponding part of the iteration matrix becomes regular, since it reads*

$$\tilde{F}_{\tilde{x}} = \begin{pmatrix} 1 & 0 & 0 \\ 0 & 0 & 1 \\ 0 & -1 & 0 \end{pmatrix}$$

for $\tilde{x} = (u_1, u_2, dotu_1)$ [24]. *In contrast, a break statement prescribing a value for* u_2 *would lead to the singular matrix*

$$\tilde{F}_{\tilde{x}} = \begin{pmatrix} 0 & 1 & 0 \\ 0 & 0 & 0 \\ 0 & -1 & 1 \end{pmatrix}$$

for $\tilde{x} = (u_1, u_2, dotu_2)$ [25].

Note that in the above example the difficulty arises from the fact that we try to compute a DC-point for an ODE that does not correspond to a classical DC-solvable electrical circuit. Of course, in general this problem becomes very complex since variables may be intertwined in an arbitrary manner. Finally, we want to recall that many difficulties can also appear due to the fact that, at least theoretically, we may have to cope with arbitrary DAEs. As a consequence, the resulting initial value may not be consistent from the DAE point of view. This may also lead to the failure of integration (cf. [4] and the references therein).

4. Conclusions

In this article we have tried to show that circuit simulation still provides challenging mathematical problems. We hope that mathematical research groups will find interest in some of the presented problems and that the eventually joint research becomes beneficial for both industry and academia.

---

[24] $dotu_1$ is a variable introduced with the break statement. It represents the value of the derivative of u_1.

[25] $dotu_2$ is a variable introduced with the break statement. It represents the value of the derivative of u_2.

For the implementation particular emphasis should be placed on the following issues:

- The strong impact of modelling on numerics should a priori be taken into consideration.
- No one single numerical algorithm can cope with all kinds of problems. Hence, more flexible numerical kernels and smart coupling of numerical methods are required.
- Diagnostic steps are of growing importance; these may be composed of checking structural and *local* numerical criteria.

References

[1] Chua, L.O., Lin, P.-M. Computer-aided analysis of electronic circuits. Prentice Hall, Englewood Cliffs (1975).

[2] Christen, E., Bakalar, K.: VHDL-AMS – A hardware description language for analog and mixed-signal applications. IEEE Trans. CAS II, 46, 1263–1272 (1999).

[3] Engl, W.L., Laur, R., Dirks, H.K.: MEDUSA – A simulator for modular circuits. IEEE Trans. CAD 1, 85–93 (1982).

[4] Estévez Schwarz, D.: Consistent initialization for index-2 differential algebraic equations and its application to circuit simulation. Humboldt-Univ. Berlin, PhD Thesis (2000).

[5] Estévez Schwarz, D., Feldmann, U., März, R., Sturtzel, S., Tischendorf, C.: Finding beneficial DAE structures in circuit simulation. In Mathematics – Key Technology for the Future. Joint Projects between Universities and Industry. Jäger, W., Krebs, H.-J. (Eds.), Springer-Verlag (2003).

[6] Estévez Schwarz, D., Tischendorf, C.: Structural analysis of electric circuits and consequences for MNA. Int. J. Circ. Theor. Appl. **(28)**, 131–162 (2000).

[7] Guenther, M., Feldmann, U.: CAD-based electric-circuit modeling in industry II. Surv. Math. Ind. 8, 131–157 (1999).

[8] Hedrich, L., Hartong. W.: Approaches to formal verification of analog circuits. In Low-power design techniques and CAD tools for analog and RF integrated circuits. Wambacq, P. (Ed.), Kluwer Academic Publishers (2001).

[9] Website of the HiSIM model: http://home.hiroshima-u.ac.jp/usdl/HiSIM.shtml.

[10] Mades, J.: Strukturelle Konsistenz und Regularisierung von VHDL-AMS-Modellen. PhD. Thesis, in preparation.

[11] Mades, J., Glesner, M., Estévez Schwarz, D.: A discrete algorithm for the regularization of hierarchical VHDL-AMS models. Proceedings of the IEEE International Symposium on Circuits and Systems ISCAS 2002, Scottsdale, Arizona, USA, May 26–29 (2002).

[12] Miura-Mattausch, M., Feldmann, U., Rahm, A., Bollu, M., Savignac, D.: Unified complete MOSFET model for analysis of digital and analog circuits. IEEE Trans. CAD 15, 1–7 (1996).

[13] Nagel, W.: SPICE2: A computer program to simulate semiconductor circuits, Tech. Rep. MEMO ERL-M520, Univ. of California (1975).

[14] Pantelides, C.C.: The consistent initialization of differential-algebraic systems. SIAM J. Sci. Statist.Comput. 9, 213–231 (1988).

[15] Reißig, G., Feldmann, U.: A simple and general method for detecting structural inconsistencies in large electrical networks. To be published in IEEE Trans. CAS I.

[16] Reißig, G., Martinson, W.S., Barton, P.I.: Differential-algebraic equations of index 1 may have an arbitrarily high structural index. SIAM J. Sci. Comput. 21, 1987–1990 (2000).

[17] Tischendorf, C., Estévez Schwarz, D.: Mathematical problems in circuit simulation. Mathematical and Computer Modelling of Dynamical Systems. 7(2), 215–223 (2001).

Acknowledgment

We specially want to thank M. Chaudhuri, J. Mades, G. Reißig, S. Sturtzel, and C. Tischendorf for many discussions and for implementing several parts of the diagnostic software. Furthermore, we gratefully acknowledge funding of parts of the work reported here by German government in the BMBF program "Neue Mathematische Verfahren in Industrie und Dienstleistungen".

Diana Estévez Schwarz
Infineon Technologies
Memory Products
Balanstr. 73
D-81541 München, Germany
email: diana.estevez-schwarz@infineon.com

Uwe Feldmann
Infineon Technologies
Memory Products
Balanstr. 73
D-81541 München, Germany
email: uwe.feldmann@infineon.com

Modeling, Simulation and Optimization of Integrated Circuits
(Eds.) K. Antreich, R. Bulirsch, A. Gilg and P. Rentrop

International Series of Numerical Mathematics Vol. 146, 101–119

Periodic Steady State Computation with the Poincaré-map Method

S.H.M.J. Houben

Abstract. A new method for the computation of the periodic steady state of autonomous oscillators is developed. The method is robust and has super-linear convergence behaviour. The method has been implemented in Pstar, the in-house circuit simulator of Philips, and has been tested on realistic oscillator circuits. Numerical results are provided.

1. Introduction

This paper focuses on determining the periodic steady-state (PSS) of an autonomous oscillator circuit. The available solution methods for both autonomous and non-autonomous types of oscillators are based on frequence-domain, time-domain or hybrid circuit-state representations (see [13]). One of the first papers in the field was [1].

For non-autonomous oscillators many efficient solution methods exist. For an overview, see [11], for a more recent overview see [14] or [7]. Harmonic Balance, which is a frequency-domain method, performs well for circuit-states which contain a few low harmonics; however, it is too expensive for states which contain a large amount of harmonics. In order to overcome this problem, hybrid methods (see [10]) and pure time-domain methods have been proposed. For a general overview of time-domain methods, see [7], which deals with shooting and finite difference methods.

For autonomous oscillators the situation is less satisfactory. Here, the period T is an additional degree of freedom, which makes the resulting system under-determined. Most methods proposed for the autonomous case are been based on methods for non-autonomous oscillators. Typically, this is done by considering the period T as an additional degree of freedom. Unfortunately, those methods are usually very sensitive with respect to the initial guess T_0 for the circuit's period T. They converge only for T_0 in a small neighbourhood of T. An example of this phenomenon is shown in this paper; see [15] for another example.

In this paper, two novel methods for computing the solution of autonomous oscillators are described. Applied to a non-autonomous circuit, where T is a priori known, our method reduces to a method similar to the one introduced in [8].

However, our method differs significantly from that in [8] in that we do not need the period T to be given or estimated before starting the algorithm. Instead, we compute T along with the solution.

2. Periodic steady-states

In this section, a mathematical formulation of the boundary value problem to be solved will be provided. We discuss a number of concepts such as *periodic steady-state*, *limit cycle* and *stability of a PSS*.

We assume that the evolution of the circuit state is described by the differential-algebraic equation

$$\frac{\partial \boldsymbol{q}(t, \boldsymbol{x})}{\partial t} + \boldsymbol{j}(t, \boldsymbol{x}) = \mathbf{0}. \tag{1}$$

We assume that (1) has index 1; see [9] for sufficient conditions for this to be the case.

Definition 2.1. *Consider an autonomous DAE of the form* (1). *A function* $\boldsymbol{x} : \mathbb{R} \to \mathbb{R}^N$ *is called a* periodic steady-state *(PSS) of* (1) *if*

- $\boldsymbol{x}$ *is a solution to* (1).
- $\boldsymbol{x}$ *is periodic, i.e., there is a* $T > 0$ *such that for all* $t \in \mathbb{R}$, $\boldsymbol{x}(t) = \boldsymbol{x}(t + T)$.

Note that according to this definition, a stationary solution, i.e., a solution of the form $\boldsymbol{x}(t) \equiv \boldsymbol{x}_0$, is also a PSS. A periodic steady-state can also be described as a solution to the following two-point periodic boundary value problem (BVP).

$$\frac{d\boldsymbol{q}(\boldsymbol{x}(t))}{dt} + \boldsymbol{j}(\boldsymbol{x}(t)) = \mathbf{0}, \qquad \text{for } 0 \le t \le T, \tag{2a}$$

$$\boldsymbol{x}(0) = \boldsymbol{x}(T), \tag{2b}$$

$\boldsymbol{x} : [0, T] \to \mathbb{R}^N$, $T > 0$ unknown.

Definition 2.2. *The* limit cycle $\mathbb{C}(\boldsymbol{x})$ *of a PSS* $\boldsymbol{x}$ *is the range of the function* $\boldsymbol{x}(t)$, *i.e.,*

$$\mathbb{C}(\boldsymbol{x}) = \{\boldsymbol{x}(t) \mid t \in \mathbb{R}\}. \tag{3}$$

A set $\mathbb{C}$ *is called a limit cycle of* (1) *if there is a PSS* $\boldsymbol{x}$ *of* (1) *so that* $\mathbb{C} = \mathbb{C}(\boldsymbol{x})$.

Definition 2.3. *A function* $\boldsymbol{x} : \mathbb{R} \to \mathbb{R}^N$ *is said to converge to a set* $S \subseteq \mathbb{R}^N$ *if and only if*

$$\lim_{t \to \infty} d(\boldsymbol{x}(t), S) = 0. \tag{4}$$

Here, $d(\boldsymbol{x}, S)$ *denotes the distance between the point* $\boldsymbol{x}$ *and the set* S, *defined as*

$$d(\boldsymbol{x}, S) := \inf_{\boldsymbol{y} \in S} \|\boldsymbol{x} - \boldsymbol{y}\|. \tag{5}$$

The concept or orbital stability will play a key role in this paper.

Definition 2.4. *A limit cycle* $\mathbb{C}$ *exhibits* orbital stability *if there is a* $\delta > 0$ *so that the following holds:*
For every solution $\tilde{\boldsymbol{x}}$ *to* (1) *with* $d(\tilde{\boldsymbol{x}}(0), \mathbb{C}) < \delta$, $\tilde{\boldsymbol{x}}$ *converges to* $\mathbb{C}$.
A PSS is orbitally stable when its limit cycle is orbitally stable.

We will concentrate on methods for finding an orbitally stable periodic steady-state. Periodic steady-states that are not orbitally stable are not interesting for the IC designer, since they do not correspond to any physical behaviour of the modelled circuit. As we will see further on, the circuit equations may have both orbitally stable and orbitally unstable periodic steady states. In that case, only the orbitally stable states correspond to physical solutions. Therefore, we want to actively avoid non-stable periodic steady-states in our solution method.

Several remarks can be made on (2).

- The condition that the PSS is stable is not implied in (2).
- For every solution $\boldsymbol{x}$ of (2), we can obtain a whole continuum of solutions $\boldsymbol{x}_\tau$, $0 \le \tau \le T$, by defining
$$\boldsymbol{x}_\tau(t) = \boldsymbol{x}((t+\tau) \mod T). \tag{6}$$
This means that a solution $\boldsymbol{x}$ can be locally unique, only if $\boldsymbol{x}$ is stationary, i.e., $\boldsymbol{x}(t) \equiv \boldsymbol{x}_0$.
- For $T = 0$, every consistent value $\boldsymbol{x}$ is a solution. For this reason, we require $T > 0$; however, we cannot always avoid that a numerical method converges to $T = 0$ anyhow.

From the above is should become clear that a solution to (2) is far from unique, and that there is in fact a large number of undesirable solutions to (2) which are to be avoided by the simulator.

3. Some example circuits

As mentioned earlier, free-running oscillators have the property that they do not have any time-dependent input signals. This implies that they can be described mathematically by an autonomous DAE. A simple example of an oscillator that can be described by an ODE is given in Figure 1. The equations for this particular circuit are
$$\frac{dv}{dt} = \frac{1}{C}i, \qquad \frac{di}{dt} = \frac{-1}{L}v. \tag{7}$$
Note that these equations are linear. For initial conditions $v(0) = v_0$, $i(0) = i_0$, the solution to (7) is given by

$$v(t) = v_0 \cos \omega t + \sqrt{\frac{L}{C}} i_0 \sin \omega t, \tag{8a}$$

$$i(t) = i_0 \cos \omega t - \sqrt{\frac{C}{L}} v_0 \sin \omega t, \tag{8b}$$

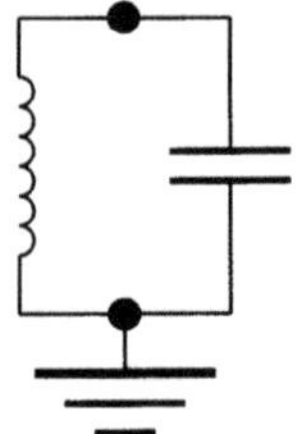

FIGURE 1. The LC-ring is a very simple free-running oscillator. It consists of a capacitor and an inductor.

where $\omega = 1/\sqrt{CL}$. In this case, it is easy to find a PSS, since *every* solution $\boldsymbol{x}$ of (7) is periodic and hence a PSS. However, none of these PSS is stable. Every PSS of (7) can be described uniquely by giving the initial values v_0 and i_0 and the period T. In Figure 2, we have sketched the set of parameter values i, v and T for which solutions exist. A nonlinear example of a free-running oscillator is Colpitt's Oscillator; its network schematic are shown in Figure 3. Colpitt's oscillator converges slowly to a periodic steady-state. The simulation results of the transient behaviour of this circuit is shown in Figure 4. Though Colpitt's oscillator is rather simple compared to industrial relevant circuits, it has no known analytical solution and thus is still too complicated for our analysis. For that reason we introduce the

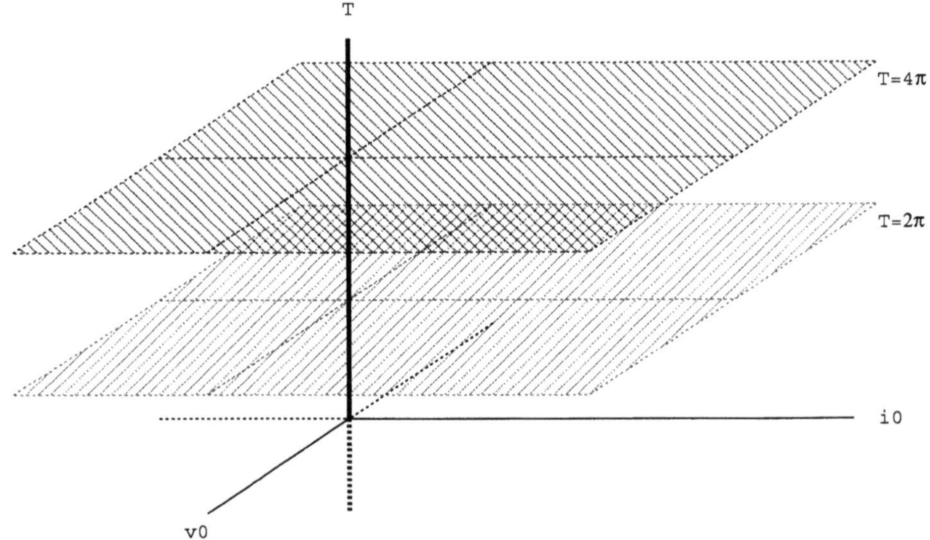

FIGURE 2. The solutions of (7) in (v_0, i_0, T)-space.

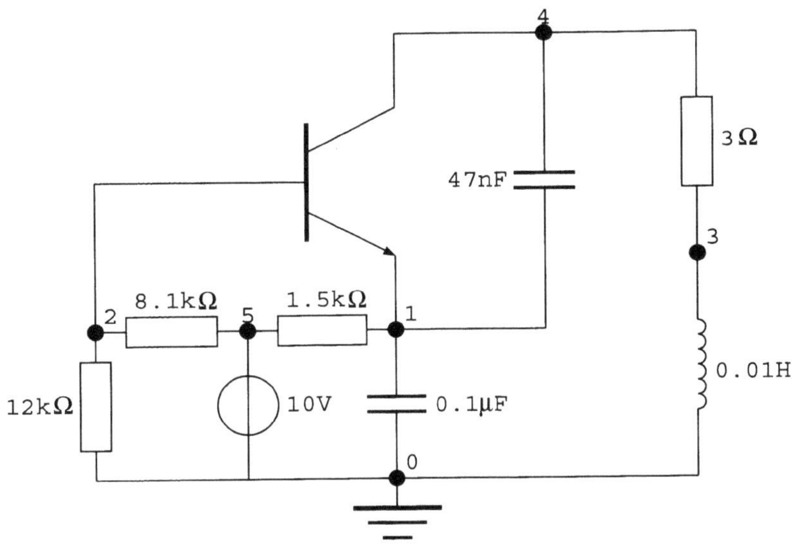

FIGURE 3. Colpitt's Oscillator.

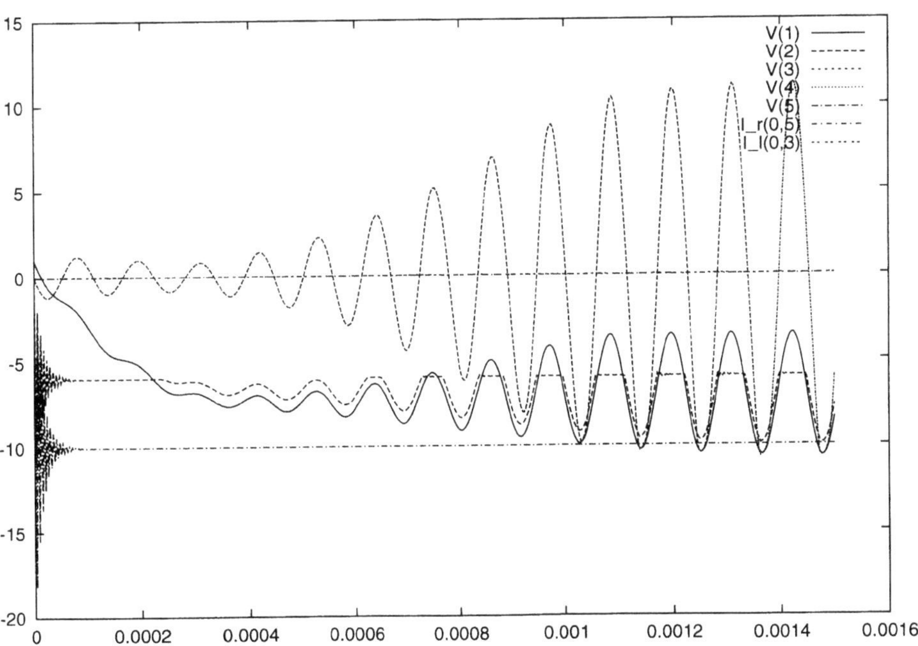

FIGURE 4. Initial transient behaviour of circuit 3.

following simple model, which is a nonlinear extension of (7) with $L = C = 1$; see also [3]. This model problem is used to assess methods for simulating free-running oscillators.

First, let $h : \mathbb{R} \to \mathbb{R}$ be a function that satisfies the conditions

- h continuous and differentiable.
- $h(0) > 0$.
- there are several points $r_k > 0$ so that $h(r_k) = 0$.

Then we define the following model ODE.

$$\frac{dx}{dt} = y + h(\sqrt{x^2 + y^2})x, \tag{9a}$$

$$\frac{dy}{dt} = -x + h(\sqrt{x^2 + y^2})y, \tag{9b}$$

The ODE (9) has the following properties

- It has at least one PSS solution, namely the stationary state with $\boldsymbol{x} = \mathbf{0}$. However, this solution is unstable.
- For every $r_k > 0$ satisfying $h(r_k) = 0$, we see that the circle described by $x^2 + y^2 = r_k^2$ is a limit cycle. Moreover, if $h'(r_k) < 0$, then the limit cycle is stable.
- As $h'(r_k) \to 0$ from below, the limit cycle $x^2+y^2 = r_k^2$ becomes an increasingly weaker attractor for nearby solutions of (7).

Some interesting choices for h are

- For $h(r) = \cos r$ there is an infinite number of stable and unstable limit cycles. Every zero of h corresponds to a limit cycle of (9). This choice of h shows that it is possible for several limit cycles to exist.
- For $h(r) = \varepsilon(1 - r)$, there is exactly one stable limit cycle, namely the unit circle. However, the solution $\boldsymbol{x}(t) \equiv \mathbf{0}$ is an unstable stationary solution. The single limit cycle is stable; convergence speed towards this limit cycle is determined by the parameter $\varepsilon > 0$. The closer ε approaches 0, the slower convergence becomes. The fact that we can tune convergence speed with ε makes this particular problem a suitable benchmark problem.

In Figure 5, we have sketched the set of parameter values i, v and T for which solutions to (9) exist.

As we will see later, finding a stable PSS becomes more difficult when the PSS behaves only as a weak attractor, i.e., the convergence towards the PSS is very slow.

4. Other approaches

Several methods have been proposed for finding a stable PSS. These methods include

- Shooting methods
- Finite Difference methods

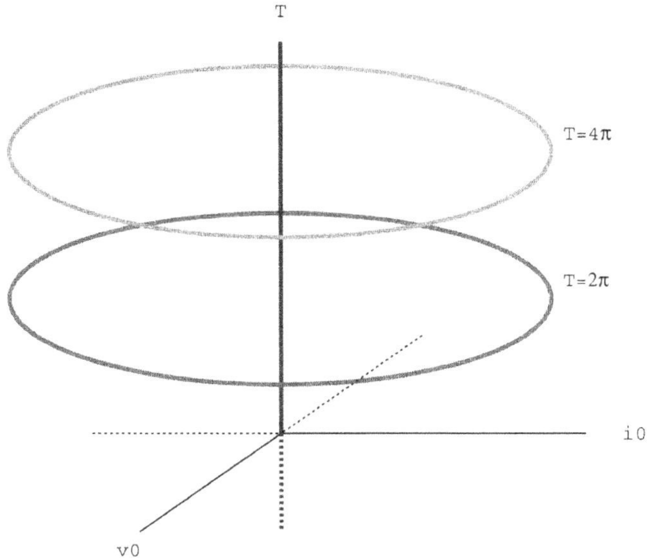

FIGURE 5. The solutions of (9) with $h(r) = \varepsilon(1 - r)$ in (v_0, i_0, T)-space.

- Harmonic Balance techniques

These methods were designed for solving Boundary Value Problems with a priori known boundaries. In order to use them for the free-running oscillator problem, where the right boundary $t = T$ is unknown a priori, it is common to scale the a priori unknown interval $[0, T]$ back to the interval $[0, 1]$, in the following way.

$$\frac{1}{T}\frac{d\boldsymbol{q}(\hat{\boldsymbol{x}}(\tau))}{dt} + \boldsymbol{j}(\hat{\boldsymbol{x}}(\tau)) = \mathbf{0}, \qquad \text{for } 0 \le \tau \le 1, \tag{10a}$$

$$\frac{d}{d\tau}T = 0, \tag{10b}$$

with the boundary conditions

$$\hat{\boldsymbol{x}}(0) = \hat{\boldsymbol{x}}(1), \tag{10c}$$

$$s(\hat{\boldsymbol{x}}(0)) := (\hat{\boldsymbol{x}}(0), \boldsymbol{n}) - \alpha = 0. \tag{10d}$$

Equation (10d) is an additional condition that is needed to make the solution locally unique. It is commonly called a *switch condition* or *phase condition*. The exact form of s needs to be provided by the circuit designer. There are no general rules to find functions s that work for an arbitrary circuit. However, for circuits in which the autonomous oscillation is generated by an LC loop, typical choices for $\alpha = 0$ and $\boldsymbol{n}$ are the following.

- Take $\boldsymbol{n}$ so that $(\boldsymbol{x}, \boldsymbol{n})$ is the voltage difference over the capacitor(s) in the LC loop. Take α to be this voltage difference in the DC state.

- Take $\boldsymbol{n}$ so that $(\boldsymbol{x}, \boldsymbol{n})$ is the current through the inductor(s) in the LC loop. Take α to be this current in the DC state.

We now discuss a method for solving (10) based on the shooting method. The well-known shooting method is described in [2], Chapter 4. Consider the function $\phi : \mathbb{R}^N \times \mathbb{R} \to \mathbb{R}$, which is defined as

$$\phi\colon (\tilde{\boldsymbol{x}}, T^*) \mapsto \hat{\boldsymbol{x}}(1), \tag{11}$$

where $\hat{\boldsymbol{x}}$ is the solution to (10a) with $\hat{\boldsymbol{x}}(0) = \tilde{\boldsymbol{x}}$ and $T = T^*$. The problem (10) can now be reformulated as follows. Find a $T > 0$ and a $\boldsymbol{x} \in \mathbb{R}^N$ such that

$$\phi(\boldsymbol{x}, T) - \boldsymbol{x} = \mathbf{0} \quad \text{and} \quad s(\boldsymbol{x}) = 0. \tag{12}$$

In essence we want to find a zero of the function $\psi(\boldsymbol{x}, T) := [(\phi(\boldsymbol{x}, T) - \boldsymbol{x})^T, s(\boldsymbol{x})]^T$. The system (12) can now be solved using a (damped) Newton method. However, this process can be extremely sensitive to bad initial guesses for T.

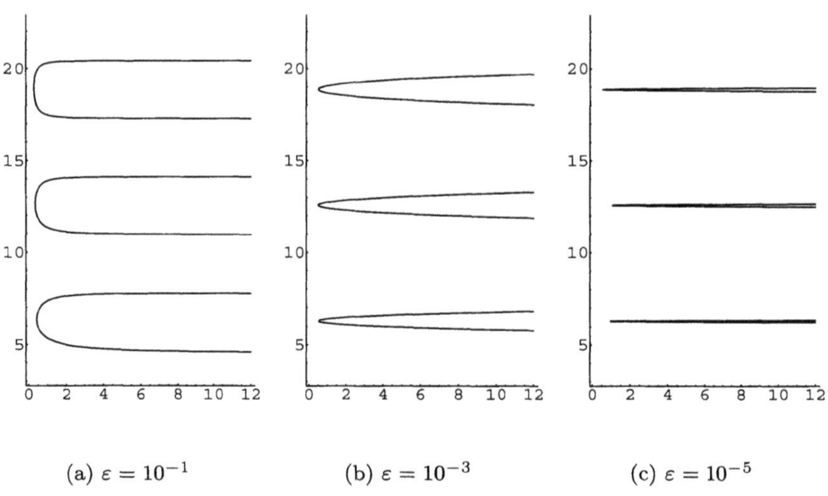

(a) $\varepsilon = 10^{-1}$ (b) $\varepsilon = 10^{-3}$ (c) $\varepsilon = 10^{-5}$

FIGURE 6. Values of (v_0, T) for which the Newton matrix becomes singular, for various values of ε.

Example 4.1. Consider problem (9) with $h(r) = \varepsilon(1-r)$, solved by single shooting. As a phase condition we will take $s(\boldsymbol{x}) := i_0$. This means that we now have two variables left, namely v_0 and T, which must be determined by a Newton process. The system of equations to be solved is

$$v(T) - v_0 = 0, \quad \text{and} \quad i(T) = 0. \tag{13}$$

It is possible to plot the points (v_0, T) for which the Newton matrix becomes singular. This has been done for $\varepsilon = 10^{-1}$, 10^{-3}, and 10^{-5} in Figure 6. For a sufficiently damped Newton method, the curves of singularity in these figures form the boundary of the areas of convergence for the various solutions to (9). It is clear that for $\varepsilon \to 0$, we have to find an increasingly accurate initial guess for the period T to ensure convergence towards the non-DC solution. In order to get a better feel

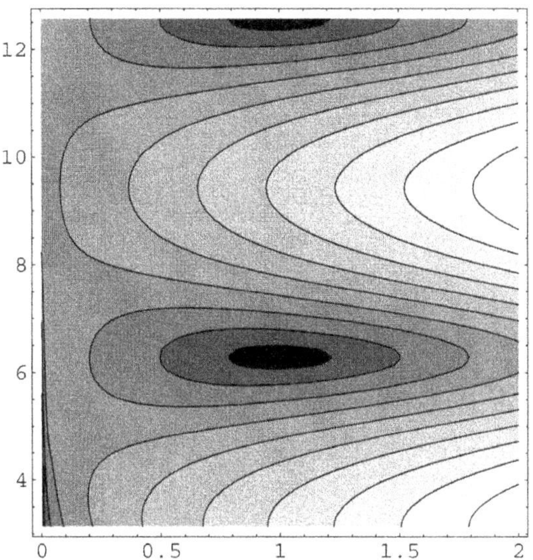

FIGURE 7. The residual $R(v_0, T)$ for $\varepsilon = 1$. Darker means a lower value. The isolated solution at $(1, 2\pi)$ is clearly visible.

for the convergence behaviour, define the residual R as

$$R(v_0, T) := \sqrt{(v(T) - v_0)^2 + i(T)^2}. \tag{14}$$

The residual R can be plotted as function of v_0 and T. This has been done in Figure 7 and 8 for $\varepsilon = 1$ and $\varepsilon = 10^{-2}$, respectively. A detail of Figure 8 is shown in Figure 9. It is also possible to plot the norm of the Newton update. This has been done in Figure 10 for $\varepsilon = 10^{-2}$ in a detail around the solution. In the white area in Figure 10, the Newton update becomes arbitrarily large. This layer separates the area of convergence towards the isolated solution $(1, 2\pi)$ from the area of convergence towards the stationary (DC) solution. $\triangle$

5. The Poincaré-map method

The Poincaré-map method is based on the following observation. Provided we start sufficiently close to a stable limit cycle $\mathbb{C}$, a transient simulation will eventually converge towards $\mathbb{C}$. After all, this is implied by the definition of a stable limit cycle. Therefore, we can simply find the PSS by starting at some point $\boldsymbol{x}_0$ and then performing a transient simulation until the solution $\boldsymbol{x}(t)$ has approached the

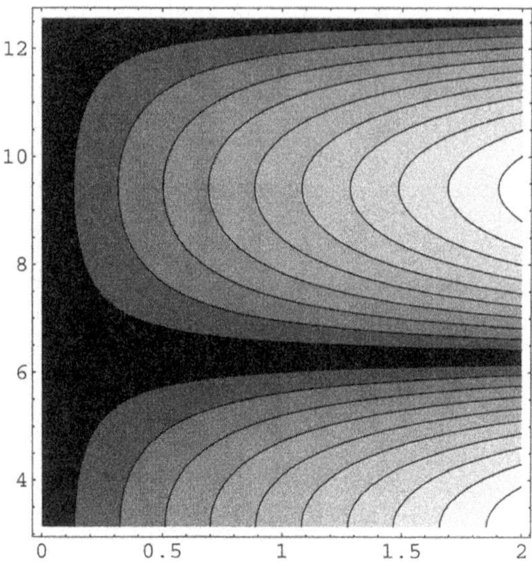

FIGURE 8. The residual $R(v_0, T)$ for $\varepsilon = 10^{-2}$. Darker means a lower value. There is an isolated solution at $(1, 2\pi)$, but this is not visible.

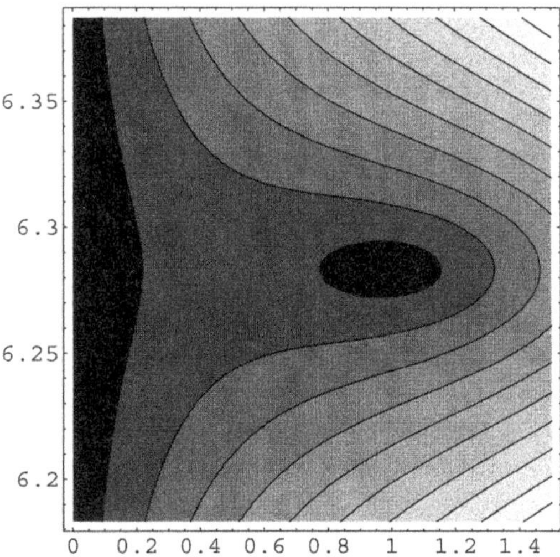

FIGURE 9. Detail of Figure 8 around $(1, 2\pi)$.

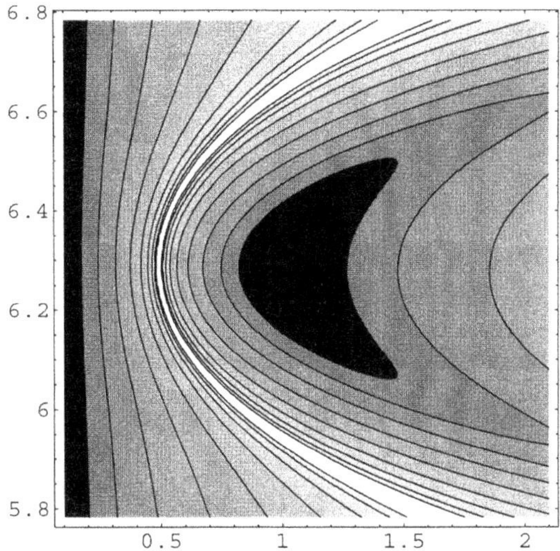

FIGURE 10. Norm of the Newton update as function of (v_0, T) for $\varepsilon = 10^{-2}$.

stable limit cycle sufficiently close. There are, however, two disadvantages to this approach:

- We have to find a way to detect when we have approached the stable limit cycle "close enough". If T is known, a "running window" can be used, i.e., the value $\boldsymbol{x}(t)$ at the current integration time t is compared to the value at $\boldsymbol{x}(t-T)$. However, T is an unknown in the autonomous case.
- Convergence will be linear at best, which means that excessive computing time is needed to arrive at the solution.

In this section we will address the first problem. The method which we derive will still be hampered by the second problem; therefore it will be rather slow. However, in the next section we shall show how we can accelerate the method.

First we note that the length of the period can be estimated by looking for periodically recurring features in the computed circuit behaviour. A possible recurring feature is the point at which a specific condition is satisfied. This is equivalent to carrying out a Poincaré-map iteration, see [5], Section I.16. The idea is to cut the transient solution $\boldsymbol{x}(t)$ by a hyperplane. The hyperplane is defined by an affine equation of the form $(\boldsymbol{x}(t), \boldsymbol{n}) = \alpha$, for some vector $\boldsymbol{n}$ and scalar α. This equation is, in fact, the phase condition. The situation is visualised in Figure 11. The resulting method will be called the *Poincaré-map method*; it can be described as follows.

Algorithm 5.1. *Let an approximate solution $\boldsymbol{x}_0$ and a required accuracy tolerance $\varepsilon > 0$ be given. Then do the following.*

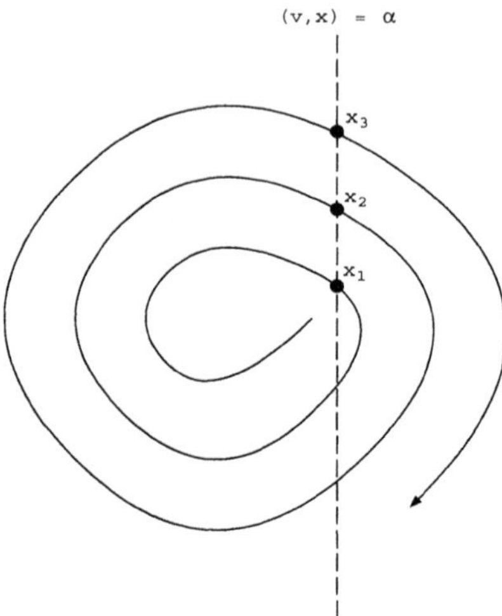

FIGURE 11. The trajectory of a solution $\boldsymbol{x}(t)$ to the Initial Value Problem. The points $\boldsymbol{x}_n$ are chosen so that they satisfy $(\boldsymbol{x}, \boldsymbol{n}) = \alpha$, for some given $\boldsymbol{v}$ and α.

$i := 0$, $t_0 := 0$
$\boldsymbol{x}_0 :=$ *some initial guess for* $\boldsymbol{x}$
repeat
  Starting with $t = t_i$, $\boldsymbol{x}(t_i) = \boldsymbol{x}_i$, *integrate* (1) *until* $(\boldsymbol{x}(t), \boldsymbol{n}) = \alpha$ *and* $d(\boldsymbol{x}(t), \boldsymbol{n})/dt > 0$.
  $\boldsymbol{x}_{i+1} := \boldsymbol{x}(t)$, $t_{i+1} := t$
  $\delta_i := \|\boldsymbol{x}_{i+1} - \boldsymbol{x}_i\|$
  $i := i + 1$
until $i \geq 2 \wedge \delta_{i-2}^2 \leq \varepsilon(\delta_{i-1} - \delta_{i-2})$
$T := t_i - t_{i-1}$
$\tilde{\boldsymbol{x}} := \boldsymbol{x}_i - \boldsymbol{x}_{i-1}$

If for a specific initial state $\boldsymbol{x}_0$ the physical circuit converges to a PSS, then the Poincaré-map method, started at that state $\boldsymbol{x}_0$, will also converge to a PSS, provided that the mathematical model is accurate and a suitable time integrator is used. Typically, the designer of the circuit will know a plausible initial state for the circuit, so it is not hard to find a good $\boldsymbol{x}_0$. Moreover, oscillator circuits are typically designed so that they will converge to the stable PSS for a large set of initial conditions. For such a circuit, the Poincaré-map method will also converge to this PSS.

Summarised, the Poincaré-map method consists of finding a fixed point of a function $F : \mathbb{R}^n \to \mathbb{R}^n$, which is defined as

$$F(\boldsymbol{x}_0) := \boldsymbol{x}(T). \tag{15}$$

Here, $\boldsymbol{x}(t)$ is the solution to (1) with $\boldsymbol{x}(0) = \boldsymbol{x}_0$, and T is the smallest $t > 0$ such that $(\boldsymbol{x}(t), \boldsymbol{n}) = \alpha$ and $d(\boldsymbol{x}(t), \boldsymbol{n})/dt > 0$.

For given $\boldsymbol{x}_0$, $F(\boldsymbol{x})$ is computed by Algorithm 5.1, i.e., by applying the ordinary Poincaré-map method. The successive approximations of the Poincaré-map method satisfy the recursion relation

$$\boldsymbol{x}_{n+1} = F(\boldsymbol{x}_n) \tag{16}$$

The period T does not explicitely enter into (16).

6. The MPE accelerated Poincaré-map method

Let $\tilde{\boldsymbol{x}}$ be a fixed point of F, i.e.,

$$F(\tilde{\boldsymbol{x}}) = \tilde{\boldsymbol{x}}. \tag{17}$$

Suppose that F is continuous and differentiable in a neighbourhood of $\tilde{\boldsymbol{x}}$. If we start sufficiently close to $\tilde{\boldsymbol{x}}$, the sequence (16) converges linearly to $\tilde{\boldsymbol{x}}$. We look for a way to accelerate this process. We will consider an *acceleration method* which operates on the first k vectors of a sequence $\{\boldsymbol{x}_n\}$, and produces an approximation $\boldsymbol{y}$ to the limit of $\{\boldsymbol{x}_n\}$. This approximation can then be used to restart (16) and generate the beginning of a new sequence $\boldsymbol{y}_0, \boldsymbol{y}_1, \boldsymbol{y}_2, \ldots$. Again, the acceleration method can be applied to this new sequence, resulting in a new approximation $\boldsymbol{z}$ of the limit. The idea is that the sequence $\boldsymbol{x}, \boldsymbol{y}, \boldsymbol{z}, \ldots$ converges much faster to the limit of $\{\boldsymbol{x}_n\}$ than the sequence $\{\boldsymbol{x}_n\}$ itself. Typically, if $\{\boldsymbol{x}_n\}$ converges linearly, then $\{\boldsymbol{x}, \boldsymbol{y}, \boldsymbol{z}, \ldots\}$ converges super-linearly. A well-known acceleration method is minimal polynomial extrapolation (MPE). A good overview of MPE, comparing it with competing techniques such as Reduced Rank Extrapolation (RRE) and Epsilon Extrapolation, is given in [12]. A more recent comparison of these methods is given in [4] and in [6], which also gives numerical comparisons. According to [6], MPE compares favourably to the other methods mentioned. As has been shown in [12], it is also numerically more stable. Therefore, we decided to base our implementation on MPE.

Consider again the function F defined as in (15). We can linearise F around $\tilde{\boldsymbol{x}}$, which gives us

$$\tilde{F}(\tilde{\boldsymbol{x}} + \boldsymbol{e}) = \tilde{\boldsymbol{x}} + \Phi \boldsymbol{e} + O(\|\boldsymbol{e}\|^2), \quad \text{with } \Phi := \frac{dF(\tilde{\boldsymbol{x}})}{d\boldsymbol{x}}. \tag{18}$$

Now consider a sequence $\{\boldsymbol{x}_n\}$, $n = 0, 1, 2, \ldots$, generated by the recursion relation $\boldsymbol{x}_{n+1} = F(\boldsymbol{x}_n)$. We can define the sequence of errors $\{\boldsymbol{e}_n\}_n$ as

$$\boldsymbol{e}_n := \boldsymbol{x}_n - \tilde{\boldsymbol{x}}, \quad \text{for } n = 0, 1, 2, \ldots. \tag{19}$$

Using (18), we find the following recurrence relation for $(\boldsymbol{e}_n)_n$

$$\boldsymbol{e}_{n+1} = \Phi \boldsymbol{e}_n + O(\|\boldsymbol{e}_n\|^2). \tag{20}$$

From this it follows that

$$\|\boldsymbol{e}_{n+1}\|^2 = \left\|\Phi \boldsymbol{e}_n + O(\|\boldsymbol{e}_n\|^2)\right\|^2 = O(\|\boldsymbol{e}_n\|^2) \tag{21}$$

Define the sequence of *differences* by

$$\boldsymbol{d}_n := \boldsymbol{x}_{n+1} - \boldsymbol{x}_n. \tag{22}$$

It is easy to show that $\boldsymbol{d}_n = \boldsymbol{x}_{n+1} - \boldsymbol{x}_n = \boldsymbol{e}_{n+1} + \tilde{\boldsymbol{x}} - \boldsymbol{e}_n - \tilde{\boldsymbol{x}} = \boldsymbol{e}_{n+1} - \boldsymbol{e}_n$. So we have for $n \geq 0$

$$\boldsymbol{d}_{n+1} = \boldsymbol{e}_{n+2} - \boldsymbol{e}_{n+1}, \tag{23a}$$

$$= \Phi \boldsymbol{e}_{n+1} - \Phi \boldsymbol{e}_n + O(\|\boldsymbol{e}_n\|^2) \tag{23b}$$

$$= \Phi(\boldsymbol{d}_n) + O(\|\boldsymbol{e}_n\|^2). \tag{23c}$$

Let k be the smallest natural number such that $\boldsymbol{d}_k \in \langle \boldsymbol{d}_0, \ldots, \boldsymbol{d}_{k-1} \rangle$. This means that there are scalars $\sigma_0, \ldots, \sigma_k$, $\sigma_0 \neq 0$, $\sigma_k \neq 0$, such that

$$\sum_{p=0}^{k} \sigma_p \boldsymbol{d}_p = \boldsymbol{0}, \quad \text{and thus} \quad \sum_{p=0}^{k} \sigma_p \Phi^p \boldsymbol{d}_0 = O(\|\boldsymbol{e}_0\|^2) \approx \boldsymbol{0}. \tag{24}$$

Once we have found these σ_p's, we can show

$$\begin{aligned}
&\sum_{p=0}^{k} \sigma_p \boldsymbol{x}_p \\
&= \sum_{p=0}^{k} \sigma_p (\tilde{\boldsymbol{x}} + \boldsymbol{e}_p) \\
&= \left(\sum_{p=0}^{k} \sigma_p\right) \tilde{\boldsymbol{x}} + \sum_{p=0}^{k} \sigma_p \Phi^p \boldsymbol{e}_0 + O(\|\boldsymbol{e}_0\|^2) \\
&= \left(\sum_{p=0}^{k} \sigma_p\right) \tilde{\boldsymbol{x}} + (\Phi - I) \sum_{p=0}^{k} \sigma_p \Phi^p \boldsymbol{d}_0 + O(\|\boldsymbol{e}_0\|^2) \\
&= \left(\sum_{p=0}^{k} \sigma_p\right) \tilde{\boldsymbol{x}} + O(\|\boldsymbol{e}_0\|^2) \\
&\approx \left(\sum_{p=0}^{k} \sigma_p\right) \tilde{\boldsymbol{x}}.
\end{aligned} \tag{25}$$

Hence an approximation of $\tilde{\boldsymbol{x}}$, called $\tilde{\boldsymbol{x}}_{\text{approx}}$, is given by

$$\tilde{\boldsymbol{x}}_{\text{approx}} := \left(\sum_{p=0}^{k} \sigma_p\right)^{-1} \sum_{p=0}^{k} \sigma_p \boldsymbol{x}_p. \tag{26}$$

We can now restart the process by taking $\boldsymbol{x}_0 = \tilde{\boldsymbol{x}}_{\text{approx}}$. By iterating this process, we obtain a sequence of approximations $\{\boldsymbol{y}_n\}$. This sequence will converge super-linearly to $\tilde{\boldsymbol{x}}$ provided we start sufficiently close to $\tilde{\boldsymbol{x}}$, i.e., $\|\boldsymbol{e}_0\|$ is sufficiently small.

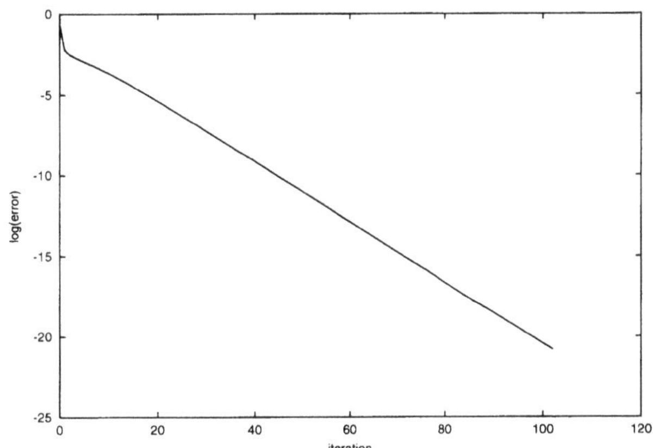

FIGURE 12. log(error) after each iteration for the Poincaré-map method applied to (9) with $h(r) := \varepsilon(1-r), \varepsilon = 3 \cdot 10^{-2}$.

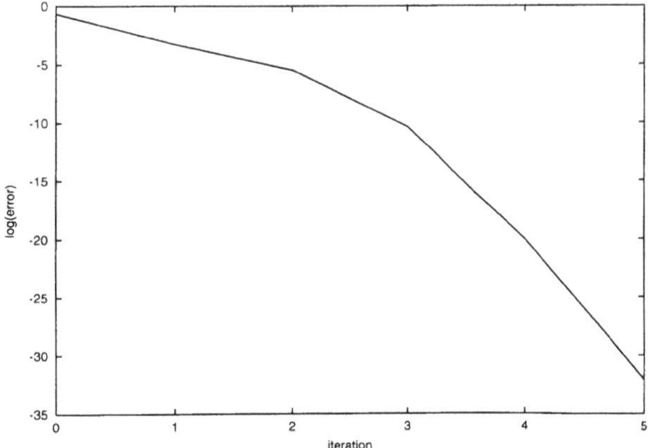

FIGURE 13. log(error) after each outer loop iteration for the accelerated Poincaré-map method applied to (9) with $h(r) := \varepsilon(1-r), \varepsilon = 3 \cdot 10^{-2}$.

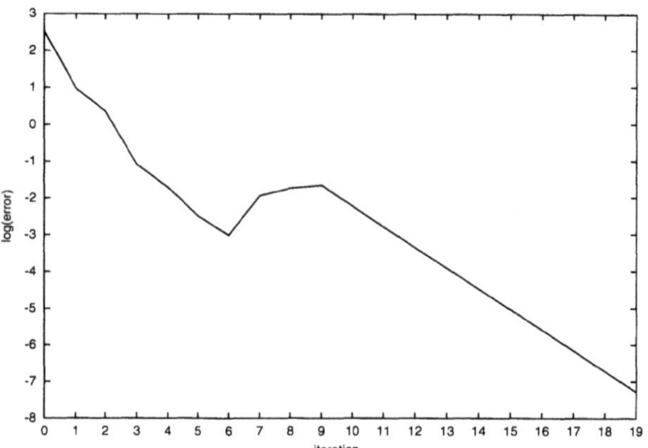

FIGURE 14. log(error) after each iteration for the Poincaré-map method applied to (3).

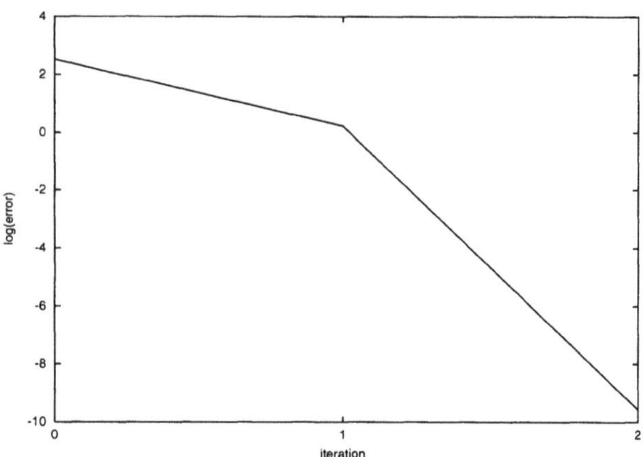

FIGURE 15. log(error) after each outer loop iteration for the accelerated Poincaré-map method applied to (3).

7. Numerical results

In this section we compare the ordinary Poincaré-map method to the MPE accelerated Poincaré-map method. They are applied to problem (9), where h was chosen as

$$h(r) := \varepsilon(1 - r). \tag{27}$$

For all $\varepsilon > 0$, the resulting problem has exactly one stable limit cycle, namely the unit circle. The parameter $\varepsilon > 0$ affects the speed of convergence towards the limit cycle; as ε approaches 0, convergence becomes increasingly slower.

ε	$1 \cdot 10^0$	$1 \cdot 10^{-1}$	$3 \cdot 10^{-2}$	$1 \cdot 10^{-2}$	$3 \cdot 10^{-3}$	$1 \cdot 10^{-3}$
Poincaré	6	35	104	290	899	2516
Accel. Poincaré	7	11	17	17	20	23

TABLE 1. The numbers of iterations (i.e., the number of evaluations of F) needed by both methods for decreasing values of ε.

The number of iterations needed for decreasing values of ε is shown in Table 1. From this it is easy to see that the Poincaré-map method becomes impractical when ε approaches 0. On the other hand, the MPE accelerated Poincaré-map method performs well even for very small values of ε.

For $\varepsilon = 3 \cdot 10^{-2}$, the errors for both methods after each iteration have been plotted in Figures 12 and 13. In Figure 13, only the error after each outer loop iteration are plotted, whereas in Table 1 the total number of iterations are listed. From Figures 12 and 13, it is clear that the Poincaré-map method gives linear convergence, whereas MPE accelerated Poincaré gives super-linear convergence for this test problem. Numerical experiments have also been performed on Colpitt's Oscillator, as shown in Figure 3. The errors after each iteration for both methods have been plotted in Figures 14 and 15. Again note that in Figure 15, only the error after each outer loop iteration are plotted.

The accelerated Poincaré-map method has been implemented in Pstar, the circuit simulator of Philips. The implementation has been tested on several larger benchmark circuits.

Example 7.1. In Figure 16, a power voltage controlled oscillator (VCO) is shown. This is a realistic implementation of an autonomous oscillator, i.e., with non-ideal inductors and capacitors. The simulation results with the Poincaré-map method in Pstar have been shown in Figure 17. For this result, Pstar required 4 outer iterations and 16 inner iterations of the accelerated Poincaré-map algorithm. △

To conclude, we observe that a considerable amount of speed-up can be accomplished by using the accelerated Poincaré-map method. The accelerated method does not require initial estimates for the period T. This makes it considerably more robust than other methods, which require such an initial guess and are very sensitive to bad guesses.

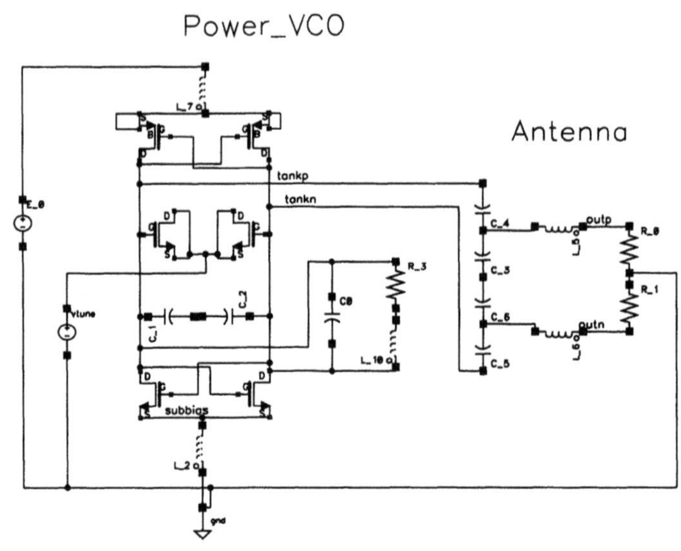

FIGURE 16. Power voltage controlled oscillator.

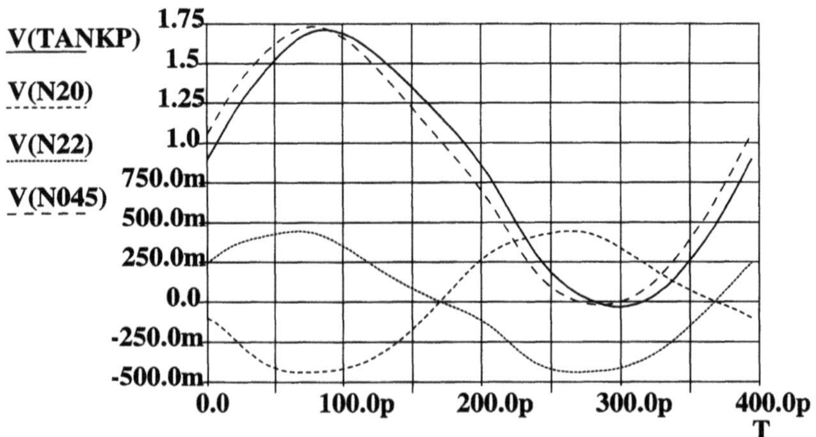

FIGURE 17. Simulation results for the power VCO.

References

[1] T.J. Aprille and T.N. Trick. Steady state analysis of nonlinear circuits with periodic inputs. *Proceedings IEEE*, 60(1):108–114, 1972.

[2] U.M. Ascher, R.M.M. Mattheij, and R.D. Russell. *Numerical solution of Boundary Value Problems for Ordinary Differential Equations.* Prentice Hall-Inc., Englewood Cliffs, New Jersey, USA, 1988.

[3] Angelo Brambilla, Dario D'Amore, and Mauro Santomauro. Simulation of autonomous circuits in the time domain. In *ECCTD'95 European Conference on Circuit Theory & Design*, pages 399–402, 1995.

[4] C. Brezinski. Convergence acceleration during the 20th century. *Journal of Computational and Applied Mathematics*, 122:1–21, 2000.

[5] E. Hairer, S.P. Nørsett, and G. Wanner. *Solving Ordinary Differential Equations I.* Springer, 1987.

[6] K. Jbilou and H. Sadok. Vector extrapolation methods. Applications and numerical comparison. *Journal of Computational and Applied Mathematics*, 122:149–165, 2000.

[7] Ken Kundert. Simulation methods for RF integrated circuits. In *Proceedings of ICCAD'97*, 1997.

[8] Linda R. Petzold. An efficient numerical method for highly oscillatory ordinary differential equations. *SIAM Journal on Numerical Analysis*, 18:455–479, 1981.

[9] D. Estévez Schwarz and C. Tischendorf. Structural analysis for electric circuits and consequences for MNA. *International Journal of Circuit Theory and Applications*, 28:131–162, 2000.

[10] A. Semlyen and A. Medina. Computation of the periodic steady state in systems with nonlinear components using a hybrid time and frequency domain method. *IEEE Transactions on Power Systems*, 10(3):1498–1504, 1995.

[11] Stig Skelboe. Time-domain steady-state analysis of nonlinear electrical systems. *Proceedings of the IEEE*, 70(10):1210–1228, 1982.

[12] David A. Smith, William F. Ford, and Avram Sidi. Extrapolation methods for vector sequences. *SIAM Review*, 29(2), 1987.

[13] R. Telichevesky, K. Kundert, I. Elfadel, and J. White. Fast simulation algorithms for RF circuits. In *Proceedings of the IEEE 1996 Custom Integrated Circuits Conference*, pages 437–444. IEEE, New York,USA, 1996.

[14] E.J.W. ter Maten. Numerical methods for frequency domain analysis of electronic circuits. *Survey on Mathematics for Industry*, 8:171–185, 1999.

[15] G. Welsh, H.G. Brachtendorf, C. Sabelhaus, and R. Laur. Minimization of the error in the calculation of the steady-state by shooting methods. Technical report, Institute of Electromagnetic Theory and Microelectronics, University of Bremen, Germany.

S.H.M.J. Houben
email: stephanh@win.tue.nl

Modeling, Simulation and Optimization of Integrated Circuits
(Eds.) K. Antreich, R. Bulirsch, A. Gilg and P. Rentrop

International Series of Numerical Mathematics Vol. 146, 121–134

Periodic AC and Periodic Noise in RF Simulation for Electronic Circuit Design

E.J.W. ter Maten, J.G. Fijnvandraat, C. Lin, and J.M.F. Peters

Abstract. This overview highlights the main steps to come to Periodic AC and Periodic Noise simulation in RF simulation. A distinction is made between forced oscillators and free oscillators. Noisy elements contribute in various ways to the overall noise effect. When dealing with free oscillators, phase noise is most important. In the case of white and coloured stochastic noise sources, the spectrum of the (stationary) autocorrelation of the phase-shifted function can be determined by summing the effects of all independent noise sources. In general one has to take the nonlinear way, in which the phase shift function changes with time, into account.

1. RF simulation methods

The newly developed simulation methods for radio frequency (RF) problems all somehow exploit the 'sparsity' of the signal spectra. The basic method is that of determining the (time) periodic steady-state (PSS) solution of a circuit. Conceptually this can be seen as a generalisation of the well-known DC operating point (the steady-state or Direct Current solution): for baseband circuits the spectral content around 0 Hz (the DC point) is important. For RF circuits the (narrow) spectral content around specific frequencies (of the PSS solution) is of interest. This PSS solution can be obtained in the frequency domain (e.g., by applying the harmonic balance method [16, 17]) or in the time domain (by methods, like shooting, based on transient simulation [7, 9, 10, 13, 18]). With baseband simulation, after determining the DC point, additional simulations like AC, noise, etc. can be done to obtain more information about the circuit. Similarly, based on the PSS solution several other simulations can be done like periodic AC, periodic noise, etc., [1, 19, 20, 21, 22]. In view of the RF circuit and signal characteristics, the PSS solution determines the nonlinear behaviour of the circuit while the periodic AC, etc. deals with the frequency shift.

The main difference between the time domain and frequency domain methods to obtain the PSS solution is that the former can easily deal with strongly nonlinear

---

2000 *Mathematics Subject Classification.* Primary 65L10; Secondary 65L80.
Key words and phrases. circuit simulation, radio frequency, RF, periodic steady state, PSS, periodic AC, periodic noise, phase noise, perturbation analysis, linear time-varying, LTV, differential-algebraic equations, autonomous systems, forced oscillators, free oscillators.

circuits and discontinuities and have good convergence properties while the latter deal naturally with components characterised in the frequency domain. Over the years combinations of both basic methods were developed resulting in mixed time-frequency domain approaches each with their own advantages and drawbacks.

A *two-step approach* appears to be powerful as well as practical for simulating *RF mixing noise*:

- Determine the *noiseless Periodic Steady-State (PSS) solution* as large-signal solution. This can be done in the time domain, the frequency domain or by using mixed time-frequency methods. The time-domain representation is a time-varying solution.

  Of course, a noiseless PSS analysis (with or without determining the oscillation frequency), has value on its own for RF simulations.
- Apply a linearisation around the PSS solution and study noise as a *small signal perturbation*. The noise sources may contain frequencies that are different from the PSS solution.

For simulating *RF phase noise* or *timing jitter* (i.e., shifts in zero crossings of the solution) of free oscillators, applying as second step a linearisation around the PSS solution and study noise as small signal perturbations is of limited use [4]. In fact, the results are only useful for small t, because the resulting deviations may grow large with time.

The *nonlinear perturbation analysis*, proposed in [4], is an alternative to the second step. Also in this approach, the first step is necessary. The nonlinear perturbation analysis results in a correct phase deviation. For the orbital deviation, again a linearisation around the PSS solution (but including phase deviation) can be used. This implies that periodicity of the coefficients of the linear time-varying differential equation can not be assumed. It also implies that, in general, the phase deviation is a time-varying function.

This paper shortly describes the PSS problem and concentrates on aspects for RF AC and RF Noise analysis.

2. The PSS problem

The Periodic Steady-State (PSS) problem for one overall period $T > 0$ is defined as:

$$\frac{\mathrm{d}}{\mathrm{d}t} q(t, x) + j(t, x) = 0 \in \mathbb{R}^N, \qquad (2.1)$$

$$x(0) - x(T) = 0, \qquad (2.2)$$

where $q(t, x)$ denotes charges (assembled at the respective nodes) and fluxes, and where $j(t, x)$ covers the static part (and sources as well). In general (2.1) is a system of Differential-Algebraic Equations (DAE). Clearly, for all $t \in \mathbb{R}$, $x(t) = x(t + T)$. A function $x : \mathbb{R} \to \mathbb{R}^n$ is called a *Periodic Steady-State Solution* if there is a $T > 0$ such that x satisfies (2.1)–(2.2). Note that according to this definition, a stationary solution (called the DC, direct current, solution), i.e., a solution of the form $x(t) \equiv x_0$, is also a PSS solution.

To define the PSS problem precisely, we have to introduce the concept of *limit cycles* and define *stability* for PSS solutions and limit cycles.

The *limit cycle* $\mathcal{C}(x)$ of a PSS solution x is the range of the function $x(t)$, i.e.,

$$\mathcal{C}(x) = \{x(t) \mid t \in \mathbb{R}\}. \tag{2.3}$$

A set $\mathcal{C}$ is called a limit cycle of (2.1) if there is a PSS solution x of (2.1) so that $\mathcal{C} = \mathcal{C}(x)$.

A PSS solution x is called *stable* (some authors prefer the term *strongly stable*) if there is a $\delta > 0$ such that for every solution x^* of (2.1) that has the property

$$\exists_{\tau_1>0} \, ||x^*(0) - x(\tau_1)|| < \delta, \tag{2.4}$$

there exists a $\tau_2 > 0$ such that

$$\lim_{t\to\infty} ||x^*(t) - x(t + \tau_2)|| = 0. \tag{2.5}$$

A limit cycle is called stable when all of its periodic steady states are stable.

Periodic steady-state solutions that are not stable are not of interest to the IC designer, since they do not correspond to any physical behaviour of the modelled circuit. In fact, we want to actively avoid non-stable periodic steady-state solutions for this reason.

An exception to the above might be the DC solution, which, in the free oscillating case, is the most well-known unstable solution. Also numerically the DC solution is of interest because it provides a way to find (approximate, initial) periodically time-varying stable solutions, by perturbing the DC solution.

For forced, or driven, (i.e., non-autonomous) problems all explicit time-dependent coefficients and sources are periodic with a common (known) period T. When dealing with autonomous circuits (also called free-running oscillator circuits) the functions q and j do not explicitly depend on time and j does not have time-dependent external sources. The PSS problem is defined by

$$\begin{aligned} \frac{\mathrm{d}}{\mathrm{d}t} q(x) + j(x) &= 0 \in \mathbb{R}^N, \\ x(0) - x(T) &= 0. \end{aligned} \tag{2.6}$$

In spite of this, a time-varying periodic steady-state solution may exist for some particular value of T. We will call this solution the oscillation solution and ω_{osc} and f_{osc}, given by $\omega_{\text{osc}} = 2\pi f_{\text{osc}} = \frac{2\pi}{T}$, the angular and "normal" oscillation frequency, respectively. In the autonomous case, solution and oscillation frequency have to be determined both. Mathematically, the problem is a nonlinear eigenproblem.

In the autonomous case, it is clear that when $x(t)$ is a solution of (2.6), another solution can simply be constructed by making a time shift: $\tilde{x}(t) = x(t-t_0)$.

To make the problem unique, in practice one gauges the solution by requiring that

$$e_i^\top x(t_0) = c \tag{2.7}$$

(for some coordinate i and constant c) [clearly c should be determined in the range of x, but not equal to a DC-value], or by imposing a condition on the time-derivative

$$e_i^\top x'(t_0) = c. \tag{2.8}$$

Now the system (2.6) and (2.7), resp. (2.8), defines a nonlinear problem with a "unique" solution for the unknowns x, T: small time shifts are excluded.

Rescaling the time by writing $t = sT$, with $s \in [0,1]$, we have

$$\begin{aligned} \frac{\mathrm{d}}{\mathrm{d}t} q(x(t)) + j(x(t)) &= \frac{1}{T}\frac{\mathrm{d}}{\mathrm{d}s} q(x(sT)) + j(x(sT)) \\ &= \frac{1}{T}\frac{\mathrm{d}}{\mathrm{d}s} q(\hat{x}(s)) + j(\hat{x}(s)), \end{aligned} \tag{2.9}$$

where $\hat{x}(s) = x(sT)$. Note that $\hat{x}(1) = x(T)$. Hence, the problem (2.6) can also be studied on the unit interval for the function $\hat{x}(s)$ after scaling the s-derivative by a factor $1/T$.

In fact, T can be nicely added to the system as well

$$\frac{1}{T}\frac{\mathrm{d}}{\mathrm{d}s} q(\hat{x}(s)) + j(\hat{x}(s)) = 0 \tag{2.10}$$

$$\frac{\mathrm{d}}{\mathrm{d}s} T = 0 \tag{2.11}$$

$$\hat{x}(0) = \hat{x}(1) \tag{2.12}$$

$$e_i^\top \hat{x}(0) = c \tag{2.13}$$

(Clearly, T automatically fulfills the periodicity condition.)

3. Perturbation analysis

We assume that the PSS solution has been determined by some numerical procedure [7, 9, 10, 13, 18]. In this section we will consider the problem for a subsequent perturbation analysis. The PSS solution of (2.1) will be denoted by x_{PSS}. It will also be called the noiseless time-varying large signal solution. Now we perturb the left-hand side of (2.1) by adding some small (noise) function n

$$\frac{\mathrm{d}}{\mathrm{d}t} q(t,x) + j(t,x) + n(t) = 0 \in \mathbb{R}^N, \tag{3.1}$$

which results in a solution of the form

$$x(t) = x_{PSS}(t + \alpha(t)) + x_n(t), \tag{3.2}$$

where the phase shift function $\alpha(t)$ still has to be prescribed and $x_n(t)$ remains small.

3.1. Linear perturbation analysis for forced systems

Linearising (3.1) around x_{PSS} (i.e., considering the case $\alpha(t) = 0$), results in a Linear Time-Varying (LTV) differential equation for x_n

$$\frac{\mathrm{d}}{\mathrm{d}t}(C(t)x_n) + G(t)x_n + n(t) = 0 \in \mathbb{R}^N, \tag{3.3}$$

$$C(t) = \left.\frac{\partial q(t,x)}{\partial x}\right|_{x_{\mathrm{PSS}}}, \quad G(t) = \left.\frac{\partial j(t,x)}{\partial x}\right|_{x_{\mathrm{PSS}}}. \tag{3.4}$$

In practical applications, a basic noise term has the form

$$n(t) = B(t)b(t), \tag{3.5}$$

$$B(t) = B(x_{\mathrm{PSS}}(t)), \tag{3.6}$$

that consists of a normalized (scalar) perturbation function $b(t)$, that is modulated by the periodic (vector) function $B(t) = B(x_{PSS}(t))$. Here $b(t)$ may be defined most conveniently in the frequency domain, while $B(x_{PSS}(t))$ is defined by expressions in the time domain. The validity of this approach was established in [4]. In noise simulations, the noise sources are included automatically by placing current sources parallel to the noisy element, which action does not disturb the topology.

For forced systems the perturbed solution $x(t)$ can be approximated by (3.2) with α being identically zero and x_n the solution of (3.3). However, when dealing with free oscillators a non-trivial choice for the phase-shift function $\alpha(t)$ has to be made.

We note that the coefficients in (3.3) are periodic in t with period T. Thus, they can be expanded in exponentials $e^{i\omega_k t}$, in which $\omega_k = 2\pi k/T$. It is instructive to consider the case for a simple sine-wave source, i.e., when

$$n(t) = Ue^{i\nu t}, \tag{3.7}$$

in which U does not depend on time, and $\nu = 2\pi f_n$; f_n may be interpreted as a noise frequency. Here ν may be different from ω_k. Introducing $y_n(t) \equiv e^{-i\nu t}x_n(t)$ results in a linear DAE of which source term and (complex) coefficients (that depend on the parameter ν) are periodical with period T

$$\frac{\mathrm{d}}{\mathrm{d}t}(C(t)y_n) + [G(t) + i\nu C(t)]y_n + U = 0 \in \mathbb{R}^N. \tag{3.8}$$

When $x_{PSS}(t) \equiv x_{DC}$, and $[G(t) + i\nu C(t)]$ is regular (and time-independent), the solution y_n is time-independent and simply equals the well-known AC solution. For the general case, we find that y_n and x_n have expansions of the form (see also [20, 21])

$$y_n(t) = \sum_{k=-\infty}^{\infty} y_{n,k}^{(\nu)} e^{i\omega_k t}, \tag{3.9}$$

$$x_n(t) = \sum_{k=-\infty}^{\infty} y_{n,k}^{(\nu)} e^{i(\nu+\omega_k)t}. \tag{3.10}$$

Because of the periodic coefficients in (3.3) and (3.8), the determination of the $y_{n,k}^{(\nu)}$ is called Periodic AC (PAC) analysis. The expansion of $x_n(t)$ implies that

$$\begin{aligned} x_n(t+T) &= \beta(\nu)x_n(t), \text{ or} &(3.11)\\ x_n(0) &= \beta(-\nu)x_n(T), \text{ where} &(3.12)\\ \beta(\nu) &= e^{i\nu T}. &(3.13)\end{aligned}$$

It is clear that, for a single input frequency ν, the solution $x_n(t)$ contains frequencies of the form $(\nu + \omega_k)$, i.e., frequency folding occurs. In practice one considers so-called wide-band input signals with input frequencies $(\nu + \omega_m)$. If we allow for several input frequencies ν_i, we can also say that a certain output frequency might originate from a large number of possible input frequencies. Hence, noise components at a certain frequency might end up in a different frequency band. This is why, for example, $1/f$ noise which has its main energy at low frequencies, still plays an important role in RF circuits.

It is important to note that we described a *linear* perturbation analysis and we will not find contributions containing for example $(\nu_1+\nu_2+\omega_k)$, $(\nu_1+2\nu_2+\omega_k)$ etc. This assumption is in general not a severe limitation when simulating noise in RF circuits.

In [20, 21] one considers the integration of (3.3) in which case the factor β easily allows adaptive re-usage of linear algebra used for solving the PSS problem (see also [2]). However, the integration of (3.8) gives rise to even more elegant algorithms. Note that now the result for x is obtained by an exact factor $e^{i\nu t}$.

3.2. Perturbation analysis for free oscillating systems

When dealing with perturbed *oscillatory* systems

$$\frac{\mathrm{d}}{\mathrm{d}t}q(x) + j(x) + n(t) = 0 \in \mathbb{R}^N, \tag{3.14}$$

it is no longer possible to assume that small perturbations $n(t)$ lead to small deviations in $x_{PSS}(t)$. [An instructive example is provided by considering $y'(t) + \cos(t)y(t) - 1 = 0$, of which the inhomogeneous solution is not periodic at all; however, note that $y(t + 2\pi)$ still satisfies the differential equation.] The main reason is that the *period* of the large signal solution is influenced by $n(t)$. This can lead to large (momentary) frequency deviations such that the difference between the noiseless and noisy solution (at the same time t) can no longer be considered to be small.

3.2.1. FLOQUET THEORY We start with the necessary background of Floquet Theory when applied to oscillatory problems and which provides a way to a proper perturbation approach [4, 5, 14, 15]. We start by noting that $x'_{\mathrm{PSS}}(t)$ satisfies the homogeneous part of (3.3)

$$\frac{\mathrm{d}}{\mathrm{d}t}(C(t)x) + G(t)x = 0, \tag{3.15}$$

and assume the case of index-1 DAEs. For literature on the higher index cases we refer to [7]. Let

$$\mathbf{S}(t) := \{z \in \mathbb{R}^N \mid (G(t) + \frac{\mathrm{d}}{\mathrm{d}t}C(t))z \in \mathrm{Im}(C(t))\}, \qquad \mathbf{N}(t) := \mathrm{Ker}(C(t)). \tag{3.16}$$

Then one has

$$\mathbf{S}(t) \cap \mathbf{N}(t) = 0, \qquad \mathbf{S}(t) \oplus \mathbf{N}(t) = \mathbb{R}^N, \tag{3.17}$$

in the index-1 case. We assume that $\mathbf{S}(t)$ is m-dimensional. There are N independent solutions of the homogeneous problem: $u_1(t)e^{\mu_1 t}, \ldots, u_m(t)e^{\mu_m t}, u_{m+1}(t), \ldots, u_N(t)$. The first $u_1(t), \ldots, u_m(t)$ are a basis of $\mathbf{S}(t)$; the last, $u_{m+1}(t), \ldots, u_N(t)$, are a basis of $\mathbf{N}(t)$. The $\mu_1, \ldots, \mu_m$ are so-called Floquet exponents; the $e^{\mu_1 T}, \ldots, e^{\mu_m T}$ are Floquet multipliers. For a stable autonomous index-1 problem we can assume that $\mu_1 = 0$ and that $\mathrm{Re}(\mu_i) < 0$ for $i = 2, \ldots, m$. In this case we can choose $u_1(t) = x'_{\mathrm{PSS}}(t)$.

The homogeneous adjoint (or dual) system corresponding to (3.3) is

$$C^\top(t)\frac{\mathrm{d}}{\mathrm{d}t}y - G^\top(t)y = 0. \tag{3.18}$$

Similar to the non-adjoint case we introduce

$$\mathbf{S}^\top(t) := \{z \in \mathbb{R}^n \mid G^\top(t)z \in \mathrm{Im}(C^\top(t))\}, \qquad \mathbf{N}^\top(t) := \mathrm{Ker}(C^\top(t)), \tag{3.19}$$

with the properties

$$\mathbf{S}^\top(t) \cap \mathbf{N}^\top(t) = 0, \qquad \mathbf{S}^\top(t) \oplus \mathbf{N}^\top(t) = \mathbb{R}^N. \tag{3.20}$$

Also $\mathbf{S}^\top$ is m-dimensional. The adjoint problem has N independent solutions: $v_1(t)e^{-\mu_1 t}, \ldots, v_m(t)e^{-\mu_m t}, v_{m+1}(t), \ldots, v_N(t)$, where $v_1(t), \ldots, v_m(t)$ are a basis of $\mathbf{S}^\top(t)$ and the last, $v_{m+1}(t), \ldots, v_N(t)$, are a basis of $\mathbf{N}^\top(t)$.
It easy to verify that if x and y are solutions of (3.3) and (3.15), respectively, the inner-products $y^\top(t)C(t)x(t)$ are constant, thus $y^\top(t)C(t)x(t) = y^\top(0)C(0)x(0)$, for all $t \geq 0$. More specifically, the bases $u_1(t), \ldots, u_N(t)$ and $v_1(t), \ldots, v_N(t)$ can be chosen such that, the $N \times N$ matrix $U(t)$ with as columns the $u_i(t)$ and the $N \times N$ matrix $V(t)$ with as rows the $v_i(t)$ satisfy a bi-orthogonality relation w.r.t. $C(t)$ and a nearly one w.r.t. $G(t)$

$$V(t)C(t)U(t) = \begin{pmatrix} I_m & 0 \\ 0 & 0 \end{pmatrix}, \qquad V(t)G(t)U(t) = \begin{pmatrix} J_m^1 & 0 \\ J_m^2 & J_m^3 \end{pmatrix}. \tag{3.21}$$

Here I_m is an $m \times m$ identity matrix. J_m^1 is an $m \times m$ block matrix. J_m^2 and J_m^3 are suitable block matrices.

The monodromy matrix is the state-transition matrix after one period, i.e., $\Phi(T) = U(T)D(T)V(0)C(0)$, where $D(T) = \mathrm{diag}(e^{\mu_1 T}, \ldots, e^{\mu_m T}, 0, \ldots, 0)$ (this matrix arises in natural way when one considers shooting methods or applies Floquet theory to analyze stability of a limit cycle). Because of the periodicity of the u_i, we see that the $u_i(0)$, for $i = 1, \ldots, m$, are eigenvectors of the monodromy matrix with corresponding eigenvalues $e^{\mu_i T}$, and that the remaining $u_i(0)$, for $i = m+1, \ldots, N$, are eigenvectors for the $(N - (m-1))$-fold eigenvalue 0. Similar to the non-adjoint case, the $v_i(0)$ are eigenvectors of the associated monodromy matrix $\Psi(T) = V^\top(T)D(-T)U^\top(0)C^\top(0)$.

3.2.2. Phase noise by nonlinear perturbation analysis We will take $u_1(t) = x'_{\rm PSS}(t)$. Here u_1 is the tangent to the orbit. Let $\alpha(t)$ be a (sufficiently smooth) phase- or time-shift function and let $s = t + \alpha(t)$ be the shifted time. If $x_{\rm PSS}(t)$ is the PSS solution of (2.6) then the phase-shifted function $y(t) \equiv x_{\rm PSS}(s) = x_{\rm PSS}(t + \alpha(t))$ satisfies

$$\begin{aligned} \frac{\mathrm{d}}{\mathrm{d}t}q(y) + j(y) &= \frac{\mathrm{d}}{\mathrm{d}s}q(x_{\rm PSS}(s)).\frac{\mathrm{d}s}{\mathrm{d}t} + j(x_{\rm PSS}(s)) \qquad (3.22)\\ &= \frac{\mathrm{d}}{\mathrm{d}x_{\rm PSS}}q(x_{\rm PSS}(s))\frac{\mathrm{d}x_{\rm PSS}}{\mathrm{d}s}\alpha'(t)\\ &= C(t+\alpha(t))u_1(t+\alpha(t))\alpha'(t). \end{aligned}$$

Hence, the phase-shifted function y satisfies a perturbed DAE in which the right-hand side has a particular form.

We now consider perturbations of the form $B(x(t))b(t)$ (cf. also (3.3)) to the original DAE (2.6)

$$\frac{\mathrm{d}}{\mathrm{d}t}q(x) + j(x) + B(x(t))b(t) = 0, \qquad (3.23)$$

and express $B(x(t+\alpha(t)))b(t)$ into its components using the basis $\{C(t+\alpha(t))u_1(t+\alpha(t)), \ldots, C(t+\alpha(t))u_m(t+\alpha(t)), G(t+\alpha(t))u_{m+1}(t+\alpha(t)), \ldots, G(t+\alpha(t))u_N(t+\alpha(t))\}$

$$\begin{aligned} B(x(t+\alpha(t)))b(t) &= \sum_{i=1}^{m} c_i(x,\alpha(t),t)C(t+\alpha(t))u_i(t+\alpha(t)) \qquad (3.24)\\ &\quad + \sum_{i=m+1}^{N} c_i(x,\alpha(t),t)G(t+\alpha(t))u_i(t+\alpha(t)),\\ c_i(x,\alpha(t),t) &= \tilde{v}_i(t+\alpha(t))b(t), \quad \text{where} \qquad (3.25)\\ \tilde{v}_i(t) &= v_i^\top(t)B(x(t)). \qquad (3.26) \end{aligned}$$

Here the scalar functions $\tilde{v}_i(t)$ are periodical in t with period T.
The first component of $B(x(t+\alpha(t)))b(t)$ will be used to determine $\alpha(t)$. We define $\alpha(t)$ to satisfy the nonlinear, scalar, differential equation

$$\begin{aligned} \alpha'(t) &= -v_1^\top(t+\alpha(t))B(x_{\rm PSS}(t+\alpha(t)))b(t), \quad \alpha(0) = 0 \qquad (3.27)\\ &= -\tilde{v}_1(t+\alpha(t))b(t), \quad \alpha(0) = 0. \end{aligned}$$

(See also [11, 12] where a first start was made to treat the phase noise problem in the time domain.) Note that if $b(t) = 0$, for $t \geq t_0$, then α becomes a constant phase shift, and the phase-shifted function $y(t)$ solves exactly (3.23) for $t \geq t_0$. Also note that the explicit form of $B(x(t))$ is never used. If we take $B(x(t))b(t) =$

$C(x(t))u_1(t)b(t)$, with $b(t)$ scalar, bi-orthogonality implies

$$\alpha'(t) = -b(t), \quad \alpha(0) = 0, \quad \text{hence} \tag{3.28}$$

$$\alpha(t) = -\int_0^t b(t)\mathrm{d}t. \tag{3.29}$$

Clearly, for this case, nonlinearity does not occur. For $b(t) = \sin(\omega t)$ we find $\alpha(t) = \frac{1}{\omega}\cos(\omega t)$, which shows a $1/\omega$ behaviour. In general, however, (3.28) is nonlinear and this effect is really needed when modelling Phase Locked Loop oscillators. A linear model, like in [8], is of limited use. Also even for small b, the phase shift function $\alpha(t)$ may increase with time (Clearly, if $N = 1, B \equiv 1, b(t) \equiv \varepsilon$, and $v_1^\top(t) \equiv \kappa$, then $\alpha(t) = \kappa\varepsilon t$).

To find α, we clearly have to know v_1. In [6] this crucial vector is called *Perturbation Projection Vector*, or *PPV*. It represents a transfer between the perturbation of the DAE and the resulting phase shift.

In [4] v_1 is determined by performing first an eigenvalue/eigenvector analysis of the monodromy matrix of the adjoint problem to obtain $v_1(0)$, followed by time integration (backward in time). To discern the proper initial value $v_1(0)$ from other eigenvectors that have eigenvalues close to 0, one can exploit the bi-orthogonality relation (3.21), because $v_1(0)$ must have a non-trivial C-inner-product with $u_1(0)$. Another, direct, approach is found in [6]. It nicely fits a Finite Difference Method approach and again exploits the bi-orthogonality relation (3.21) in an elegant way.

For deterministic perturbations one has to integrate (3.28). Because of this action in the time domain, all Fourier components of $n(t)$ are treated in a combined way. However, for stochastic noise, such a detail is not necessary. In [4, 5] expressions for the power due to the noise are derived that depend on the asymptotic behaviour (i.e., for large t) of the variance var$[\alpha(t)]$. The authors derive power spectrum expressions that depend on the Fourier components of the PSS solution x_{PSS}, on the DC component of $v_1(t)$, and on the power spectrum of b. The power of the j-th harmonic of x_{PSS} is preserved in the power of the 'asymptotic' j-th harmonic of y (i.e., the shifted x_{PSS}). Consequently, by summing over j, we see that also the total power is preserved. In case of stochastically independent sources, one can determine the contributions to the variance var$[\alpha(t)]$ quite elegantly.

More specifically, let

$$x_{\mathrm{PSS}}(t) = \sum_{j=-\infty}^{\infty} X_j e^{i\omega_j t}, \tag{3.30}$$

where $\omega_j = 2\pi j f_{\mathrm{osc}} = 2\pi j/T$. The stationary autocorrelation of the phase-shifted function $y(t)$ can be expressed as an asymptotic stochastic expectation of the autocorrelation

$$R_y(\tau) \equiv \lim_{t\to\infty} R_y(t,\tau), \quad \text{where} \tag{3.31}$$

$$R_y(t,\tau) = E[y(t)y^*(t+\tau)].$$

The spectral density that corresponds to $R_y(\tau)$ can be expressed as

$$S_y(\omega) = \sum_{j=-\infty}^{\infty} X_j X_j^* S_j(\omega + \omega_j), \tag{3.32}$$

where each S_j satisfies $\int_{\infty}^{\infty} S_j(2\pi f)\mathrm{d}f = 1$. The expressions for S_j can be related to the spectra for the local noise sources $b(t)$ and involve the DC component of the PPV transfer $\tilde{v}_1(t)$ [4, 5].

In practice, noise criteria are based on the two-sided phase noise power spectral density in V^2/Hz

$$\begin{aligned} S_{\text{pp}}^k(\omega) &= S_y(\omega + \omega_k), \\ &= \sum_{j=-\infty}^{\infty} X_j X_j^* S_j(\omega + \omega_k + \omega_j), \\ &= \sum_{j=-\infty}^{\infty} X_j X_j^* S_j(\omega + \omega_{k+j}). \end{aligned} \tag{3.33}$$

Summation is performed for all j for which both j and $j+k$ are in the range of calculated harmonics. Other well-known representations are the power spectral density in dB/Hz

$$dBS_{\text{pp}}^k = 10\log_{10}(S_{\text{pp}}^k), \tag{3.34}$$

and the power spectral density with respect to the carrier in dBc/Hz

$$dBcS_{\text{pp}}^k = 10\log_{10}\left(\frac{S_{\text{pp}}^k}{|X_k|^2}\right). \tag{3.35}$$

3.2.3. Orbital deviation The solution x of (3.23) may be written as $x(t) = y(t) + z(t)$, where z is the orbital deviation function, which can be analysed by a proper linear perturbation analysis (but with linearised equations which now have non-periodic coefficients!). Because n also affects the phase-shifted function, around which one linearises to study the orbital devations, there is no simple summation formula known for cumulative noise contributions. In [4, 5] it is argued that, in first order, $z(t)$ satisfies $||z||_\infty <$ Const.$||b||_\infty$ (and even $z(t) \to 0$ $(t \to \infty)$). We will assume that b is a scalar function, for simplicity. Then, in first order, z satisfies a linear differential equation

$$\frac{\mathrm{d}}{\mathrm{d}t}[C_y(t)z(t)] + [G_y(t) + B_y(t)b(t)]z(t) + B_1(y(t))b(t) = 0, \tag{3.36}$$

where

$$\begin{aligned} C_y(t) &= \frac{\mathrm{d}}{\mathrm{d}y}q(y(t)), \qquad G_y(t) = \frac{\mathrm{d}}{\mathrm{d}y}j(y(t)), \\ B_y(t) &= \frac{\mathrm{d}}{\mathrm{d}y}B(y(t)), \qquad B_1(y(t))b(t) = [B(y(t)) - B_2(y(t))]b(t), \\ B_2(y(t)) &= v_1^\top(t+\alpha(t))B(y(t))C(t+\alpha(t))u_1(t+\alpha(t)), \end{aligned}$$

where B_2 is the component of B in the C-inner-product along u_1. If we write $z(t) = \tilde{z}(t + \alpha(t)) = \tilde{z}(s)$, (3.36) can be written more in terms of s

$$\begin{aligned} -\frac{\mathrm{d}}{\mathrm{d}s}[C(x_{\mathrm{PSS}}(s))\tilde{z}(s)]v_1^{\top}(s)B(x_{\mathrm{PSS}}(s))b(t) & \\ +[G(x_{\mathrm{PSS}}(s)) + B_x(x_{\mathrm{PSS}}(s))b(t)]\tilde{z}(s) + B_1(x_{\mathrm{PSS}}(s))b(t) &= 0. \end{aligned} \tag{3.37}$$

We note that if $b(t) \equiv 0$ for $t \geq t_0$, $\tilde{z}(s) = 0$ for some $s \geq s_0$ (assuming that $G(x_{\mathrm{PSS}}(s))$ has full rank). If $b(t) \equiv c$ for some constant c all coefficients in (3.37) are periodic in s.

For small b this linear differential equation in $\tilde{z}(s)$ is singularly perturbed and inhomogeneous. The coefficients depend on b, which implies that the transfer function depends on the (group of) sources as well.

If we assume that $\tilde{z}(s)$ and $\tilde{z}'(s)$ are both of order $\mathcal{O}(||b||)$, (3.37) implies the first order linear system

$$\begin{aligned} G(x_{\mathrm{PSS}}(s))\tilde{z}(s) &= -B_1(x_{\mathrm{PSS}}(s))b(t), \\ &= -[B(x_{\mathrm{PSS}}(s)) - B_2(x_{\mathrm{PSS}}(s))]b(t), \\ &= -[B(x_{\mathrm{PSS}}(s)) - v_1^{\top}(s)B(x_{\mathrm{PSS}}(s))C(x_{\mathrm{PSS}}(s))u_1(s)]b(t), \\ &= -[B(x_{\mathrm{PSS}}(s)) - v_1^{\top}(s)B(x_{\mathrm{PSS}}(s))C(x_{\mathrm{PSS}}(s))u_1(s)]\tilde{b}(s), \end{aligned} \tag{3.38}$$

where $\tilde{b}(s)$ is defined by $\tilde{b}(s) = b(t)$. We note that for the system (3.38), at the right-hand side, the term between the brackets can completely be determined in the s-variables; note that all quantities are periodic in s. The factor $\tilde{b}(s)$ has to be determined in the frequency domain.

4. Simple RLC oscillator example

We consider a simple RLC oscillator to simulate phase noise. The schematic is shown in Fig. 1 (Top-Left). Here $\mathrm{L}_0 = 1\mu\mathrm{H}$, $\mathrm{C}_1 = 1\mu\mathrm{F}$, $\mathrm{E}_0 = 5.0\mathrm{V}$, $\mathrm{R}_3 = 1m\Omega$, $\mathrm{R}_l = 50\Omega$, $\mathrm{R}_s = 50\Omega$. An oscillator is "controlledly unstable". The instability is caused by the negative resistance, $\mathrm{R}_0 : i = 0.4\tanh(-5v)$, the nonlinearity provides control in excluding unlimited growth. The value 0.4 determines the height of the oscillation amplitude and the $\tanh(-5v)$ factor determines the time moment when the oscillation will start.

The basic RLC-components are $\mathrm{L}=\mathrm{L}_0$, $\mathrm{C}=\mathrm{C}_1$ and $\mathrm{R}=\mathrm{R}_3$. The series resistor R_3 models the effect of a non-ideal coil. The resonant angular frequency ω_0 of an RLC oscillator, with a non-ideal coil, can be determined by $\omega_0 = \frac{1}{\sqrt{LC}}\sqrt{1 - \frac{R^2C}{4L}} \approx \frac{1}{\sqrt{LC}} = 10^6$ because $\frac{R^2C}{4L} \ll 1$. This corresponds to a frequency $f_0 = 159.24\mathrm{kHz}$. In general one will estimate the frequency by a pole-zero analysis, or by applying the Kurokawa condition [16], or from an initial transient analysis, or as part of a PSS analysis [9, 10]. The intended oscillation can be observed at the nodal voltage $v_n(1)$. From the PSS analysis one derives $f_0 = 159.11\mathrm{kHz}$ as PSS frequency. In Fig. 1 results are shown. At the Top-Right the amplitude of the Fourier Transform

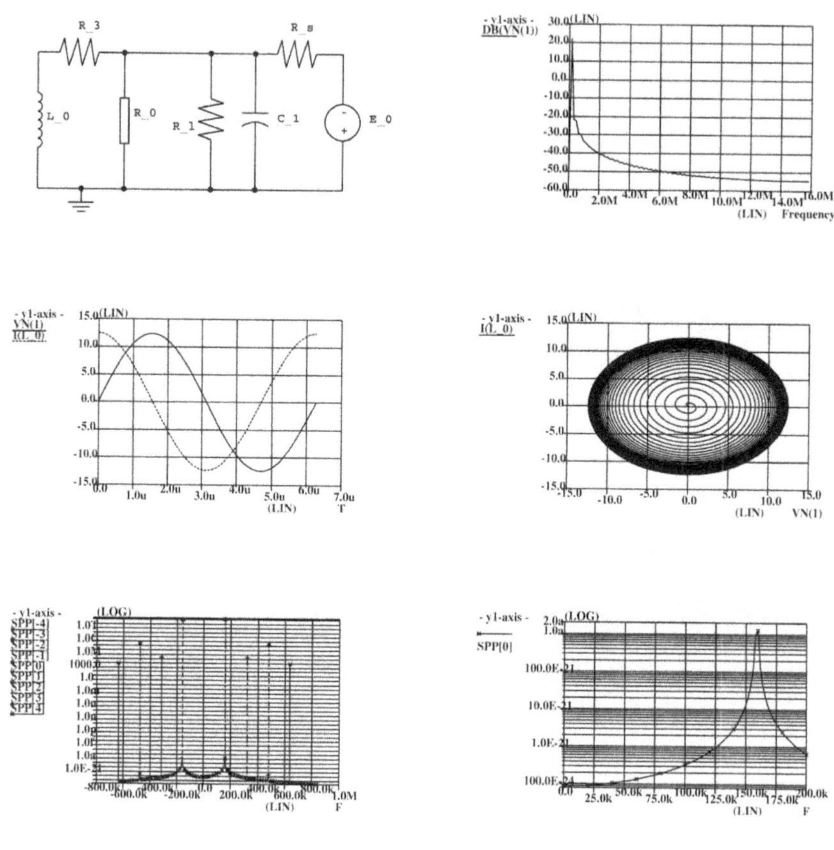

FIGURE 1. RLC Oscillator: *Top-Left:* Schematic, *Top-Right:* Amplitude of Fourier Transform of $v_n^{\mathrm{PSS}}(1)$, *Center-Left:* Time signal of $v_n^{\mathrm{PSS}}(1)$, *Center-Right:* $i(L_0)$ is plotted against $v_n(1)$, *Bottom-Left:* Output phase noise voltage spectrum S_{pp}^k, *Bottom-Right:* Output phase noise voltage spectrum S_{pp}^0.

of $v_n(1)$ is shown, which has magnitude about 21.9 dBV, which compares well to the amplitude of the PSS time signal, shown at Center-Left. In the Center-Right figure, the current $i(L_0)$ is plotted against $v_n(1)$. Sweeping the "noise" frequency f_{sw} from 0 to 200kHz in a Periodic AC analysis leads to output at frequencies $f_{\mathrm{out}} = f_{\mathrm{sw}} + k f_0$. At the Bottom-Right the output phase noise S_{pp}^k voltage spectrum in V^2/Hz is displayed for the whole circuit. S_{pp}^0 gives the phase noise around the oscillation frequency. The smaller the integrated area, the less phase noise is present.

References

[1] W. Anzill, F.X. Kärtner, P. Russer: *Simulation of the phase noise of oscillators in the frequency domain*, Int. J. Electron. Commun. (AEÜ) **48-1** (1994), 45–50.

[2] W. Bomhof, H.A. van der Vorst (2001): *A parallelizable GMRES-type method for p-cyclic matrices with applications in circuit simulation*, in U. van Rienen, M. Günther, D. Hecht (Eds): *Scientific computing in electrical engineering*, Proc. SCEE-2000, Warnemünde, Springer, Berlin (2001), 293–300.

[3] A. Demir, A. Sangiovanni-Vincentelli: *Analysis and Simulation of noise in nonlinear electronic circuits and systems*, Kluwer Academic Publ., Boston, USA, 1998.

[4] A. Demir, A. Mehrotra, J. Roychowdhury: *Phase noise in oscillators: a unifying theory and numerical methods for characterisation*, IEEE Trans. on Circuits and Systems – I: Fund. Theory and Applics. **47-5** (2000), 655–674.

[5] A. Demir: *Phase noise in oscillators: DAEs and coloured noise sources*, Proc. ICCAD'98, Int. Conf. on Computer Aided Design, San José, CA, USA (1998), 170–177.

[6] A. Demir, D. Long, J. Roychowdhury: *Computing phase noise eigenfunctions directly from Harmonic Balance/Shooting matrices*, Proc. ICCAD 2000, Int. Conf. on Computer Aided Design, San José, CA, USA (2000), 283–288.

[7] M. Günther, U. Feldmann, J. ter Maten: *Discretization of circuit problems*, to appear in *Numerical Analysis of Electromagnetics*, Series Handbook of Numerical Analysis, Elsevier Science.

[8] A. Hajimiri, T.H. Lee: *A general theory of phase noise in electrical oscillators*, IEEE J. Solid-State Circ., **33-2** (1998), 179–194.

[9] S.H.M.J. Houben, E.J.W. ter Maten, J.M. Maubach, J.M.F. Peters: *Novel time-domain methods for free-running oscillators*, In: V. Porra, M. Valtonen, I. Hartimo, M. Ilmonen, O. Simula, T. Veijola (Eds): *ECCTD'01 – Proceedings of the 15TH European Conference on Circuit Theory and Design*, Helsinki University of Technology (ISBN 951-22-5571-5), Finland (2001), III-393–III-396.

[10] S.H.M.J. Houben: *Time-domain simulation techniques for finding the PSS of electric oscillators*, these Proceedings.

[11] F.X. Kärtner: *Untersuchung des Rauschverhaltens von Oszillatoren*, PhD-Thesis TU München, 1989.

[12] F.X. Kärtner: *Analysis of white and $f^{-\alpha}$ noise in electrical oscillators*, Int. J. Circuit Theory Appl. **18** (1990), 485–519.

[13] T.A.M. Kevenaar: *Periodic Steady State Analysis using Shooting and Waveform-Newton*, Int. J. Circuit Theory and Applics. **22** (1994), 51–60.

[14] R. Lamour, R. März, R. Winkler: *How Floquet Theory applies to Index 1 differential algebraic equations*, J. of Math. Analysis and Applics. **217** (1998), 372–394.

[15] R. Lamour: *Floquet-Theory for differential-algebraic equations (DAE)*, ZAMM **78-3** (1998), S989–S990.

[16] S.L̃ampe, H.G. Brachtendorf, E.J.W. ter Maten, S.P. Onneweer, R. Laur: *Robust limit cycle calculations of oscillators*, In U. van Rienen, M. Günther and D. Hecht (Eds.): *Scientific computing in electrical engineering*, Proc. SCEE-2000, Warnemünde/Rostock, LNCSE 18, Springer Verlag, Berlin (2001), 233–240.

[17] E.J.W. ter Maten: *Numerical methods for frequency domain analysis of electronic circuits*, Surv. Math. Ind. **8** (1999), 171–185.

[18] K. Mayaram, D.C. Lee, S. Moinian, D. Rich, J. Roychowdhury: *Overview of computer-aided analysis tools for RFIC: algorithms, features, and limitations*, IEEE 1997 Custom Integrated Circuit Conference, Santa Clara, CA, USA (1997), 505–512.

[19] M. Okumura, T. Sugawara, H. Tanimoto: *An efficient small signal frequency analysis method for nonlinear circuits with two frequency excitations*, IEEE Trans. of Comp.-Aided Design of Integrated Circuits and Systems **9-3** (1990), 225–235.

[20] M. Okumura, H. Tanimoto, T. Itakura, T. Sugawara: *Numerical noise analysis for nonlinear circuits with a periodic large signal excitation including cyclostationary noise sources*, IEEE Trans. on Circuits and Systems – I: Fund. Theory and Applics. **40-9** (1993), 581–590.

[21] R. Telichevesky, K.S. Kundert, J.K. White: *Efficient Steady-State Analysis based on matrix-free Krylov-Subspace methods*, Proc. DAC'95, 1995.

[22] R. Telichevesky, K. Kundert: *Efficient AC and noise analysis of two-tone RF circuits*, Proc. DAC'96, Las Vegas, 1996.

Acknowledgment

We thank all involved in organising the inspiring 2001 Oberwolfach Meeting 48 on *Modellierung, Simulation und Optimierung integrierter Schaltkreise*.

Joint address:
Philips Research Laboratories and
Eindhoven University of Technology
c/o Prof. Holstlaan 4, Bldng WAY
NL-5656 AA Eindhoven, The Netherlands
email: `Jan.ter.Maten@philips.com`

Modeling, Simulation and Optimization of Integrated Circuits
(Eds.) K. Antreich, R. Bulirsch, A. Gilg and P. Rentrop

International Series of Numerical Mathematics Vol. 146, 135–151

Differential Algebraic Systems with Properly Stated Leading Term and MNA Equations

R. März

Abstract. Differential algebraic equations with properly stated leading term are equations of the form $A(x(t),t)(d(x(t),t))' + b(x(t),t) = 0$ with in some sense well-matched coefficients. Systems resulting from the modified nodal analysis (MNA) in circuit simulation promptly fit into this form.
Recent results concerning solvability and numerical treatment of those equations are discussed. An index notion that works via linearization is given. This allows for index criteria just in terms of the coefficients A, d, b and their first partial derivatives, no further derivative arrays are used.

1. Introduction

Differential algebraic equations (DAEs) with properly stated leading term are equations of the form

$$A(x(t),t)(d(x(t),t))' + b(x(t),t) = 0, \tag{1.1}$$

with in some sense well-matched coefficients A and d. Roughly speaking, there is no gap but also no overlap between A and d, and the seam in between is solution-independent.

By means of the DAE (1.1), more information on the process to be modelled is preserved than by a standard form DAE

$$E(x(t),t), x'(t) + f(x(t),t) = 0 \tag{1.2}$$

Namely, it is precisely figured out which derivatives of the unknown function are involved and in what way. In contrast to this, equation (1.2) with singular $E(x,t)$ leaves this question open and suggests that all components of the solution are involved together with their derivatives.

Stronger solvability results, and a better performance of numerical methods can be realized as benefits of the more precise model. For instance, as it is well

---

2000 *Mathematics Subject Classification.* 34A09, 34A34, 65L80.
Key words and phrases. differential algebraic equations, tractability index, index monitor, modified nodal analysis.

known, the famous index two DAE (cf. [BrCaPe], page 46)

$$\begin{pmatrix} 0 & 0 \\ 1 & \eta t \end{pmatrix} x'(t) + \begin{pmatrix} 1 & \eta t \\ 0 & 1+\eta \end{pmatrix} x(t) = \begin{pmatrix} g(t) \\ 0 \end{pmatrix}, \tag{1.3}$$

which has the solution $x_1(t) = g(t) + \eta t g'(t)$, $x_2(t) = -g'(t)$, cannot be solved by the implicit Euler method if $\eta < \frac{1}{2}$. However, rewriting (1.3) with properly stated leading term as

$$\begin{pmatrix} 0 \\ 1 \end{pmatrix} (x_1(t) + \eta t x_2(t))' + \begin{pmatrix} 1 & \eta t \\ 0 & 1 \end{pmatrix} x(t) = \begin{pmatrix} g(t) \\ 0 \end{pmatrix}, \tag{1.4}$$

and then applying the implicit Euler method yields $x_{1n} = g(t_n) + \eta t_n \frac{1}{h_n}(g(t_n) - g(t_{n-1}))$, $x_{2n} = -\frac{1}{h_n}(g(t_n) - g(t_{n-1}))$, i.e., in this version the implicit Euler method provides good approximations.

The idea to consequently formulate the leading term in a DAE by means of two well-matched coefficients was born in [BaMä]. First, the reason was only the desire for more symmetry in DAE theory, in particular, a uniform treatment of linear DAEs and their adjoints. While a linear standard form DAE $E(t)x'(t) + F(t)x(t) = q(t)$ and its adjoint $-(E(t)^*y(t))' + F(t)^*y(t) = r(t)$ are equations of completely different type, a DAE with proper leading term

$$A(t)(D(t)x(t))' + B(t) = q(t) \tag{1.5}$$

has the adjoint equation

$$-D(t)^*(A(t)^*y(t))' + B(t)^*y(t) = r(t), \tag{1.6}$$

which, in turn, has a properly stated leading term. The solutions $x(.)$ and $y(.)$ satisfy a generalized Lagrange identity, and there are interesting relations among the characteristic subspaces corresponding to (1.5) and (1.6), respectively ([BaMä]). Positive consequences for the optimal control can be expected here.

Numerical integration methods for lower index DAEs with properly stated leading term are considered in [HiMä], [HiMäTi], [HiMäTi2], [Te], [Fl].

A completely different aspect, which emphasizes the importance of equations of the form (1.1), is the fact that the modified nodal analysis (MNA) used in circuit simulation provides DAEs promptly fitting into the form (1.1) provided that the structured description is used [EsTi], cf. Section 2 below). From this point of view the results on the DAE (1.1) and their numerical treatment give the theoretical background for many things that are already common practice in circuit simulation due to practical experience and intuition.

At present, one of the challenging tasks in circuit simulation is monitoring the DAE index before or during the simulation. In this context let us remark that, unfortunately, the so-called structural index of linear constant coefficients, which can be calculated relatively easily by means of the Pantelides algorithm, is not relevant here (cf. [ReMaBa]). The development of a derivative array ([BrCaPe]) is impossible because of the missing smoothness on the one hand as well as because of the huge dimensions in circuit simulation on the other hand.

Basing on the tractability index, an index monitor that can detect whether the given DAE has index one, two or greater than two has been developed for the simulation package TITAN (cf. [Es et al.]). This index monitor uses topological methods as well as numerical criteria. Now this gives rise to the explicit wish for exactly determining higher indices, too. This paper aims at developing appropriate index criteria for that. We hope that these criteria can later be applied by using structural properties by means of topological methods.

The paper is organized as follows. In Section 2, structured MNA equations due to [EsTi] are described. Section 3 contains the basics on DAEs with properly stated leading term. Results on index one DAEs are collected in Section 4. Section 5 concerns linearizations, and the tractability index is introduced for linear DAEs. Finally, in Section 6, the tractability index of nonlinear DAEs is given in such a way that the corresponding notion for linear DAEs appears to be a special case, and, furthermore, all admissible linearizations of a nonlinear index μ DAE have index μ, too. Let us emphasize here that just the first partial derivatives of A, d, b are used, but no higher derivatives and derivative arrays. We finish with proving further necessary index μ criteria concerning the so-called local matrix pencil.

2. MNA equations

The modified nodal analysis (MNA), which is one of the most applied modelling techniques in circuit simulation packages, provides equation systems of the form (cf. [EsTi])

$$
\begin{aligned}
A_C(q(A_C^T e(t),t))' \; & + A_R r(A_R^T e(t),t) + A_L j_L(t) + A_V j_V(t) \\
& + A_I i(\tilde{A}^T e(t), j_L(t), j_V(t), t) && = 0, && (2.1)\\
(\phi(j_L(t),t))' \; & - A_L^T e(t) && = 0, && (2.2)\\
& A_V^T e(t) - v(\tilde{A}^T e(t), j_L(t), j_V(t)) && = 0. && (2.3)
\end{aligned}
$$

The unknown functions are $e(t), j_L(t)$ and $j_V(t)$, where $e(t)$ consists of the node potentials (excepting the datum node), $j_L(t)$ and $j_V(t)$ combine the currents of inductances and voltage sources, respectively. A_C, A_L, A_R, A_V and A_I are the element-related incidence matrices, they describe the branch-current relations for the capacitive branches, inductive branches, resistive branches, branches of voltage sources, and branches of current sources.

$\tilde{A} := (A_C, A_L, A_R, A_V, A_I)$ is the (reduced) incidence matrix of the whole system, its entries are just from $\{-1, 0, 1\}$. The functions $q(u,t)$ and $\phi(j,t)$ describe charges and fluxes, and the partial Jacobians

$$C(u,t) := q_u(u,t), \quad L(j,t) := \phi_j(j,t)$$

are positive-definite.

Choosing

$$x := \begin{pmatrix} e \\ j_L \\ j_V \end{pmatrix}, \quad A := \begin{pmatrix} A_C & 0 \\ 0 & I \\ 0 & 0 \end{pmatrix}, \quad d(x,t) := \begin{pmatrix} A_C^+ A_C q(A_C^T e, t) \\ \phi(j_L, t) \end{pmatrix},$$

with the Moore-Penrose Inverse A_C^+ of A_C,

$$b(x,t) := \begin{pmatrix} A_R r(A_R^T e, t) + A_L j_L + A_V j_V + A_I i(\tilde{A}^T e, j_L, j_V, t) \\ -A_L^T e \\ A_V^T e - v(\tilde{A}^T e, j_L, j_V, t) \end{pmatrix},$$

we rewrite the MNA system (2.1)–(2.3) as

$$A(d(x(t),t))' + b(x(t),t) = 0. \tag{2.4}$$

The partial Jacobian $D(x,t) := d_x(x,t)$ is now of the special form

$$D(x,t) = \begin{pmatrix} A_C^+ AC(A_C^T e, t) A_C^T & 0 \\ 0 & L(j_L, t) \end{pmatrix} = \begin{pmatrix} A_C^+ AC(A_C^T e, t) & \\ & L(j_L, t) \end{pmatrix} A^T. \tag{2.5}$$

We observe that

$$AD(x,t) = A \begin{pmatrix} C(A_C^T e, t) & 0 \\ 0 & L(j_L, t) \end{pmatrix} A^T,$$

where the inner factor on the right-hand side is a positive-definite $n \times n$ matrix. Recall that A is a large rectangular matrix. It follows that (cf. [EsTi], Lemma 2.4)

$$\ker AD(x,t) = \ker A^T, \quad \operatorname{im} AD(x,t) = \operatorname{im} A \tag{2.6}$$

is valid independently of x and t. Then we also have the relations

$$ker A \cap im D(x,t) = 0, \quad im D(x,t) = im A^+ A = (ker A)^\perp, \tag{2.7}$$

since $Az = 0, z = D(x,t)w$ yield $AD(x,t)w = 0$, thus $D(x,t)w = 0$, i.e., $z = 0$. It comes out that we are given the decomposition

$$\mathbb{R}^n = ker A \oplus im D(x,t). \tag{2.8}$$

Obviously, the constant projector $R := A^+ A = \begin{pmatrix} A_C^+ A_C & 0 \\ 0 & I \end{pmatrix}$ realizes the decomposition (2.8).

At this place we emphasize that (2.4) is, formally, slightly different from (2.1)–(2.3). Namely, while the first term in (2.1) is $A_C(q(A_C^T e(t), t)' = A_C A_C^+ A_C (q(A_C^T e(t), t))'$, the corresponding one in (2.4) is $A_C(A_C^+ A_C q(A_C^T e(t), t))'$. But, since we are allowed to move the constant matrix factor $A_C^+ A_C$ from outside into the inner of the derivative, we know these terms to coincide in fact.

Notice that the factor $A_C^+ A_C$ is introduced to obtain a constant subspace $im D(x,t)$. As we shall see below, there is no need for using this factor $A_C^+ A_C$ in practical computations.

Theoretically, the orthoprojector $A_C^+ A_C$ may be replaced by any projector $P_{A_C} \in L(I\!R^n)$ with $ker P_{A_C} = ker A_C$ if the subspace $im\, C(A_C^T e, t) A_C^T$ is independent of x and t a priori, the factor $A_C^+ A_C$ in the definition of $d(x,t)$ can be dropped completely.
Since the MNA system (2.1)–(2.3) is not in the standard form $f(x'(t), x(t), t) = 0$, which is mostly used in DAE theory as well as for numerical methods, one turns to the so-called conventional MNA or to the so-called charge-oriented MNA (cf. [EsTi]). Both of them are in standard DAE form. In contrast to this we shall treat the equations (2.1)–(2.3) resp. (2.4) as they are.

3. DAEs with properly stated leading term

We consider equations

$$A(x(t), t)(d(x(t), t))' + b(x(t), t) = 0 \tag{3.1}$$

with coefficients $A(x,t) \in L(I\!R^n, I\!R^m)$, $d(x,t) \in I\!R^n$, $b(x,t) \in I\!R^m$ given for $x \in \mathcal{D}$, $t \in \mathcal{I}$. $\mathcal{D} \subseteq I\!R^m$ is open and connected, $\mathcal{I} \subseteq I\!R$ is an interval. A, b, d are assumed to be continuous with continuous partial derivatives $A_x, b_x, d_x, d_t, d_{xt}, d_{xx}$. Denote

$$D(x,t) := d_x(x,t).$$

A continuous function $x : \mathcal{I}_x \subseteq \mathcal{I} \to I\!R^m$ is said to be a solution of equation (3.1) if $x(t) \in \mathcal{D}$, for $t \in \mathcal{I}_x$, $d(x(t), t)$ is continuously differentiable with respect to t, and (3.1) is satisfied for $t \in \mathcal{I}_x$.

In this context, the coefficients are not supposed to have higher derivatives and the solution is not expected to belong to C^1, as it is usually the case in standard DAE theory [RaRh]. Of course, if a solution is C^1, then

$$A(x(t), t) D(x(t), t) x'(t) + b(x(t), t) + A(x(t), t) d_t(x(t), t) = 0 \tag{3.2}$$

is satisfied. Turning from (3.1) to (3.2) corresponds to the solution of the MNA system (2.1)–(2.3) by means of the conventional MNA.

Definition 3.1. *Equation* (3.1) *is a DAE with properly stated leading term if the decomposition*

$$ker A(x,t) \oplus im D(x,t) = I\!R^n, \quad x \in \mathcal{D},\ t \in \mathcal{I}, \tag{3.3}$$

is valid, $d(x,t) \in im D(x,t), x \in \mathcal{D}, t \in \mathcal{I}$, is satisfied, and if the subspaces $ker A(x,t)$ and $im D(x,t)$ are independent of x but have bases that are continuously differentiable in t.

In the following, the leading term in (3.1) is supposed to be stated properly. In particular, $A(x,t)$ and $D(x,t)$ have common constant rank then. Denote by $R(t) \in L(I\!R^n)$ the projector matrix that realizes the decomposition (3.3), i.e., $R(t)^2 = R(t)$, $im D(x,t) = im R(t)$, $ker A(x,t) = ker R(t)$, $t \in \mathcal{I}$. Since the subspaces $im R(t)$, $ker R(t)$ have continuously differentiable bases, the projector function R is continuously differentiable. Further it holds that $d(x,t) = R(t) d(x,t)$. Obviously, Definition 3.1 coincides with [HiMä], Definition 5.1.

The MNA system (2.1)–(2.3) forms a particular class of DAEs with properly stated leading term. Moreover, so-called semi-explicit DAEs

$$\begin{aligned} x_1'(t) + b_1(x_1(t), x_2(t), t) &= 0, \\ b_2(x_1(t), x_2(t), t) &= 0 \end{aligned}$$

have properly stated leading terms by simply choosing

$$A = \begin{pmatrix} I \\ 0 \end{pmatrix}, \; d(x,t) = x_1, \; D(x,t) = (I, 0).$$

A properly formulated leading term contains more precise information on which part of the derivative of the unknown function is actually involved. This information gets lost if one turns to standard form DAEs like (3.2).

Sometimes, the augmented system

$$A(x(t),t)(R(t)y(t))' + b(x(t),t) = 0 \tag{3.4}$$

$$y(t) - d(x(t),t) = 0 \tag{3.5}$$

is easier to handle. It has a properly stated leading term, too.

A solution of (3.4), (3.5) is a pair of continuous functions $x : \mathcal{I}_x \to I\!R^m, y : \mathcal{I}_x \to I\!R^n, x(t) \in \mathcal{D}, t \in \mathcal{I}_x$, such that $R(t)y(t)$ is continuously differentiable in t, and the equations (3.4), (3.5) are satisfied. Then, also $y(t)$ is continuously differentiable in t, since $y(t) = d(x((t),t) = R(t)d(x(t),t) = R(t)y(t)$. Consequently, the DAE (3.1) and its augmented form (3.4), (3.5) are equivalent via $y(.) = d(x(.),.)$.

Notice that the augmented form (3.4), (3.5) of the MNA system (2.1)–(2.3) corresponds to the charge-oriented MNA.

Obviously, if $x : \mathcal{I}_x \to I\!R^m$ is a solution of the DAE (3.1), then $x(t) \in \mathcal{M}_0(t)$, $t \in \mathcal{I}_x$, must hold, where

$$\mathcal{M}_0(t) := \{x \in \mathcal{D} : b(x,t) \in imA(x,t)\}, \quad t \in \mathcal{I}, \tag{3.6}$$

is the so-called constraint set. The flow of the DAE is restricted to $\mathcal{M}_0(t)$.

For given $t_0 \in \mathcal{I}, x_0 \in \mathcal{M}_0(t)$, there is a uniquely determined $y_0 \in I\!R^n$ such that

$$A(x_0,t_0)y_0 + b(x_0,t_0) = 0, \quad y_0 = R(t_0)y_0. \tag{3.7}$$

Definition 3.2. *The value $x_0 \in \mathcal{M}_0(t)$ is said to be consistent if there is a solution passing through (x_0,t_0). Then, with y_0 from (3.7), the pair (y_0,x_0) is a consistent initialization at t_0.*

In case of DAEs (3.1) with index 1 (cf. Section 4 below), the flow generated by the DAEs is restricted to $\mathcal{M}_0(t)$, but $\mathcal{M}_0(t)$ is completely filled by this flow, thus each $x_0 \in \mathcal{M}_0(t)$ is consistent. The flow of higher index DAEs is additionally restricted by so-called hidden constraints, and the set of consistent values at t_0 is a proper subset of $\mathcal{M}_0(t)$ only.

Introduce the matrix functions

$$G_0(x,t) := A(x,t)D(x,t), \tag{3.8}$$

$$B_0(y,x,t) := b_x(x,t) + A(x,t)y)_x \tag{3.9}$$

and the subspaces

$$N_0(x,t) := kerG_0(x,t) = kerD(x,t), \tag{3.10}$$
$$S_0(y,x,t) := \{z \in \mathbb{R}^m : B_0(y,x,t)z \in imG_0(x,t)\}, \tag{3.11}$$

for $x \in \mathcal{D}, t \in \mathcal{I}, y \in \mathbb{R}^n$.

Because of the identity $A(x,t) \equiv A(x,t)R(t)$ it holds that $B_0(y,x,t) \equiv B_0(R(t)y,x,t)$ and $S_0(y,x,t) \equiv S_0(R(t)y,x,t)$.

If the tangent space $T_{x_0}\mathcal{M}_0(t)$ is well defined for $x_0 \in \mathcal{M}_0(t)$, then it coincides with $S_0(y_0,x_0,t_0)$, y_0 from (3.7).

4. Index-1 DAEs

In this section we collect certain relevant results obtained recently.

Definition 4.1. *([HiMä]) The DAE* (3.1) *has tractability index* $\mu = 1$ *if*

$$N_0(x,t) \cap S_0(y,x,t) = 0, \quad x \in \mathcal{D},\ t \in \mathcal{I},\ y \in \mathbb{R}^n, \tag{4.1}$$

is satisfied, i.e., these subspaces intersect transversally.

It is well known that (4.1) is equivalent to

$$N_0(x,t) \oplus S_0(y,x,t) = \mathbb{R}^m, \quad x \in \mathcal{D},\ t \in \mathcal{I},\ y \in \mathbb{R}^n.$$

In particular, if $T_x\mathcal{M}_0(t)$ is the tangent space to $\mathcal{M}_0(t)$ at $x \in \mathcal{M}_0(t)$, then $N_0(x,t) \oplus T_x\mathcal{M}_0(t) = \mathbb{R}^m$ must be true for $x \in \mathcal{M}_0(t)$.

Theorem 4.2. *([HiMä]) Let the DAE* (3.1) *have tractability index* $\mu = 1$.

(i) *Then, for each* $t_0 \in \mathcal{I}, x_0 \in \mathcal{M}_0(t_0)$, *there is exactly one solution passing through* (x_0,t_0).

(ii) *If* $x_* : \mathcal{I}_* \to \mathbb{R}^m$ *is a solution of* (3.1), $\mathcal{I}_* \subseteq \mathcal{I}$ *compact, then the perturbed initial value problems*

$$A(x(t),t)(d(x(t),t))' + b(x(t),t) = q(t), \tag{4.2}$$
$$d(x(t_0),t_0) = y^0, \tag{4.3}$$

with arbitrary $q \in C(\mathcal{I}_*,\mathbb{R}^m)$, $y^0 \in imR(t_0)$, $\|q\|_\infty$ *and* $|d(x_*(t_0),t_0) - y^0|$ *sufficiently small, are uniquely solvable on* $\mathcal{I}_*$, *and the solution* $x : \mathcal{I}_* \to \mathbb{R}^m$ *satisfies the inequality*

$$\|x - x_*\|_\infty \le K(\|q\|_\infty + |d(x_*(t_0),t_0) - y^0|). \tag{4.4}$$

Theorem 4.3. *Let the DAE* (3.1) *have tractability index* $\mu = 1, t_0 \in \mathcal{I}, y^0 \in imR(t_0)$. *Then, supposed the system*

$$A(x_0,t_0)y_0 + b(x_0,t_0) = 0, \tag{4.5}$$
$$(I - R(t_0))y_0 + d(x_0,t_0) - y^0 = 0, \tag{4.6}$$

is solvable for x_0, y_0, *it is locally uniquely solvable and provides a consistent initialization.*

Proof. The Jacobian of the system (4.5), (4.6) is of the form

$$J = \begin{bmatrix} B_0 & A \\ D & I-R \end{bmatrix}.$$

$J\begin{pmatrix} z_1 \\ z_2 \end{pmatrix} = 0$ implies $z_2 = Rz_2,\ z_1 \in kerD,\ B_0 z_1 + Az_2 = 0$, hence $z_1 \in kerD = N_0,\ B_0 z_1 = -Az_2 = -ADD^- z_2$, thus $z_1 \in N_0 \cap S_0$, i.e., $z_1 = 0$, further $z_2 = Rz_2,\ Az_2 = 0$, which leads to $z_2 = 0$. Consequently, J is nonsingular. □

Naturally, a variable stepsize variable order (up to k) BDF applied to (3.1) reads

$$A(x_n, t_n)\frac{1}{h_n}\sum_{j=0}^{k} \alpha_{nj}\ d(x_{n-j}, t_{n-j}) + b(x_n, t_n) = 0. \tag{4.7}$$

A stiffly accurate s-stage Runge-Kutta method applied to (3.1) is of the form

$$x_n = X_{ns},$$

$$A(X_{ni}, t_{ni})\frac{1}{h_n}\sum_{j=1}^{s} \alpha_{ij}(d(X_{nj}, t_{nj}) - d(x_{n-1}, t_{n-1})) + b(X_{ni}, t_{ni}) = 0, \tag{4.8}$$

$$i = 1, \ldots, s,$$

where X_{ni} are the stage approximations, $t_{ni} = t_{n-1} + c_i h_n$ are the stages and $\alpha_{ij} = (\mathcal{A}^{-1})_{ij}$ are the entries of the inverse of the Runge Kutta matrix $\mathcal{A}$.

These methods are discussed in [HiMä], [HiMäTi]. For index-1 DAEs (3.1), stability, consistency and convergence is proved. Roughly speaking, for sufficiently small stepsizes these integration methods work well as expected. However, caused by subspaces $imR(t) = imD(x,t)$ varying with t, certain extra stepsize restrictions arise. Conversely, if the matrix $D(x,t)$ has a constant image-space, then, by means of (4.7) resp. (4.8), the so-called inherent regular ODE of (3.1) is automatically integrated by the same method. No conversion, e.g., of the implicit Euler method into the explicit one, may happen. Contractivity and dissipativity properties are preserved (cf. [HiMäTi]).

Definition 4.4. *(*[HiMäTi] *The DAE* (3.1) *with tractability index* $\mu = 1$ *is* numerically qualified *if the subspace* $imD(x,t)$ *is constant.*

The implementations of the BDF in [Fl] and of Radau IIA in [Te] confirm the discussed theoretical aspects. It is evident that one should try for numerically qualified models from the very beginning. If the nullspace $kerA(x,t)$ is constant, but $imD(x,t)$ is not, we may put $\tilde{R} := A(x,t)^+ A(x,t)$ and reformulate (3.1) as

$$A(x(t),t)(\tilde{R}d(x(t),t))' + b(x(t),t) = 0 \tag{4.9}$$

which is now numerically qualified. Due to $A(x,t) \equiv A(x,t)\tilde{R}$ the integration methods (4.7) resp. (4.8) are invariant under this procedure. Therefore, there is no need for computing $\tilde{R}$ in practice. In this sense also DAEs (3.1) with constant $kerA(x,t)$ are numerically qualified.

Remark 4.5. Fortunately, as it was realized in Section 2, the MNA system (2.1)–(2.3) generates just a constant subspace $imD(x,t)$, namely (cf. (2.7)) $imD(x,t) = (ker A)^{\perp}$. Hence, the MNA system is, if it has tractability index 1, just numerically qualified. Furthermore, there is no need for any practical computation of $A_C^+ A_C$.

Notice that the conventional MNA system resulting from (2.1)–(2.3) (cf. (3.2))

$$AD(x(t),t)x'(t) + b(x(t),t) + Ad_t(x(t),t) = 0$$

may be rewritten in its turn as (cf. (2.6))

$$AD(x(t),t)(AA^+x(t))' + b(x(t),t) + Ad_t(x(t),t) = 0, \tag{4.10}$$

which has a properly stated leading term and which is numerically qualified, too. Again, when realizing numerical integration methods for (4.10) there is no need to calculate AA^+ in fact. However, as it has been well known in circuit simulation for a long time, when applying numerical integration methods to (4.10) one is confronted with the need of practical computations of the second derivatives d_{xt}, d_{xx}, whereas, in (4.7) and (4.8) one can do with $D = d_x$.

Remark 4.6. In essence, the investigations of DAEs with properly stated leading terms in the index 1 case confirm long, practical experience in circuit simulation. However, what is of interest less for practical simulation but for its mathematical backing up are the now simpler proofs and investigations. Namely, in [Gü] a loop way via an artificially enlarged standard form index-two DAE was still expedient. Having a closer look at the assumptions in [Gü] reveals that we just have numerically qualified DAEs (3.1) with tractability index 1 there.

5. Linearization

Observe the MNA equation (2.4) to be just a special case of the general DAE (3.1). Namely, the matrix function $A(x,t)$ and the nullspace $N_0(x,t)$ ($N_0 = ker A^T$, cf. (2.6)) are invariant at all of x and t. In this case we have

$$A(x,t) \equiv A, \quad N_0(x,t) \equiv N_0, \quad G_0(x,t) \equiv AD(x,t),$$
$$B_0(y,x,t) \equiv b_x(x,t) =: B_0(x,t).$$

Further, if $Q_0 \in L(\mathbb{R}^m))$ denotes a projector onto N_0 (i.e., $Q_0^2 = Q_0$, $imQ_0 = N_0$), $P_0 := I - Q_0$, we derive

$$d(x,t) - d(P_0x,t) = \int_0^1 D(sx + (1-s)P_0x, t)Q_0x ds = 0,$$

for all $x \in \mathcal{D}, t \in \mathcal{I}, [x, P_0x] \subset \mathcal{D}$. Recall that $\mathcal{D} \times \mathcal{I} \subseteq \mathbb{R}^m \times \mathbb{R}$ was introduced in S3 to denote the definition domain of the functions d and b.

In accordance with this, and to be more transparent, we are now going to consider more special DAEs of the form

$$A(d(x)t,t))' + b(x(t),t) = 0, \tag{5.1}$$

with properly stated leading term, constant subspace $N_0 = kerAD(x,t)$ and a domain $\mathcal{D}$ that includes $[x, P_0x]$ together with x. Let us stress that all semi-explicit systems

$$x_1'(t) + b_1(x_1(t), x_2(t), t) = 0, \tag{5.2}$$
$$b_2(x_1(t), x_2(t), t) = 0, \tag{5.3}$$

and, in particular, all Hessenberg form DAEs, have this simpler form, where

$$A = \begin{pmatrix} I \\ 0 \end{pmatrix}, \ d(x,t) \equiv x_1, \ D(x,t) = (I, 0), \ N_0 = O \times \mathbb{R}^{m_2}.$$

Consider a continuous function $x_* : \mathcal{I}_* \to \mathbb{R}^m$, $\mathcal{I}_* \subseteq \mathcal{I}$, with values in $\mathcal{D}$, i.e., $x_*(t) \in \mathcal{D}$, $t \in \mathcal{I}_*$. Introduce the accompanying function $y_* : \mathcal{I}_* \to \mathbb{R}^n$, $y_*(t) := d(x_*(t), t), t \in \mathcal{I}_*$.

A priori, y_* is continuous. However, is it also continuously differentiable?

Lemma 5.1. *If the component P_0x_* is continuously differentiable, then y_* is so, and vice versa.*

Proof. Supposed $P_0x_* \in C^1(\mathcal{I}, \mathbb{R}^m)$, the function $y_* = d(x_*(.),.) = d(P_0x_*(.),.)$ is continuously differentiable as a superposition of C^1 functions. Conversely, assume $y_* \in C^1(\mathcal{I}, \mathbb{R}^n)$, and consider the equation $F(y,x,t) = 0$, where the map $F : \mathbb{R}^n \times (\mathcal{D} \cap imP_0) \times \mathcal{I} \to imA$ is given by $F(y,x,t) := Ay - Ad(x,t)$. The derivative $F_x(y,x,t) = -AD(x,t) = -AD(x,t)P_0$ acts bijectively between imP_0 and imA. Due to $F(y_*(t), P_0x_*(t), t) = 0$, $t \in \mathcal{I}_*$, the Implicit Function Theorem provides a continuously differentiable function $x = \xi(y,t)$, which solves the equation $F(y,x,t) = 0$. In particular, $P_0x_*(t) = \xi(y_*(t), t), t \in \mathcal{I}_*$, holds true. Thus, P_0x_* is continuously differentiable. □

The function space

$$C^1_{N_0}(\mathcal{I}_*, \mathbb{R}^m) := \{x \in C(\mathcal{I}_*, \mathbb{R}^m) : P_0x \in C^1(\mathcal{I}_*, \mathbb{R}^m)\}$$

is independent of the special choice of the projector P_0. Namely, if $P_0, \tilde{P}_0$ are two projectors, $kerP_0 = ker\tilde{P}_0 = N_0$, then it holds that $P_0 = P_0\tilde{P}_0$, $\tilde{P}_0 = \tilde{P}_0P_0$.

Now we are well prepared to define the map $\mathcal{F} : \mathcal{D}_\mathcal{F} \to C(\mathcal{I}_*, \mathbb{R}^m)$,

$$(\mathcal{F}x)(t) := A(d(x(t),t))' + b(x(t),t), \quad t \in \mathcal{I}_*, \ x \in \mathcal{D}_\mathcal{F}, \tag{5.4}$$

with the definition domain

$$\mathcal{D}_\mathcal{F} := \{x \in C^1_{N_0}(\mathcal{I}_*, \mathbb{R}^m) : x(t) \in \mathcal{D}, \ t \in \mathcal{I}_*\}.$$

The operator equation $\mathcal{F}x = 0$ represents the DAE (5.1) on the interval $\mathcal{I}_* \subseteq \mathcal{I}$.

For fixed $x_* \in \mathcal{D}_\mathcal{F}$ being not necessarily a solution of the DAE we form the coefficients

$$D_*(t) := D(x_*(t), t), \quad B_*(t) := b_x(x_*(t), t), \quad t \in \mathcal{I}_*. \tag{5.5}$$

If $\mathcal{I}_*$ is compact, the map $\mathcal{F}$ is Fréchet differentiable, and we obtain

$$\mathcal{F}'(x_*)x = A(D_*x)' + B_*x, \quad x \in C^1_{N_0}(\mathcal{I}_*, \mathbb{R}^m).$$

Hence, the operator equation linearized at x_*, $\mathcal{F}'(x_*)x = q$, is just the abstract form of the so-called *DAE linearized along the function* x_*

$$A(D_*x)' + B_*x = q. \tag{5.6}$$

Coming from the linear DAE theory (e.g., [Mä1, Mä2]) one would consider the function space

$$C^1_{D_*}(\mathcal{I}_*, I\!R^m) := \{x \in C(\mathcal{I}_*, I\!R^m) : D_*x \in C^1(\mathcal{I}_*, I\!R^n)\}$$

as the solution space for (5.6). Does it depend on D_*? And is the leading term in (5.6) also properly stated? These questions are answered by the next lemma.

Lemma 5.2.

(i) *The linearized DAE* (5.6) *has a properly stated leading term with the same projector function* $R \in C^1(\mathcal{I}_*, L(I\!R^n))$ *as given for the nonlinear DAE* (5.1).

(ii) *The function spaces* $C^1_{D_*}(\mathcal{I}_*, I\!R^m)$ *and* $C^1_{N_0}(\mathcal{I}_*, I\!R^m)$ *coincide.*

Proof. (i) The properties of the leading term (cf. Definition 3.1) apply simply to the linearized DAE (5.6) via $D_*(t) = D(x_*(t), t)$.
(ii) Let the generalized inverse D_*^- be determined (pointwise for $t \in \mathcal{I}_*$) by the four conditions

$$D_*^- D_* D_*^- = D_*^-, \quad D_* D_*^- D_* = D_*, \quad D_* D_*^- = R, \quad D_*^- D_* = P_0. \tag{5.7}$$

Since $D_* = D(P_0 x_*(.), .)$ belongs to the class C^1, so does D_*^-. Therefore, because of $P_0 x = D_*^- D_* x$, $D_* x = D_* P_0 x$, the above function spaces coincide in fact. □

To apply the index notion proposed for DAEs in [Mä1, Mä2] we construct the following sequence of matrix functions and possibly time-dependent subspaces. We begin with

$$\begin{aligned} G_{*0} &:= AD_*, \quad B_{*0} := B_*, \\ N_{*0} &:= kerG_{*0} = N_0, \quad Q_{*0} := Q_0, \quad P_{*0} := I - Q_{*0}, \end{aligned}$$

and introduce D_*^- as in Lemma 5.2 by (5.7).
Then, for $i \geq 0$, we define

$$\begin{aligned} G_{*i+1} &:= G_{*i} + B_{*i}Q_{*i}, \qquad (5.8)\\ N_{*i+1} &:= kerG_{*i+1}, \\ Q_{*i+1} &= Q^2_{*i+1}, \; imQ_{*i+1} = N_{*i+1}, \; P_{*i+1} := I - Q_{*i+1}, \\ B_{*i+1} &:= B_{*i}P_{*i} - G_{*i+1}D_*^-(D_*P_{*0}\cdots P_{*i+1}D_*^-)'D_*P_{*0}\cdots P_{*i}. \qquad (5.9) \end{aligned}$$

All expressions are meant pointwise for $t \in \mathcal{I}_*$. When realizing this sequence, one has to take care for the existence of the derivative in B_{*i+1}. As discussed in [Mä1, Mä2], this derivative corresponds to a consecutive decomposition of the *smooth* subspace imR into further such subspaces. This is closely related to solvability properties of the linear DAE (5.6).

Definition 5.3. *([Mä2]) The linear DAE* (5.6) *has tractability index* μ *if there is a sequence* (5.8), (5.9) *such that, for* $i \geq 1$,

(i) $G_{*i}(t)$ *has constant rank* r_{*i} *on* $\mathcal{I}_*$,
(ii) $N_{*i} \oplus \cdots \oplus N_{*i-1} \subseteq kerQ_{*i}$,
(iii) $Q_{*i} \in C(\mathcal{I}_*, L(\mathbb{R}^m))$, $D_* P_{*0} \cdots P_{*i} D_*^- \in C^1(\mathcal{I}_*, L(\mathbb{R}^n))$
hold true, and further $r_{*\mu-1} < r_{*\mu} = m$.

Remark 5.4. The flexible part within the sequence (5.8), (5.9) is given by the projector functions Q_{*i}. The index μ as well as the characteristic values $r_{*0}, \ldots, r_{*\mu-1}$ are independent of the choice of these projector functions in Definition 5.3 ([Mä2]). The index is also invariant under regular transformations of the unknown function and refactorizations of the leading term ([Mä1]).

It is a common practice to define properties of nonlinear differentiable maps as $\mathcal{F}$ is one via linearizations. In accordance with this, in the next section, we try to construct a pointwise matrix function sequence for the nonlinear DAE (5.1) and to give a constructive index definition of tractability index μ in such a way that all admissible linearizations have tractability index μ, too. With this we intend to propose rather simple index criteria, which can be checked by linear algebra tools.

If, reversely, the desired properties of the pointwise nonlinear matrix function sequence are always given in case all admissible linearizations have tractability index μ, is not clear yet. Clarifying this here would go far beyond the scope of this paper. Thus, we do with the following conjecture, which will be proved for the case of $\mu = 1$ in Theorem 5.6.

Conjecture 5.5. *It is a mere supposition that the nonlinear DAE* (5.1) *has tractability index* μ *(in the sense of Definition* 6.1 *below) if all linearizations at* $x_* \in \mathcal{D}_{\mathcal{F}} \cap C^{(\mu-1)}(\mathcal{I}_*, \mathbb{R}^m)$ *with* $\mathcal{I}_* \subseteq \mathcal{I}$ *an interval, have tractability index* μ.

Clearly, the index notion below shall cover Definition 4.1 in the index-1 case.

Theorem 5.6. *If the DAE* (5.1) *has tractability index* $\mu = 1$, *then, for each* $x_* \in \mathcal{D}_{\mathcal{F}}, \mathcal{I}_* \subseteq \mathcal{I}$, *the linearized DAE* (5.6) *has tractability index* $\mu = 1$. *Conversely, if all these linearizations have tractability index* $\mu = 1$, *then the nonlinear DAE* (5.1) *has tractability index* $\mu = 1$.

Proof. Let the DAE (5.1) have tractability index $\mu = 1$. Then, due to (4.1), the matrix function $G_1(x,t) := G_0(x,t) + B_0(x,t)Q_0$ remains nonsingular for all $x \in \mathcal{D}, t \in \mathcal{I}$. Consider the linearization (5.6) for fixed $x_* \in \mathcal{D}_{\mathcal{F}}$. Choosing $Q_{*0} = Q_0$ we find $G_{*1}(t) = AD_*(t) + B_*(t)Q_{*0} = G_1(x_*(t), t)$, $t \in \mathcal{I}_*$. Thus, the linear DAE (5.6) has tractability index $\mu = 1$. Conversely, let all linearizations have index $\mu = 1$. Consider the function $G_1(x,t)$ given above and fix an arbitrary $(\bar{x}, \bar{t}) \in \mathcal{D} \times \mathcal{I}$. Take any function $x_* \in \mathcal{D}_{\mathcal{F}}$ passing through $(\bar{x}, \bar{t})$. Since the linearization along x_* has index $\mu = 1$, $AD_*(\bar{t}) + B_*(\bar{t}) = G_1(\bar{x}, \bar{t})$ must be nonsingular. □

6. Index criteria for nonlinear DAEs

We continue to investigate the nonlinear DAE (5.1) as we agreed upon in Section 5. For $x \in \mathcal{D}, t \in \mathcal{I}$, we apply once more the matrices

$$G_0(x,t) := AD(x,t), \quad B_0(x,t) := b_x(x,t),$$

as well as the subspaces $N_0 := kerG_0(x,t)$. $Q_0 \in L(I\!R^m)$ is a projector onto N_0, $P_0 := I - Q_0$.

Let $D(x,t)^-$ denote the generalized inverse of $D(x,t)$, which satisfies the four conditions $DD^-D = D$, $D^-DD^- = D^-$, $DD^- = R$, $D^-D = P_0$ pointwise on $\mathcal{D} \times \mathcal{I}$. Since $D(x,t)$ depends continuously differentiably on x,t and has constant rank, $D(x,t)^-$ does so, too.

Introduce further, for $x \in \mathcal{D}, t \in \mathcal{I}$,

$$\begin{aligned} G_1(x,t) &:= G_0(x,t) + B_0(x,t)Q_0, \\ N_1(x,t) &:= kerG_1(x,t), \\ Q_1(x,t) &\in L(I\!R^m) \quad \text{a projector onto } N_1(x,t), \\ P_1(x,t) &:= I - P_1(x,t). \end{aligned} \tag{6.1}$$

Then we construct, for $x \in \mathcal{D}, t \in \mathcal{I}, x^1 \in I\!R^m$,

$$Diff_1(x^1,x,t) := (DP_0P_1D^-)_x(x,t)x^1 + (DP_0P_1D^-)_t(x,t),$$

$$B_1(x^1,x,t) := B_0(x,t)P_0 - G_1(x,t)D(x,t)^- Diff_1(x^1,x,t)D(x,t). \tag{6.2}$$

Of course, when using these expression we have to take care of the existence of the involved partial derivatives. For $i \geq 1$, we form the sequence

$$\begin{aligned} G_{i+1} &:= G_i + B_iQ_i, &(6.3)\\ N_{i+1} &:= kerG_{i+1}, \\ Q_{i+1} &= Q_{i+1}^2, \; imQ_{i+1} = N_{i+1}, \; P_{i+1} := I - Q_{i+1}, \\ B_{i+1} &:= B_iP_i - G_{i+1}D^- Diff_{i+1} DP_0 \cdots P_i &(6.4) \end{aligned}$$

pointwise for $x \in \mathcal{D}, t \in \mathcal{I}, x^1, \ldots, x^{i+1} \in I\!R^m$. Thereby, $G_{i+1}, N_{i+1}, Q_{i+1}, P_{i+1}$, $DP_0 \cdots \cdots P_{i+1}D^-$ depend on the arguments $x^i, \ldots, x^1, x, t$ while $Diff_{i+1}$ and B_{i+1} have the arguments $x^{i+1}, \ldots, x^1, x, t$. The definition

$$\begin{aligned} \text{Diff}_{i+1}(x^{i+1},\ldots,x^1x,t) := \sum_{j=1}^{i} (DP_0 \cdots P_{i+1}D^-)_{x^j}(x^i,\ldots,x^1x,t)x^{j+1} \\ + (DP_0 \cdots P_{i+1}D^-)_x(x^i,\ldots,x^1x,t)x^1 + (DP_0 \cdots P_{i+1}D^-)_t(x^i,\ldots,x^1x,t) \end{aligned} \tag{6.5}$$

shows how the new variable x^{i+1} comes into this expression at this stage.

Definition 6.1. *The DAE* (5.1) *has the tractability index* μ *if there is a sequence* (6.1)–(6.4) *such that, for* $i \geq 1$,

(i) $G_i(x^{i-1}, \ldots, x^1, x, t)$ *has constant rank* r_i *for* $x \in \mathcal{D}, t \in \mathcal{I}, x^1, \ldots, x^{i-1} \in I\!R^m$,

(ii) $N_0 \oplus \cdots \oplus N_{i-1} \subseteq kerQ_i$ *pointwise,*

(iii) Q_i *is continuous, but* $DP_0 \cdots P_iD^-$ *is continuously differentiable,*

and $r_{\mu-1} < r_\mu = m$.

Remark 6.2. The notion of the tractability index allows for an index monitoring via linear algebra tools (rank detection, projector calculations). A special numerical algorithm realizing the condition $N_0 \oplus \cdots \oplus N_{i-1} \subseteq ker Q_i$ is proposed in [La]. There are several possibilities to reduce the computational amount.

In particular, G_μ is nonsingular if and only if

$$N_{\mu-1} \cap ker((I - G_{\mu-1}G_{\mu-1}^{-})B_0) = 0, \tag{6.6}$$

hence, one can avoid to compute $B_{\mu-1}$ and G_μ. Moreover, this proves the equivalence of Definition 4.1 for the case of $\mu - 1$ in Definition 6.1.

For more details and computational experiments we refer to [La].

Remark 6.3. An index monitor to be used in TITAN for circuit simulation is reported in [Es et al.]. Its mathematical background is given by the tractability index (former version for standard DAEs, $\mu \leq 2$) and the careful and detailed structural analysis of electrical circuits [EsTi] providing the characteristic subspaces and projectors. Due to the combination of topological and numerical tools the index can be calculated (up to $\mu = 2$) even for very large circuits. It is hoped that it will become possible to update this index monitor according to the new higher index criteria given here to realize at least index three precisely, too.

Obviously, if the DAE (5.1) itself is just a linear one, then Definition 6.1 coincides with the index notion given for linear DAEs in [Mä2] (cf. Definition 5.3). In this case, the sequence (6.1)–(6.4) is simply independent of the values x and x^j, and (6.5) reads $Diff_{i+1}(t) = (DP_0 \cdots P_{i+1}D^-)_t(t)$.

Theorem 6.4. *If the DAE* (5.1) *has tractability index* μ*, then, for each* $x_* \in \mathcal{D}_\mathcal{F} \cap C^{\mu-1}(\mathcal{I}_*, \mathbb{R}^m)$*,* $\mathcal{I}_* \subseteq \mathcal{I}$*, the linearized DAE* (5.6) *has tractability index* μ*, too. All these linearizations have common characteristic values* $r_{*i} = r_i$*,* $i = 0, \ldots, \mu$*.*

Proof. Fix an $x_* \in \mathcal{D}_\mathcal{F} \cap C^{\mu-1}(\mathcal{I}_*, \mathbb{R}^m)$, $\mathcal{I}_* \subseteq \mathcal{I}$ and construct an appropriate sequence for the linearized DAE (5.6) (cf. Definition 5.3). Choose $P_{*0} := P_0, Q_{*0} := Q_0$. It results that $G_{*1}(t) := AD_*(t) + B_*(t)Q_0 = G_1(x_*(t), t)$, $t \in \mathcal{I}_*$, and $r_{*1} = r_1$. Because of $N_0 \subseteq ker Q_1(x,t), x \in \mathcal{D}, t \in \mathcal{I}$, we may choose $Q_{*1}(t) := Q_1(x_*(t), t), t \in \mathcal{I}_*$. This implies $N_0 \subseteq ker Q_{*1}(t)$. Further, Q_{*1} is obviously continuous, and $(D_* P_{*0} P_{*1} D_*^-)(t) = (DP_0P_1D^-)(x_*(t), t)$ depends continuously differentiably on t as a superposition of C^1 functions.

From $Diff_1(x_*'(t), x_*(t), t) = \frac{d}{dt}(DP_0P_1D^-)(x_*(t), t) = (D_*P_{*0}P_{*1}D_*^-)'(t)$ we derive

$$B_{*1}(t) = B_1(x_*'(t), x_*(t), t), \quad G_{*2}(t) = G_2(x_*'(t), x_*(t), t), \quad r_{*2} = r_2.$$

Next we choose $Q_{*2}(t) := Q_2(x_*'(t), x_*(t), t)$ and so on. □

Example 6.5. Consider the Hessenberg form DAE of size three (cf. [BrCaPe])

$$\begin{aligned} x_1'(t) + x_1(t) + x_4(t) &= q_1(t), \\ x_2'(t) + \alpha(x_3(t), t)x_4(t) &= q_2(t), \\ x_3'(t) + x_1(t) + x_2(t) + x_3(t) &= q_3(t), \\ x_3(t) &= q_4(t), \end{aligned} \tag{6.7}$$

α is continuously differentiable, $1+\alpha(x_3,t) \neq 0$ for all x and t. With

$$A = \begin{pmatrix} 1 & 0 & 0 \\ 0 & 1 & 0 \\ 0 & 0 & 1 \\ 0 & 0 & 0 \end{pmatrix}, \quad D = \begin{pmatrix} 1 & 0 & 0 & 0 \\ 0 & 1 & 0 & 0 \\ 0 & 0 & 1 & 0 \end{pmatrix}, \quad D^- = A, \; G_0 = \begin{pmatrix} 1 & 0 & 0 & 0 \\ 0 & 1 & 0 & 0 \\ 0 & 0 & 1 & 0 \\ 0 & 0 & 0 & 0 \end{pmatrix},$$

$$P_0 = G_0, B_0 = \begin{pmatrix} 1 & 0 & 0 & 1 \\ 0 & 0 & \gamma & \alpha \\ 1 & 1 & 1 & 0 \\ 0 & 0 & 1 & 0 \end{pmatrix}, \quad \gamma(x_3,x_4,t) := \alpha_{x_3}(x_3,t)x_4$$

we put the DAE (6.7) in the form (5.1).

Derive further

$$G_1 = \begin{pmatrix} 1 & 0 & 0 & 1 \\ 0 & 1 & 0 & \alpha \\ 0 & 0 & 1 & 0 \\ 0 & 0 & 0 & 0 \end{pmatrix}, \quad Q_1 = \begin{pmatrix} 1 & 0 & 0 & 0 \\ \alpha & 0 & 0 & 0 \\ 0 & 0 & 0 & 0 \\ -1 & 0 & 0 & 0 \end{pmatrix}, \quad DP_0P_1D^- = \begin{pmatrix} 0 & 0 & 0 \\ -\alpha & 1 & 0 \\ 0 & 0 & 1 \end{pmatrix},$$

$$\mathit{Diff}_1 = \begin{pmatrix} 0 & 0 & 0 \\ \delta & 0 & 0 \\ 0 & 0 & 0 \end{pmatrix}, \quad \delta(x_3^1, x_3, t) := \alpha_{x_3}(x_3,t)x_3^1 + \alpha_t(x_3,t),$$

$$G_2 = \begin{pmatrix} 2 & 0 & 0 & 1 \\ \delta & 1 & 0 & \alpha \\ 1+\alpha & 0 & 1 & 0 \\ 0 & 0 & 0 & 0 \end{pmatrix}, \quad Q_2 = \begin{pmatrix} 0 & 0 & -\frac{1}{1+\alpha} & 0 \\ 0 & 0 & \frac{\delta-2\alpha}{1+\alpha} & 0 \\ 0 & 0 & 1 & 0 \\ 0 & 0 & \frac{2}{1+\alpha} & 0 \end{pmatrix},$$

$$DP_0P_1P_2D^- = \begin{pmatrix} 0 & 0 & 0 \\ -\alpha & 1 & \frac{\alpha-\delta}{1+\alpha} \\ 0 & 0 & 0 \end{pmatrix},$$

and check that the intersection (cf. 6.6) $N_2 \cap ker((I - G_2G_2^-)B_0) = \{z \in \mathbb{R}^4 : z_3 = 0,\ (1+\alpha)z_1 = 0,\ \delta z_1 + z_2 + \alpha z_4 = 0,\ 2z_1 + z_4 = 0\}$ is trivial. Hence, system (6.7) has tractability index $\mu = 3$ as expected. Observe that $DP_0P_1P_2D^-$ actually depend, via δ, on x^1.

Remark 6.6. The index notion proposed in [Mä3] means in essence Definition 6.1 applied to a somewhat restricted class of DAEs. Namely, in [Mä3], the projector $DP_0P_1P_2D^-$ is supposed to depend just on x, t, but to be independent of x^1. In the consequence, Diff_2 depends just on x^1, x, t, and so do B_2 and G_3. At the next stage, $DP_0P_1P_2P_3D^-$ is supposed to depend only on x, t, and so on. These additional structural properties lead to a sequence of matrix functions $G_i(x^1, x, t),\ i \geq 2$, depending on the variables x^1, x, t only. Then, Theorem 6.4 applies to each $x_* \in \mathcal{D}_{\mathcal{F}} \cap C^1(\mathcal{I}_*, \mathbb{R}^m)$. However, now we restrict the admissible linearizations to a smoother class, which remains huge enough, and we avoid additional structural restrictions. As we have seen in Example 6.5, the projector $DP_0P_1P_2D^-$ may actually depend on x^1.

In practice it ist hard to carry out formula (6.5) or to approximate the needed derivatives. Sometimes, necessary conditions, which are much easier to handle, will do.

Corollary 6.7. *Let the DAE* (5.1) *have tractability index* μ.

(i) *Then, for each* $c \in \mathcal{D}$, *the linear DAE*

$$A(D(c,t)x(t))' + b_x(c,t)x(t) = q(t) \tag{6.8}$$

is tractable with index μ *and it has the characteristic values* $r_{ci} = r_i$, $i = 0, \ldots, \mu$.

(ii) *If the DAE* (5.1) *is autonomous, then, for each* $c \in \mathcal{D}$, *the local matrix pencil*

$$\lambda AD(c) + b_x(c)$$

is regular with the Kronecker index μ, *and* $r_i - r_{i-1}$ *is the number of nilpotent Jordan blocks of size* i *in the Kronecker canonical normal form,* $i = 1, \ldots, \mu$.

(iii) *If the subspaces* $imD(c,t)$, $im(DP_0 \cdots P_{i-1}Q_i)(0, \ldots, 0, c, t)$, $i = 1, \ldots, \mu - 1$, *do not vary at all, then the local pencil*

$$\lambda AD(c,t) + b_x(c,t)$$

is regular with the Kronecker index μ *uniformly for* $t \in \mathcal{I}$.

Proof. We linearize along constant functions $x_*(t) \equiv c, t \in \mathcal{I}$. The second part is a consequence of [GrMae], Theorem 3, whereas the third one follows from [Mä1], Theorem 4.4. □

We finish by stressing that there is some hope for realizing and exploiting constant relevant subspaces in circuit simulation (cf. [HiMäTi2], [EsTi]). This is not only important for the index monitoring, but also for a good performance of numerical integration methods.

References

[BaMä] K. Balla, R. März: A unified approach to linear differential algebraic equations and their adjoints. Zeitschrift f. Analysis und ihre Anwendungen 21(2002)3, 783–802.

[BrCaPe] K.E. Brenan, S.L. Campbell, L.R. Petzold: Numerical solution of initial value problems in differential-algebraic equations. North Holland, Amsterdam, 1989.

[Es et al.] D. Estévez Schwarz, U. Feldmann, R. März, S. Sturtzel, C. Tischendorf: Finding beneficial DAE structures in circuit simulation. W. Jäger, H.-J. Krebs (eds.) Mathematics – Key Technology for the future. Springer 2003.

[EsTi] D. Estévez Schwarz, C. Tischendorf: Structural analysis of electric circuits and consequences for MNA. Int. J. Circuit Theory and Applications 28, 131–162, 2000.

[Fl] J. Flügel: Lösung von Algebro-Differentialgleichungen mit properem Hauptterm durch die BDF.Diplomarbeit, Humboldt-Universität zu Berlin, Institut für Mathematik, 2002.

[GrMae] E. Griepentrog, R. März: Basic properties of some differential-algebraic equations. Zeitschrift für Analysis und ihre Anwendungen 8, 25–40, 1998.

[Gü] M. Günther: Simulating digital circuits numerically – a charge-oriented ROW approach. Numerische Mathematik 79, 203–212, 1998.

[HiMä] I. Higueras, R. März: Differential algebraic equations with properly stated leading terms. Computers Math. Applic. (to appear)

[HiMäTi] I. Higueras, R. März, C. Tischendorf: Numerically well-qualified index-1 DAEs. Applied Numerical Mathematics, to appear.

[HiMäTi2] I. Higueras, R. März, C. Tischendorf: Numerically qualified index-2 DAEs. Applied Numerical Mathematics, to appear.

[Mä1] R. März: The index of linear differential algebraic equations with properly stated leading terms. Results in Mathematics, to appear.

[Mä2] R. März: Solvability of linear differential algebraic equations with properly stated leading terms. Humboldt-Universität zu Berlin, Institut für Mathematik, Preprint 02-12.

[Mä3] R. März: Characterizing differential algebraic equations without the use of derivative arrays. Proc. NMCM 2002, University of Miskolc, Hungary, July 15–19, 2002.

[La] R. Lamour: Index determination and calculation of consistent initial values for DAEs. Proc. NMCM 2002, University of Miskolc, Hungary, July 15–19, 2002.

[ReMaBa] G. Reißig, W.S. Martinson, P.J. Barton: Differential-algebraic equations of index 1 may have an arbitrarily high structural index. SIAM J. Scie. Comp. 21(6), 1997–1990, 2000.

[RaRh] P.J. Rabier, W.C. Rheinboldt: Theoretical and Numerical Analysis of Differential-Algebraic Equations. Handbook of Numerical Analysis Vol. VIII, Elsevier Publ. Amsterdam 2002.

[Te] S. Tegtmeier: Numerische Lösung von Algebro-Differentialgleichungen mit proper formuliertem Hauptterm durch Runge-Kutta-Verfahren. Diplomarbeit, Humboldt-Universität zu Berlin, Institut für Mathematik, 2002.

R. März
Humboldt-University Berlin
Institute of Mathematics
Unter den Linden 6
D-10099 Berlin, Germany
email: maerz@mathematik.hu-berlin.de

Modeling, Simulation and Optimization of Integrated Circuits
(Eds.) K. Antreich, R. Bulirsch, A. Gilg and P. Rentrop

International Series of Numerical Mathematics Vol. 146, 153–166

A Parallel Finite Difference Method for Multirate Partial Differential Algebraic Equations

Roland Pulch

Abstract. In oscillatory electric circuits, the presence of several time rates causes quasi-periodic responses. A multi-dimensional signal model yields an alternative approach for numerical simulation of such circuits in case of widely separated time scales. Consequently, the time-dependent network equations change into partial differential algebraic equations. Efficient time domain methods are obtained by using the special form of information transport in the arising system. We construct an according finite difference method, which performs a discretisation along characteristic curves. In addition, this technique includes an inherent potential for parallelism. Test results verify that the parallel implementation of the finite difference technique generates good speedups.

1. Introduction

In radio frequency applications, the time scales of quasi-periodic signals often differ by several orders of magnitude. Since the fastest rate limits the integration step size, while the slowest rate determines the total time interval, numerical simulation by means of transient analysis becomes costly. A multi-dimensional model is able to decouple the time scales by using separate variables for each rate. Consequently, the describing differential algebraic equation (DAE) changes into a PDAE, the multirate partial differential algebraic equation (MPDAE), introduced by Brachtendorf et al. [2]. If efficient techniques are applied, then the MPDAE approach can be solved faster than the corresponding DAE.

The MPDAE system determines an information transport along characteristic curves, see [9]. The resulting structure allows the design of efficient time domain methods. A corresponding finite difference scheme is constructed for periodic boundary value problems in [10]. The employed characteristic grid regards the underlying information transport. Accordingly, the amount of computation

Key words and phrases. Differential Algebraic Equation, Multirate Partial Differential Algebraic Equation, Finite Difference Method, Characteristic System.

time and memory decreases drastically in comparison to finite difference methods on uniform grids, because the arising nonlinear equations are significantly less coupled.

A parallel implementation of the finite difference scheme reduces the amount of computation time even further. In this paper, we present a corresponding parallel algorithm. Thereby, the emerging nonlinear system is arranged such that its Jacobian matrix allows an efficient block Gaussian elimination in a Newton method. This scheme implies an according parallelisation, where the specific structure of the submatrices has to be considered in order to achieve a good parallel efficiency. The respective code uses multiple instruction and multiple data patterns. Furthermore, the presented technique can be generalised to a multilevel Newton method for the purpose of reducing the communication overhead, see [14].

The paper is organised as follows. Section 2 gives a brief review of the multidimensional model and the resulting MPDAE formulation. Then we describe the finite difference technique, which results from the structure of the characteristic curves. In Section 4, we introduce the corresponding parallel algorithm. Thereby, its generalisation to multilevel Newton methods is outlined. Finally, test results verify good speedups in parallel runs of the code.

2. Multi-dimensional model

Electric circuits often exhibit signals with widely separated time scales. Especially in radio frequency applications, oscillatory responses including several frequencies play an important role. A multi-dimensional model for such signals yields an alternative approach. Thereby, each separate time scale is given its own variable. A simple example demonstrates the transition

$$b(t) = \sin\left(\tfrac{2\pi}{T_1}t\right)\sin\left(\tfrac{2\pi}{T_2}t\right) \quad \longrightarrow \quad \hat{b}(t_1,t_2) = \sin\left(\tfrac{2\pi}{T_1}t_1\right)\sin\left(\tfrac{2\pi}{T_2}t_2\right). \tag{2.1}$$

Fig. 1 illustrates these two functions. If $T_1 \gg T_2$ holds, then the signal b possesses a huge number of oscillations in the interval $[0, T_1]$ and thus we require many sampling points to capture this function. On the contrary, the description $\hat{b}$ is biperiodic and we can sample it with relatively few points in the underlying rectangle $[0, T_1] \times [0, T_2]$, since only one oscillation occurs in each coordinate direction. Nevertheless, this function includes the original signal on the diagonal by $b(t) = \hat{b}(t,t)$. The representation $\hat{b}$ is called the *multivariate function* (*MVF*) of b and leads to a more efficient description of the multitone signal.

Likewise, each multitone quasi-periodic signal can be represented by a corresponding multiperiodic function. The definitions of these function types are given below. One can see the connections by expanding the multiperiodic function into a multi-dimensional Fourier series.

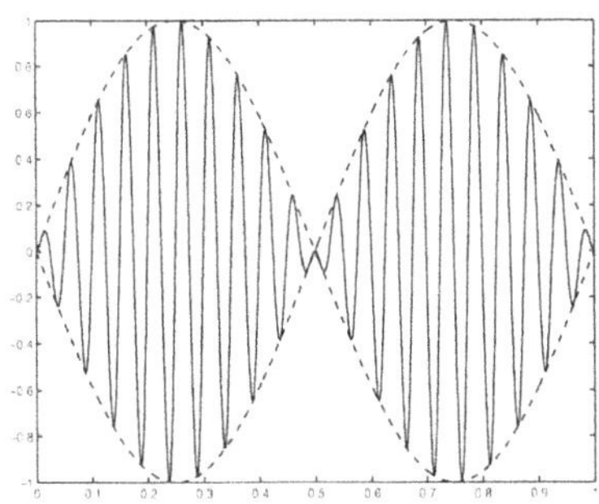

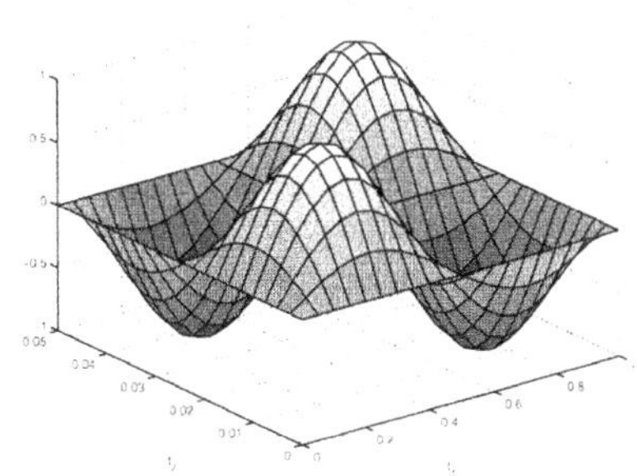

FIGURE 1. Multitone signal b (left) and corresponding multivariate function $\hat{b}$ (right).

Definition 2.1. A function $\mathbf{x} : \mathbb{R} \to \mathbb{C}^k$ is said to be m-tone quasi-periodic, if it can be expanded in the form

$$\mathbf{x}(t) = \sum_{j_1,\ldots,j_m=-\infty}^{+\infty} \mathbf{X}_{j_1,\ldots,j_m} \exp\left(i2\pi\left(\tfrac{j_1}{T_1} + \cdots + \tfrac{j_m}{T_m}\right)t\right) \tag{2.2}$$

with real- or complex-valued constant vectors $\mathbf{X}_{j_1,\ldots,j_m}$.

Definition 2.2. A function $\hat{\mathbf{x}} : \mathbb{R}^m \to \mathbb{C}^k$ is said to be m-periodic, if real constants $T_1, \ldots, T_m$ exist, which satisfy $\hat{\mathbf{x}}(t_1 + k_1T_1, \ldots, t_m + k_mT_m) = \hat{\mathbf{x}}(t_1, \ldots, t_m)$ for all $t_1, \ldots, t_m \in \mathbb{R}$ and $k_1, \ldots, k_m \in \mathbb{Z}$. Then the constant T_j is called the jth period of $\hat{\mathbf{x}}$.

In circuit simulation, the transient behaviour of all node voltages and some branch currents is calculated. Nodal analysis yields systems of *differential algebraic equations* (*DAEs*), see [6], which we write in the form

$$\tfrac{d}{dt}\mathbf{q}(\mathbf{x}(t)) = \mathbf{f}(\mathbf{b}(t), \mathbf{x}(t)). \tag{2.3}$$

Here, $\mathbf{x} : \mathbb{R} \to \mathbb{R}^k$ denotes unknown voltages and currents and the function $\mathbf{q} : \mathbb{R}^k \to \mathbb{R}^k$ represents a charge term. The right-hand side $\mathbf{f} : \mathbb{R}^p \times \mathbb{R}^k \to \mathbb{R}^k$ depends on input signals $\mathbf{b} : \mathbb{R} \to \mathbb{R}^p$. Now we assume the presence of exactly two widely separated time scales in the independent signals. However, modifications to more rates are straightforward. Consequently, we introduce the corresponding MVF $\hat{\mathbf{b}} : \mathbb{R}^2 \to \mathbb{R}^p$ with $\mathbf{b}(t) = \hat{\mathbf{b}}(t,t)$. This transition changes the DAE model of the circuit into a PDAE model, the *multirate partial differential algebraic equation* (*MPDAE*)

$$\frac{\partial \mathbf{q}(\hat{\mathbf{x}}(t_1,t_2))}{\partial t_1} + \frac{\partial \mathbf{q}(\hat{\mathbf{x}}(t_1,t_2))}{\partial t_2} = \mathbf{f}(\hat{\mathbf{b}}(t_1,t_2), \hat{\mathbf{x}}(t_1,t_2)), \tag{2.4}$$

where $\hat{\mathbf{x}} : \mathbb{R}^2 \to \mathbb{R}^k$ represents the MVF of $\mathbf{x}$. The crucial relation is that a solution $\hat{\mathbf{x}}$ of (2.4) includes a solution $\mathbf{x}$ of (2.3) via $\mathbf{x}(t) = \hat{\mathbf{x}}(t,t)$. If the input signals $\mathbf{b}$ are two-tone quasi-periodic, then we are able to construct a corresponding biperiodic MVF $\hat{\mathbf{b}}$. Accordingly, a biperiodic solution $\hat{\mathbf{x}}$ of (2.4) yields a two-tone quasi-periodic signal $\mathbf{x}$ satisfying (2.3). More details and existence theorems for multiperiodic MPDAE solutions in conjunction with multitone quasi-periodic DAE solutions are given in [12].

In the model above, the time rates are driven by the input signals. Furthermore, the MPDAE approach can also be applied in case of autonomous time rates, too, see, e.g., [3]. A mixture of driven and autonomous time scales often produces frequency modulation. In this case, the concept of warped MPDAEs, which has been introduced by Roychowdhury [13], allows to determine frequency-modulated quasi-periodic solutions.

Thus the MPDAE model provides an alternative approach for the computation of quasi-periodic responses. In the presence of widely separated time scales, the behaviour of the involved functions causes savings in computation time in comparison to attacking the DAE directly. Now we need efficient methods for solving the MPDAE system.

3. Finite difference technique

Our aim is to compute a numerical solution of the MPDAE (2.4) for biperiodic boundary conditions, where the periods are known from the input signals. We consider a finite difference technique here in view of two advantages. Firstly, this strategy yields a robust algorithm and is also suitable for strongly nonlinear differential equations. Secondly, we can easily apply the techniques to DAEs or PDAEs, respectively. A finite difference method performs a discretisation of the derivatives on a grid in time domain. For example, we consider a uniform grid in the rectangle $[0, T_1[\times[0, T_2[$, i.e.,

$$(t_{1,j}, t_{2,i}) = ((j-1)h_1, (i-1)h_2); \; h_1 = \tfrac{T_1}{n_1}; \; h_2 = \tfrac{T_2}{n_2} \tag{3.1}$$

for $j = 1, \ldots, n_1$ and $i = 1, \ldots, n_2$. The derivatives in the MPDAE are replaced with difference formulae, e.g., centred differences of second order. Hence we obtain a nonlinear system of k equations in each grid point

$$\begin{aligned} \mathbf{F}_{j,i} \equiv \; & \tfrac{1}{2h_1}\left[\mathbf{q}(\hat{\mathbf{x}}_{j+1,i}) - \mathbf{q}(\hat{\mathbf{x}}_{j-1,i})\right] + \tfrac{1}{2h_2}\left[\mathbf{q}(\hat{\mathbf{x}}_{j,i+1}) - \mathbf{q}(\hat{\mathbf{x}}_{j,i-1})\right] \\ & -\mathbf{f}(\hat{\mathbf{b}}_{j,i}, \hat{\mathbf{x}}_{j,i}) = \mathbf{0}, \end{aligned} \tag{3.2}$$

where $\hat{\mathbf{x}}_{j,i}$ is the approximation in the corresponding grid point $(t_{1,j}, t_{2,i})$. Involved points, which lie outside the grid, are identified with points inside the grid using the periodicities. Consequently, this approach produces a biperiodic numerical solution. We note that the discretisation in (3.2) owns the form of a five point star due to the derivatives in the two coordinate directions. Thus one grid point is coupled with four neighbouring points. Now we can solve the nonlinear system (3.2) by methods of Newton type. The Jacobian matrix in the resulting linear systems is sparse. However, due to the couplings in the nonlinear equations, many fill-ins are produced in the according LU-decomposition. Hence the main part of computational work consists in the LU-decompositions of the arising Jacobian

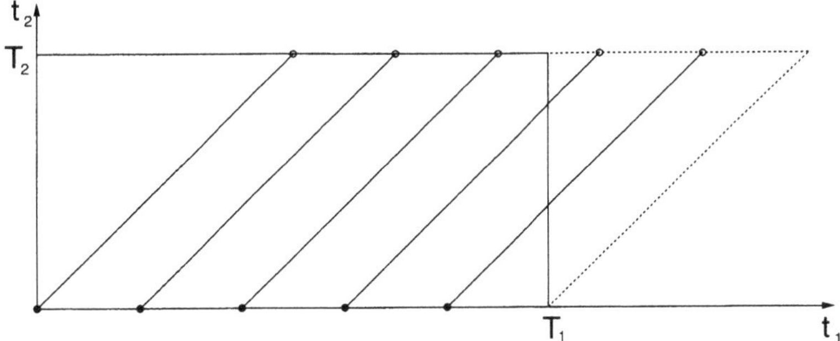

FIGURE 2. Characteristic curves in domain of dependence.

matrices. Reorderings of the matrices are able to decrease the number of fill-ins. Yet it takes also much time to compute an efficient reordering. Iterative solvers are problematic, too, because the Jacobian matrix is neither symmetric nor positive definite. Therefore we develop an alternative finite difference technique to reduce the computational effort.

A detailed analysis of the MPDAE system (2.4) exhibits an information transport along characteristic curves, as shown in [9]. These curves are just the straight lines in diagonal direction, i.e., $(t_1, t_2) = (\tau + c, \tau)$ for a parameter $\tau \in \mathbb{R}$ and any fixed $c \in \mathbb{R}$. Accordingly, the characteristic system of (2.4) is defined as

$$\tfrac{d}{d\tau}\mathbf{q}(\bar{\mathbf{x}}_c(\tau)) = \mathbf{f}(\hat{\mathbf{b}}(\tau + c, \tau), \bar{\mathbf{x}}_c(\tau)), \tag{3.3}$$

which represents a continuum of DAEs for the restrictions $\bar{\mathbf{x}}_c(\tau) := \hat{\mathbf{x}}(\tau + c, \tau)$. It follows that a function $\hat{\mathbf{x}} \in C^1(\mathbb{R}^2)$ solves the MPDAE (2.4), if and only if its restrictions $\bar{\mathbf{x}}_c$ satisfy the characteristic system (3.3) for each $c \in \mathbb{R}$, see [10].

Now we consider periodic boundary conditions in both coordinate directions and thus have to compute a biperiodic solution of (2.4) for given periods T_1 and T_2. Without loss of generality, we assume $T_1 \geq T_2$. Due to the periodicity in the first coordinate direction, we perform a discretisation by choosing n_1 equidistant characteristic curves with step size $h_1 = T_1/n_1$ via $c_j = (j-1)h_1$ for $j = 1, \ldots, n_1$. Fig. 2 outlines the situation. Consequently, a system of n_1 independent DAEs

$$\tfrac{d}{d\tau}\mathbf{q}(\tilde{\mathbf{x}}_j(\tau)) = \mathbf{f}(\hat{\mathbf{b}}(\tau + (j-1)h_1, \tau), \tilde{\mathbf{x}}_j(\tau)), \quad j = 1, \ldots, n_1 \tag{3.4}$$

is extracted from the continuum (3.3) with $\tilde{\mathbf{x}}_j(\tau) := \hat{\mathbf{x}}(\tau + (j-1)h_1, \tau)$. Therefore this strategy represents a kind of semi-discretisation. The periodicity condition in the second coordinate direction yields boundary conditions for system (3.4). The values $\tilde{\mathbf{x}}_j(T_2)$ on the line $t_2 = T_2$ can be interpolated by the values $\tilde{\mathbf{x}}_j(0)$ on the line $t_2 = 0$. Thereby, the periodicity in the first coordinate direction is also used to project required points into the rectangle of periodicities. The resulting boundary conditions are of the form

$$(\tilde{\mathbf{x}}_1(T_2), \ldots, \tilde{\mathbf{x}}_{n_1}(T_2))^T = \mathcal{B}(\tilde{\mathbf{x}}_1(0), \ldots, \tilde{\mathbf{x}}_{n_1}(0))^T \tag{3.5}$$

with a constant matrix $\mathcal{B} \in \mathbb{R}^{n_1 k \times n_1 k}$. The structure of $\mathcal{B}$ depends on the employed interpolation formula. To achieve consistent initial values of the DAEs, one can interpolate only independent components and determine the other components of the initial values by algebraic relations. Thus we obtain a system of independent DAEs, which are only coupled by the boundary conditions. Its corresponding solution yields an approximation to the biperiodic MPDAE solution.

We are able to solve the boundary value problem (3.4), (3.5) by a finite difference method, too, which means a discretisation of the DAEs by a difference formula on a corresponding grid. For simplicity, we apply the same constant step size on each characteristic curve. Thus we encounter the characteristic grid with the points

$$(t_{1,j}, t_{2,i}) = ((j-1)h_1 + (i-1)h_2, (i-1)h_2); \; h_1 = \tfrac{T_1}{n_1}; \; h_2 = \tfrac{T_2}{n_2} \tag{3.6}$$

for $j = 1, \ldots, n_1$ and $i = 1, \ldots, n_2$ in a corresponding parallelogram, see Fig. 2. Using centred differences again, the nonlinear system

$$\mathbf{G}_{j,i} \equiv \tfrac{1}{2h_2} \left[\mathbf{q}(\tilde{\mathbf{x}}_{j,i+1}) - \mathbf{q}(\tilde{\mathbf{x}}_{j,i-1})\right] - \mathbf{f}(\hat{\mathbf{b}}_{j,i}, \tilde{\mathbf{x}}_{j,i}) = \mathbf{0} \tag{3.7}$$

arises in each grid point for the according unknowns $\tilde{\mathbf{x}}_{j,i} \in \mathbb{R}^k$. The evaluations of $\mathbf{q}$ on the layer $i = n_2 + 1$, which correspond to points outside the grid, are interpolated by the values on the layer $i = 1$ applying the two periodicities. Analogously, layer $i = n_2$ yields approximations for layer $i = 0$. Thus we use extended boundary conditions for the solution of the boundary value problem (3.4), (3.5) in view of the grid point positions in the biperiodic domain of dependence. In the nonlinear system (3.7), each grid point is connected with only two neighbouring points on the same characteristic curve. Just the boundary conditions produce further couplings via the interpolation. Hence the nonlinear system (3.7) is less coupled than the nonlinear system (3.2). It follows that an LU-decomposition of a Jacobian matrix with respect to (3.7) generates significantly fewer fill-ins in a corresponding Newton method. Consequently, the amount of computational work and memory reduces drastically for the finite difference technique on the characteristic grid in comparison to the uniform grid option. These savings are analysed and compared more detailed in [10].

Furthermore, we are also able to use other discretisations for the characteristic system. There is no restriction to just replacing the derivatives by a difference formula here. An arbitrary multistep scheme, which performs an integration of (3.4), can be included analogue to the described procedure. Thus the discretisation of the characteristic system allows more flexibility than a direct finite difference method for the MPDAE. Attention has to be paid in the DAE case, where consistent solution values are required. One can apply a numerical integration formula for a semi-explicit DAE in a direct or an indirect approach, see [11]. In some integration schemes, it is necessary to perform an indirect approach for the boundary value problem (3.4), (3.5) in order to obtain consistent values.

4. Parallel algorithm

Now we construct a parallel algorithm of the finite difference method from the previous section. To obtain an advantageous block structure, we perform a permutation of the equations and the unknowns in the nonlinear system (3.7). Therefore we define

$$\begin{aligned} \mathbf{y}_j &:= (\tilde{\mathbf{x}}_{j,2}, \ldots, \tilde{\mathbf{x}}_{j,n_2-1}), \quad j = 1, \ldots, n_1 \\ \mathbf{y}_{n_1+1} &:= (\tilde{\mathbf{x}}_{1,1}, \ldots, \tilde{\mathbf{x}}_{n_1,1}, \tilde{\mathbf{x}}_{1,n_2}, \ldots, \tilde{\mathbf{x}}_{n_1,n_2}), \end{aligned} \tag{4.1}$$

where $\mathbf{y}_j$ for $j = 1, \ldots, n_1$ are unknowns along the jth characteristic curve and $\mathbf{y}_{n_1+1}$ includes all variables, which are involved in the interpolation at the boundaries. Likewise, we arrange the equations to

$$\begin{aligned} \mathbf{H}_j &:= (\mathbf{G}_{j,2}, \ldots, \mathbf{G}_{j,n_2-1}), \quad j = 1, \ldots, n_1 \\ \mathbf{H}_{n_1+1} &:= (\mathbf{G}_{1,1}, \ldots, \mathbf{G}_{n_1,1}, \mathbf{G}_{1,n_2}, \ldots, \mathbf{G}_{n_1,n_2}). \end{aligned} \tag{4.2}$$

This strategy can be interpreted as a partition of the characteristic grid in the domain of dependence. Accordingly, $\mathbf{y}_{n_1+1}$ represents coupling variables of the subsystems $\mathbf{H}_j$ for $j = 1, \ldots, n_1$ and $\mathbf{H}_{n_1+1}$ forms the interconnect system. Hence the weakly coupled nonlinear system

$$\begin{aligned} \mathbf{H}_j(\mathbf{y}_j, \mathbf{y}_{n_1+1}) &= \mathbf{0}, \quad j = 1, \ldots, n_1 \\ \mathbf{H}_{n_1+1}(\mathbf{y}_1, \ldots, \mathbf{y}_{n_1}, \mathbf{y}_{n_1+1}) &= \mathbf{0} \end{aligned} \tag{4.3}$$

has to be solved. The Newton-Raphson method yields the iteration

$$\mathbf{y}^{(\nu+1)} = \mathbf{y}^{(\nu)} - \mathbf{\Delta y}^{(\nu)} = \mathbf{y}^{(\nu)} - \left(\tfrac{\partial \mathbf{H}}{\partial \mathbf{y}}(\mathbf{y}^{(\nu)})\right)^{-1} \mathbf{H}(\mathbf{y}^{(\nu)}), \quad \nu = 0, 1, 2, \ldots \tag{4.4}$$

starting from appropriate values $\mathbf{y}^{(0)}$. Consequently, the arising Jacobian matrix has the structure

$$\frac{\partial \mathbf{H}}{\partial \mathbf{y}} = \frac{\partial(\mathbf{H}_1, \ldots, \mathbf{H}_{n_1}, \mathbf{H}_{n_1+1})}{\partial(\mathbf{y}_1, \ldots, \mathbf{y}_{n_1}, \mathbf{y}_{n_1+1})} = \begin{pmatrix} \mathcal{A}_1 & & & & \mathcal{N}_1 \\ & \mathcal{A}_2 & & & \mathcal{N}_2 \\ & & \ddots & & \vdots \\ & & & \mathcal{A}_{n_1} & \mathcal{N}_{n_1} \\ \mathcal{M}_1 & \mathcal{M}_2 & \cdots & \mathcal{M}_{n_1} & \mathcal{C} \end{pmatrix}. \tag{4.5}$$

Due to the employed difference formula, the submatrices $\mathcal{A}_j \in \mathbb{R}^{(n_2-2)k \times (n_2-2)k}$ exhibit a block tridiagonal structure. The matrix $\mathcal{C} \in \mathbb{R}^{2n_1k \times 2n_1k}$ includes all values, which result from interpolation in the interconnect system $\mathbf{H}_{n_1+1}$, and thus $\mathcal{C}$ is not sparse. Furthermore, the other submatrices $\mathcal{N}_j \in \mathbb{R}^{(n_2-2)k \times 2n_1k}$ and $\mathcal{M}_j \in \mathbb{R}^{2n_1k \times (n_2-2)k}$ contain some entries with respect to the variables $\mathbf{y}_{n_1+1}$ and the system $\mathbf{H}_{n_1+1}$. These matrices are sparse, since each of them holds only two blocks of size $k \times k$, which emerge from the difference formula.

The partition of the Jacobian matrix (4.5) allows the use of a block Gaussian elimination for the linear systems in (4.4). This strategy results in the following procedure for the solution of the linear system in each Newton step.

Algorithm 1: block Gaussian elimination

1. Decompose: $\mathcal{A}_j = \mathcal{L}_j \mathcal{U}_j$, $j = 1, \ldots, n_1$
2. Compute: $\mathcal{M}_j \mathcal{A}_j^{-1} \mathcal{N}_j$, $\mathcal{M}_j \mathcal{A}_j^{-1} \mathbf{H}_j$, $j = 1, \ldots, n_1$
3. Solve: $(\mathcal{C} - \sum_{j=1}^{n_1} \mathcal{M}_j \mathcal{A}_j^{-1} \mathcal{N}_j) \mathbf{\Delta y}_{n_1+1} = \mathbf{H}_{n_1+1} - \sum_{j=1}^{n_1} \mathcal{M}_j \mathcal{A}_j^{-1} \mathbf{H_j}$
4. Compute: $\mathbf{\Delta y}_j = \mathcal{A}_j^{-1}(\mathbf{H}_j - \mathcal{N}_j \mathbf{\Delta y}_{n_1+1})$, $j = 1, \ldots, n_1$

This algorithm suggests a parallel implementation. In step one, each process computes and decomposes some matrices $\mathcal{A}_j$, which are portioned block-wise. The same holds for the computation of the Schur complements $\mathcal{M}_j \mathcal{A}_j^{-1} \mathcal{N}_j$ and the right-hand side part $\mathcal{M}_j \mathcal{A}_j^{-1} \mathbf{H}_j$ in step two. The results are sent to the first process. Due to the structure of the matrices $\mathcal{M}_j, \mathcal{N}_j$ here, each Schur complement includes only four submatrices of the size $k \times k$. Hence the communication overhead is low in this partial stage. For step three, the first process sets up matrix $\mathcal{C}$ and subtracts the Schur complements. The arising matrix is not sparse and thus its LU-decomposition causes a relatively large computational effort. In contrast, the decompositions of the matrices $\mathcal{A}_j$ are cheap in view of their block tridiagonal structure. If the linear system of step three is solved by the first process alone, then a sequential bottleneck will occur. However, since the according LU-decomposition is dense, this linear system can be efficiently solved in parallel by all processes. For this purpose, the matrix is distributed across the processes in a two-dimensional block-cyclic manner, for example, which is the most efficient for a parallelisation. Then the processes solve the linear system together. Finally, the resulting solution is sent to all processes in order to solve the local linear systems in step four. This completes the block Gaussian elimination.

The above described algorithm has been implemented in FORTRAN77. The underlying finite difference scheme uses the discretisation with centred differences and linear interpolation at the boundaries. Hence the method is of order two. The code is based on a distributed memory pattern and thus executes message passing, where the library MPI (Message Passing Interface) is applied. More information concerning MPI can be found in [5], for example. The dense interconnect system is solved in parallel using subroutines of the ScaLAPACK (Scalable Linear Algebra PACKage) library, see [1]. Subroutines of the sequential LAPACK library yield the solution of the other subsystems, where band matrices arise. We investigate the time performance of the code in the next section.

Moreover, the block Gaussian elimination can be modified into a two-level Newton method in order to decrease the communication overhead in the whole iteration. Thereby, the parallelisation is adapted to the arising nonlinear systems.

The inner subsystems $\mathbf{H}_j$ for $j = 1, \ldots, n_1$ are more often handled than the interconnect system $\mathbf{H}_{n_1+1}$. If we perform m inner iterations, then one outer iteration reads as follows.

Algorithm 2: multilevel Newton iteration

1. Solve for $j = 1, \ldots, n_1$:

$$\mu = 0, \ldots, m-1: \quad (\mathbf{y}_j^{*(0)} = \mathbf{y}_j^{(\nu)})$$

$$\mathbf{y}_j^{*(\mu+1)} = \mathbf{y}_j^{*(\mu)} - \left(\frac{\partial \mathbf{H}_j}{\partial \mathbf{y}_j}(\mathbf{y}_j^{*(\mu)}, \mathbf{y}_{n_1+1}^{(\nu)}) \right)^{-1} \mathbf{H}_j(\mathbf{y}_j^{*(\mu)}, \mathbf{y}_{n_1+1}^{(\nu)})$$

2. Solve:

$$\left(\frac{\partial \mathbf{H}_{n_1+1}}{\partial \mathbf{y}_{n_1+1}} - \sum_{j=1}^{n_1} \frac{\partial \mathbf{H}_{n_1+1}}{\partial \mathbf{y}_j} \left(\frac{\partial \mathbf{H}_j}{\partial \mathbf{y}_j} \right)^{-1} \frac{\partial \mathbf{H}_j}{\partial \mathbf{y}_{n_1+1}} \right) \Bigg|_{\substack{\mathbf{y}_j = \mathbf{y}_j^{*(m)} \\ \mathbf{y}_{n_1+1} = \mathbf{y}_{n_1+1}^{(\nu)}}} \Delta \mathbf{y}_{n_1+1}$$

$$= \mathbf{H}_{n_1+1}(\mathbf{y}_1^{*(m)}, \ldots, \mathbf{y}_{n_1}^{*(m)}, \mathbf{y}_{n_1+1}^{(\nu)})$$

3. Compute for $j = 1, \ldots, n_1$:

$$\mathbf{y}_j^{(\nu+1)} = \mathbf{y}_j^{*(m)} + \left(\frac{\partial \mathbf{H}_j}{\partial \mathbf{y}_j}(\mathbf{y}_j^{*(m)}, \mathbf{y}_{n_1+1}^{(\nu)}) \right)^{-1} \frac{\partial \mathbf{H}_j}{\partial \mathbf{y}_{n_1+1}}(\mathbf{y}_j^{*(m)}, \mathbf{y}_{n_1+1}^{(\nu)}) \Delta \mathbf{y}_{n_1+1}$$

Set: $\mathbf{y}_{n_1+1}^{(\nu+1)} = \mathbf{y}_{n_1+1}^{(\nu)} - \Delta \mathbf{y}_{n_1+1}$

In one outer iteration, the processes perform several Newton iterations for the inner subsystems, where the coupling variables are kept constant. After solving the modified interconnect system in the second step, the current solutions of the inner subsystems are corrected in order to obtain better approximations for the next outer iteration. If a master process solves the interconnect system alone, then this approach omits communication. One hopes that the number of required outer iterations for a given accuracy is well below the number of steps in the Newton-Raphson method. Although one outer iteration is more expensive than a standard Newton iteration, since the inner subsystems are solved repeatedly, the inherent potential of parallelism leads to savings in the total run time.

Indeed, the solution of the interconnect system represents the main part of the computational work in a Newton step (4.4). Thus this procedure demands a parallelisation, too, at the cost of further communication overhead. If the number of outer iterations is smaller than the number of Newton-Raphson iterations, then the modified interconnect system has to be solved fewer times in the two-level Newton approach. Hence time savings would already emerge in a purely sequential implementation of this technique due to the large computational effort for this stage. Furthermore, the two-level Newton method reduces communication by omitting frequent parallel solutions of the interconnect system.

5. Test results

We investigate the time performance of the implemented parallel algorithm now. As test example, we consider the ring modulator circuit, which produces a multiplicative mixture of two input signals. Fig. 3 displays the circuit diagram. The corresponding mathematical model was introduced by Horneber [7]. Thereby, nodal

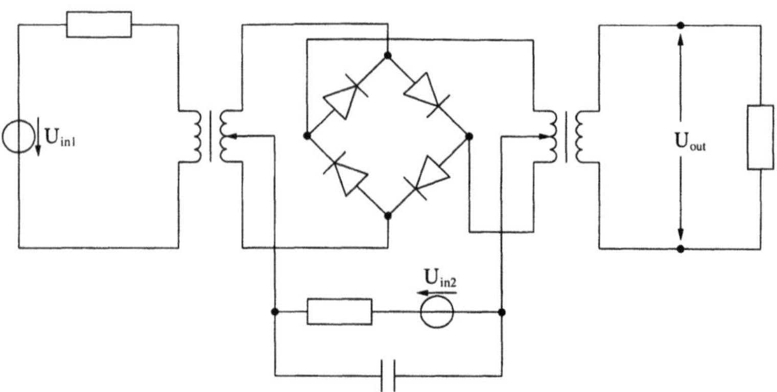

FIGURE 3. Ring modulator circuit.

analysis of an appropriate substitute circuit yields a system of fifteen equations for seven node voltages and eight branch currents. Depending on the choice of a special capacitance, this system is either an ordinary differential equation (ODE) or a DAE of index two. In the ODE case, the capacitance produces a parasitic oscillation. Thus numerical integration demands much more time steps for the ODE than for the DAE description here, see [4]. More information about this modelling and the arising system is also given in [8]. Therefore we apply the DAE model of the ring modulator in the following. In particular, we choose sinusoids with the widely separated periods $T_1 = 100\,\text{msec}$ and $T_2 = 0.1\,\text{msec}$ for the two input signals. This choice gives the desired multirate behaviour and the application of the MPDAE model is advantageous.

Accordingly, we used the parallel code to solve the arising MPDAE system on a characteristic grid. Two grids with 50 points and 100 points in each coordinate direction were selected. This yields a nonlinear system of 37500 and 150000 equations, respectively. The Newton iteration applying start values zero converged to the approximate solution. Thereby, 25 steps were needed to achieve our demanded accuracy that the Newton correction becomes smaller than a relative and absolute tolerance of 10^{-2} for each component. We realised the computations on a SUN 450 Enterprise with SPARC architecture and operating system Solaris 2.7. Four symmetric processors were available here.

Table 1 illustrates the computation times for a single Newton step, where different numbers of processors are involved. The parallel subroutine can also be run on a single processor and thus we obtain the corresponding speedups by comparing the required times with the time for one processor alone. Surprisingly, the parallel linear equation solver from ScaLAPACK is more efficient on the single processor than the according sequential solver from LAPACK on our architecture. Hence the speedups would even enlarge in a comparison with a purely sequential code here. We recognise good speedups in our test example. Moreover, the parallel

	50 × 50 grid		100 × 100 grid	
No. of proc.	Time (secs)	Speedup	Time (secs)	Speedup
1	11.2	—	64.1	—
2	6.5	1.7	36.5	1.8
3	4.9	2.3	27.0	2.4
4	4.1	2.7	22.4	2.9

TABLE 1. Time results for one Newton step in parallel code.

		50 × 50 grid		100 × 100 grid	
Part	No. of proc.	Time (secs)	Speedup	Time (secs)	Speedup
	1	4.8	—	18.9	—
I	2	2.5	1.9	10.2	1.9
	3	1.8	2.7	7.1	2.7
	4	1.5	3.2	6.0	3.2
	1	6.2	—	44.4	—
II	2	3.9	1.6	25.9	1.7
	3	3.0	2.1	19.4	2.3
	4	2.6	2.4	16.2	2.7
	1	0.19	—	0.8	—
III	2	0.09	2.0	0.4	2.0
	3	0.07	2.9	0.27	3.0
	4	0.05	3.8	0.2	4.0

TABLE 2. Time behaviour for different parts in one Newton step of parallel code.

efficiency increases in case of the finer grid due to the higher problem size, which reflects the scalability of the underlying procedures.

For a more detailed examination of the time performance, we split the single Newton step into three separate parts and specify the respective speedups. The first part (I) consists in the computation of the Jacobian matrix and the corresponding right-hand side, the LU-decompositions of the band matrices, calculation of Schur complements and their transmission to the first process. Therefore this stage represents mainly step one and two of Algorithm 1 from the previous section. The distribution of the interconnect system on the process grid and its parallel solution by LU-decomposition, i.e., step three of Algorithm 1, represents the second part (II). Finally, the third part (III) includes the solution of the other subsystems using the result from the interconnect system, which is step four of Algorithm 1. Table 2 shows the time consumptions and corresponding speedups. Thereby, we observe good results in the first stage. However, the parallel efficiency does not increase in case of the finer grid due to the large communication overhead. The second part also features satisfying accelerations and additionally indicates the

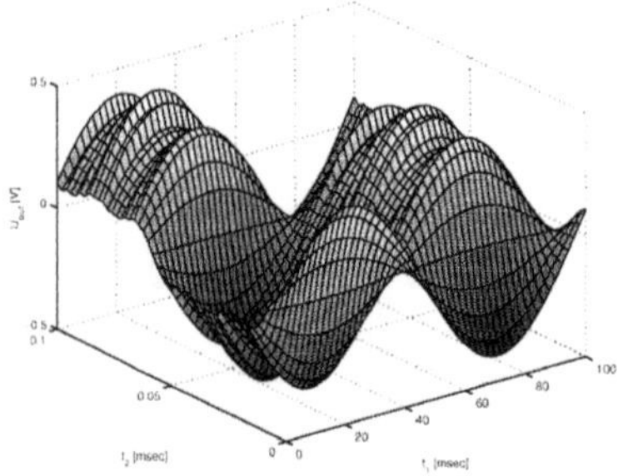
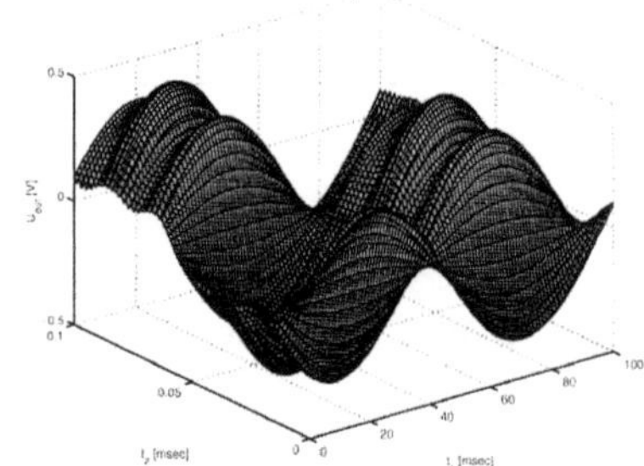

FIGURE 4. Resulting MVF of output voltage from finite difference technique on 50 × 50 grid (left) and 100 × 100 grid (right).

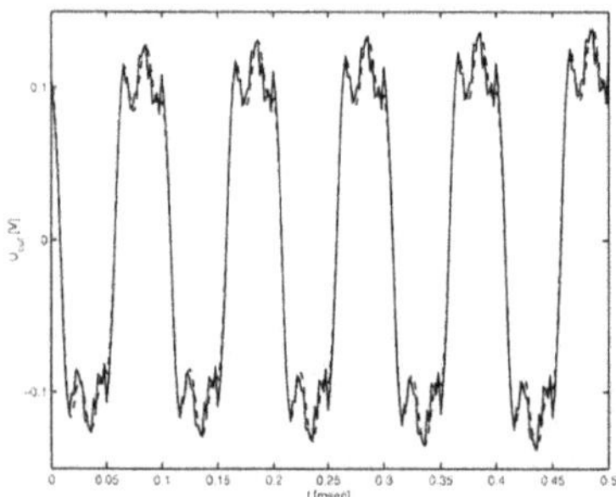
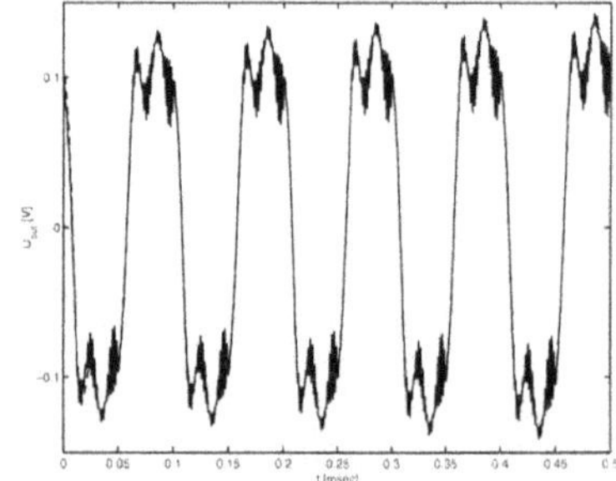

FIGURE 5. First five cycles of interpolated DAE solution from the MVF on 50 × 50 grid (left) and 100 × 100 grid (right) together with integrated DAE solution (dashed line).

scalability of this parallel procedure. Furthermore, we see that this stage includes the main part of computational work during one Newton step. The last part exhibits perfect parallelism. On the other hand, its time demand is too small to influence the total speedup significantly.

The Newton iteration converged to the desired approximate solution on the two characteristic grids. Fig. 4 illustrates the resulting biperiodic MVFs for the output voltage of the ring modulator, where the function is interpolated on a uniform grid employing the periodicities. We obtain the corresponding DAE solution by interpolating the grid functions on the diagonal with the use of the periodicities again. Fig. 5 shows the first five oscillations of these solutions. For comparison, the plots contain also a solution, which was computed by integrating the DAE directly with the RADAU5 scheme under high accuracy demands. Thereby, the MPDAE solution in $t_1 = t_2 = 0$ provided the initial values. We observe a good agreement of interpolated and integrated DAE solution in the steep gradients. However, incorrect fast oscillations within the MVFs occur near the edges, especially on the finer grid, which is typical for centred differences. Some other formulae are able to achieve better approximations in the discretisation of the characteristic system, cf. [10]. The parallel code can easily be modified for an arbitrary formula interconnecting three successive points.

A plot of the DAE solutions over the whole time rate T_1 includes thousand oscillations in the example here. Hence this plot is indistinct, since the details can not be seen. On the contrary, the MVF representation permits a compact visualisation of the behaviour, which belongs to the underlying multitone signal. Thus the concept of MVFs yields an efficient postprocessing tool.

6. Conclusions

The numerical simulation of MPDAE models can efficiently be performed by time domain methods, which apply the underlying information transport along characteristic curves. In particular, the according finite difference method represents a sophisticated and favourable technique. A corresponding parallel algorithm has been presented. This scheme is tailored to the special structure of the arising linear systems in a Newton iteration. Consequently, the parallel code achieves good speedups for practical examples, where linear systems of high orders emerge. Moreover, the developed technique can easily be transferred to a multilevel Newton method. Then the parallelisation operates on the nonlinear system levels and may produce additional savings in computation time.

References

[1] Blackford, L.S.; Choi, J.; Cleary, A.; D'Azevedo, E.; Demmel, J.; Dhillon, I.; Dongarra, J.; Hammarling, S.; Henry, G.; Petitet, A.; Stanley, K.; Walker, D.; Whaley, R.C.: ScaLAPACK Users' Guide. SIAM, Philadelphia 1997.

[2] Brachtendorf, H.G.; Welsch, G.; Laur, R.; Bunse-Gerstner, A.: Numerical steady state analysis of electronic circuits driven by multi-tone signals. Electrical Engineering **79** (1996), pp. 103–112.

[3] Brachtendorf, H.G.; Laur, R.: Multi-rate PDE methods for high Q oscillators. In: Mastorakis, N. (ed.): Problems in Modern Applied Mathematics. CSCC 2000, MCP 2000, MCME 2000 Multi-conference, Athens, July 2000. World Scientific 2000, pp. 391–398.

[4] Denk, G.; Rentrop, P.: Mathematical models in electric circuit simulation and their numerical treatment. In: Strehmel, K. (ed.): Numerical Treatment of Differential Equations. Proc. of Fifth Seminar 'NUMDIFF-5' at Halle, 1989. Teubner 1991, pp. 305–316.

[5] Gropp, W.; Lusk, E.; Skjellum, A.: Using MPI: Portable Parallel Programming with the Message-Passing Interface. (2nd Ed.) MIT Press, Cambridge 1999.

[6] Günther, M.; Feldmann, U.: CAD based electric circuit modeling in industry I: Mathematical structure and index of network equations. Surv. Math. Ind. **8** (1999), pp. 97–129.

[7] Horneber, E.H.: Analyse nichtlinearer RLCÜ-Netzwerke mit Hilfe der gemischten Potentialfunktion mit einer systematischen Darstellung der Analyse nichtlinearer dynamischer Netzwerke. PhD thesis, Universität Kaiserslautern, Kaiserslautern 1976.

[8] Kampowsky, W.; Rentrop, P.; Schmitt, W.: Classification and numerical simulation of electric circuits. Surv. Math. Ind. **2** (1992), pp. 23–65.

[9] Pulch, R.; Günther, M.: A method of characteristics for solving multirate partial differential equations in radio frequency application. Appl. Numer. Math. **42** (2002) 1, pp. 397–409.

[10] Pulch, R.: A finite difference method for solving multirate partial differential algebraic equations. Preprint 02/04, IWRMM, University Karlsruhe (2002).

[11] Rentrop, P.; Strehmel, K.; Weiner, R.: Ein Überblick über Einschrittverfahren zur numerischen Integration in der technischen Simulation. GAMM-Mitt., Berlin **19** (1996) H. 1, pp. 9–43.

[12] Roychowdhury, J.: Analysing circuits with widely-separated time scales using numerical PDE methods. IEEE Trans. CAS I **48** (2001) 5, pp. 578–594.

[13] Roychowdhury, J.: Multi-time PDEs for dynamical system analysis. In: Rienen, U. van; Günther, M.; Hecht, D. (eds.): Scientific Computing in Electrical Engineering. Lecture Notes in Computational Science and Engineering, Springer, Berlin 2001, pp. 3–14.

[14] Wever, U.; Zheng, Q.: Parallel circuit simulation on workstation clusters. In: Neunzert, H. (ed.): Progress in Industrial Mathematics at ECMI 94. Wiley/Teubner 1996, pp. 274–284.

Acknowledgements

This work has been supported by the "Bundesministerium für Bildung und Forschung" (BMBF) within the programme "Neue mathematische Verfahren in Industrie und Dienstleistungen" (Project No. 03REM3KA). The author is indebted to Prof. P. Rentrop and Priv.-Doz. Dr. M. Günther for helpful discussions.

Roland Pulch
Technische Universität München
Zentrum Mathematik – M2
Boltzmannstr. 3
85748 Garching b. München, Germany
email: pulch@mathematik.tu-muenchen.de

Modeling, Simulation and Optimization of Integrated Circuits
(Eds.) K. Antreich, R. Bulirsch, A. Gilg and P. Rentrop

International Series of Numerical Mathematics Vol. 146, 167–184

Feasibility Regions and their Significance to the Hierarchical Optimization of Analog and Mixed-Signal Systems

Guido Stehr, Helmut Graeb, and Kurt Antreich

Abstract. Physical effects have to be considered throughout the design of analog integrated circuits. The difficulties in modeling these effects at higher levels of abstraction is a major obstacle for a top-down analog design that proceeds from high to low levels of abstraction. This paper discusses a hierarchical top-down design style with feasibility regions as a means to describe physical effects at different levels of abstraction. We suggest a method to calculate linearized approximations to the feasibility regions at higher levels based on Fourier-Motzkin elimination.

1. Introduction

Analog components play important roles in modern integrated electronic systems. Telecommunication applications, for example, combine digital circuitry for signal processing and analog components such as filters, amplifiers and signal converters on a single chip. The design of such mixed analog/digital, i.e., *mixed-signal*, systems suffers from the entirely different design styles in the digital and the analog domain.

The development of digital circuits is characterized by a *top-down* design style. Starting from the system specifications, digital design successively adds implementation details. In this way the abstraction level is lowered until the final circuit implementation is found. This methodology is based on the abstract digital view, which completely hides physical details in the early design stages. The prominent challenge in digital design is a large system complexity. Hardware description languages (HDLs) such as VHDL and Verilog are the foundation of digital circuit simulation and synthesis, which results in a high development productivity.

In contrast, the properties of an analog circuit are tightly coupled with its physical implementation. Therefore, the neglection of physical effects is hardly possible even in early design steps. For that reason, analog design usually proceeds from low to high levels of abstraction in a *bottom-up* fashion.

Key words and phrases. nonlinear optimization, sizing rules, Fourier-Motzkin elimination, hierarchical optimization.

Analog circuit design consists of three steps:

- Topology selection/generation: Find a circuit structure.
- Circuit sizing: Assign physical values to designable parameters.
- Layout: Generate actual geometric structures to be implemented in silicon.

A major problem in analog design is the tight coupling of these steps, which results in a low level of automation. Therefore, most circuits are still developed manually by experienced designers. Although in electronic systems, the analog components occupy only a small fraction of the entire chip area, they increasingly constitute a design bottleneck. Emerging hardware description languages for mixed-signal systems, most prominently VHDL-AMS and Verilog-AMS, promise to stimulate the development of new design tools and methodologies.

The approach presented in this paper focuses on topology selection and sizing. It is intended as a contribution to the realization of an analog and mixed-signal top-down design flow.

For the sizing of small analog circuits, mature tools are available, which rely on circuit simulation [1]. The aim of the approach at hand is to extend their applicability to large systems by the introduction of hierarchy. The idea of hierarchical design was presented before [2, 8, 10, 14]. Yet, a core problem remained unsolved, namely the suitable formulation of physically motivated design constraints at higher levels of abstraction. We suggest a solution to this problem.

This article is organized as follows: Section 2 introduces a phase-locked loop (PLL) as an example of a mixed-signal system that our optimization approach aims at. Section 3 describes modeling and simulation at circuit and system level. Also, a hierarchical simulation approach is formulated that combines both levels of abstraction. In a similar fashion, Section 4 describes hierarchical optimization. Additionally, the concept of sizing rules is explained, which formalizes circuit design knowledge for automatic design. In hierarchical optimization, the notion of a *system feasibility region* plays a key role. Section 5 elaborates on this idea and shows how a linear approximation to this feasibility region can be obtained. In this method, the Fourier-Motzkin elimination plays a prominent role. It is described in Section 6. Finally, Section 7 concludes this article.

2. Mixed-signal systems

A phase-locked loop (PLL) as depicted in Fig. 1 is an example of a mixed-signal system. It is frequently used for clock generation in digital circuits and consists of about 1500 transistors. Its function is to synthesize a rectangular output signal that has K times the frequency of the input reference signal: $f_{out} = K \cdot f_{in}$.

The core building block is a current-controlled oscillator (CCO). Its output signal passes a frequency divider ($1/K$) and is compared to the input signal. This is done by a phase detector (PD), which generates a current based on the signal phase difference. Then, this control current passes a loop filter and is fed into the

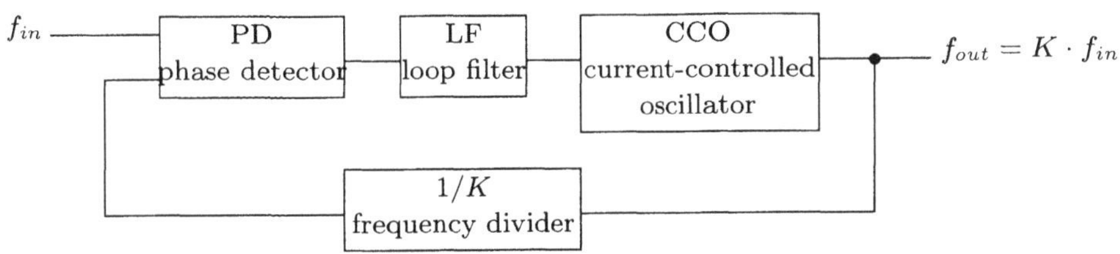

FIGURE 1. Phase-locked loop (PLL): a mixed-signal system

CCO. The oscillator and the loop filter exhibit a behavior that has to be captured in continuous analog models. The frequency divider and the phase detector are modeled as discrete digital components. Therefore, the PLL is a combined analog/digital, i.e., a *mixed-signal*, system.

Given a certain circuit structure or topology, a designer can tune the circuit properties by adjusting the attributes of the individual transistors. For CMOS technology, these designable transistor parameters $\mathbf{t}$ [1] are the transistor channel width and length. The circuit performances $\mathbf{f}$, which characterize the behavior of the PLL, are the lock time and the lock range, among others. The lock time is how long it takes the PLL to adjust to a given input frequency, and the lock range is the frequency range that the PLL can lock on. The sizing task is to choose appropriate parameter values to realize the specified performances.

3. Simulation

Simulation is the main tool in analog circuit design. In the following sections, the properties of simulation at circuit and system level will be discussed. It will also be shown how hierarchical simulation combines both levels.

3.1. Circuit-level simulation

At the circuit abstraction level, a circuit is modeled by a netlist of transistors. As mentioned before, the transistor count of a PLL is about 1500. On an average, a simulator has to solve 1 to 2 nonlinear differential-algebraic equations (DAEs) per transistor. These DAEs describe physical effects in a detailed way. There are two adjustable design parameters per transistor, namely its channel length and width. In fact, not every single transistor needs to be designed individually. Therefore, a parameter reduction is possible heuristically. Consequently, the PLL has a total complexity of about 2000 nonlinear DAEs and roughly 50 design parameters.

The PLL performances $\mathbf{f}$ depend on the transistor parameters $\mathbf{t}$. They are obtained from circuit simulation and subsequent postprocessing. These two steps can

[1] In this paper, vectors and matrices are written as bold lower case and bold upper case letters, respectively. Regular lower case letters denote scalars.

be described mathematically by the performance function at circuit level $\mathbf{f}^{\,\mathrm{circ}}(\cdot)$:

$$\mathbf{f} = \mathbf{f}^{\,\mathrm{circ}}(\mathbf{t})\,. \tag{1}$$

An advantage of the circuit-level simulation is that it is an accurate direct mapping from the design parameters to the performances. On the other hand, simulation is very time-consuming. Since the time constant of the CCO differs from the time constant of the control loop by several orders of magnitude, the PLL can oscillate numerous times before it eventually locks. Consequently, a single simulation run may take up to 50 hours. Therefore, faster simulation alternatives are required.

3.2. System-level simulation

One way to shorter simulation times is the transition to the system abstraction level. In this more abstract view, the PLL is a netlist of its 4 functional blocks (phase detector, loop filter, oscillator and frequency divider). These blocks are modeled by 5 to 10 equations embedded in a VHDL-AMS model, which is called a *behavioral model* because it does not focus on physical effects any longer. It is rather a phenomenological description of the block behavior. The oscillator, for example, can be described as a controlled rectangular voltage source. Only 3 to 5 behavioral parameters suffice to adjust such a model. The free-running frequency of the oscillator is an example of a behavioral parameter. The behavioral parameters of the entire circuit are combined in a vector $\mathbf{b}$.

In comparison to the circuit level, the total model complexity is reduced dramatically. The number of equations drops from roughly 2000 to about 30. Instead of 50 design parameters, there are only 10. A simulator can now calculate the system performances from a much simplified system-level description of the PLL:

$$\mathbf{f} = \mathbf{f}^{\,\mathrm{sys}}(\mathbf{b})\,. \tag{2}$$

The greatest advantage of this method is its fast execution speed. The simulation time drops down to less than 10 minutes, which means a speedup of more than 300 when compared to a full circuit-level simulation. Yet, the behavioral parameters $\mathbf{b}$ cannot be adjusted directly any longer. However, if the modeling guarantees that the *behavioral parameters are block performances*, which can be extracted from circuit-level simulations, we can establish a hierarchical simulation approach as discussed below.

3.3. Hierarchical simulation

Hierarchical simulation combines circuit and system simulation. Leaving the PLL example behind, a general system consists of N blocks.

In a first step, which can be called *model calibration*, for each block a circuit simulation is performed in order to obtain its block performances from its transistor parameters. *The circuit-level performances of the blocks are the system-level parameters for their behavioral models* later on:

$$\underset{\substack{\text{for all blocks } i\\ \text{of a system}}}{\forall_{i\in\{1,2,\ldots,N\}}} \quad \underset{\substack{\text{system-level}\\ \text{parameters}\\ \text{of block } i}}{\mathbf{b}_i} \quad = \quad \underset{\substack{\text{circuit-level}\\ \text{performances}\\ \text{of block } i}}{\mathbf{b}_i^{\text{circ}}} \underset{\substack{\text{circuit-level}\\ \text{parameters}\\ \text{of block } i}}{(\mathbf{t}_i)}\,. \tag{3}$$

The vectors [2]

$$\mathbf{t} = (\mathbf{t}_1^T\ \mathbf{t}_2^T\ \ldots \mathbf{t}_N^T)^T \tag{4}$$

$$\text{and} \quad \mathbf{b} = (\mathbf{b}_1^T\ \mathbf{b}_2^T\ \ldots \mathbf{b}_N^T)^T \tag{5}$$

comprise the transistor and behavioral parameters, respectively, of all N system blocks.

In a second step, the system performances are found by a system-level simulation with behavioral parameters:

$$\mathbf{f} = \mathbf{f}^{\,\text{sys}}(\mathbf{b})\,. \tag{6}$$

Since the model calibration takes only a few minutes, there is still a speedup of about 100 compared to a pure circuit-level simulation. Fig. 2 illustrates this concept graphically.

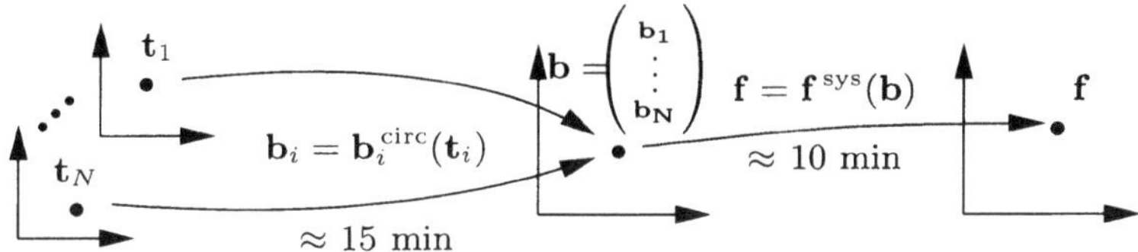

FIGURE 2. Hierarchical simulation

4. Optimization

While the task of simulation is to map parameters to performances, optimization has to propagate performance specifications onto parameter values in the opposite direction. In this section, optimization is examined from the three hierarchical points of view introduced in Section 3.

4.1. Circuit-level optimization

4.1.1. PROBLEM FORMULATION Optimization starts from a set of specified performance values $\hat{\mathbf{f}}$ and has to find corresponding parameter values $\hat{\mathbf{t}}$. This leads to a nonlinear optimization problem:

$$\hat{\mathbf{t}} = \underset{\mathbf{t}}{\operatorname{argmin}} \parallel \mathbf{f}^{\,\text{circ}}(\mathbf{t}) - \hat{\mathbf{f}} \parallel\,. \tag{7}$$

[2]The superscript ^{T} denotes a transposed vector.

The argmin operator yields the argument $\hat{\mathbf{t}}$ leading to a minimum objective value. In the formulation above, the vector norm measures the difference between the specifications $\hat{\mathbf{f}}$ and the achieved performances $\mathbf{f}^{\,\mathrm{circ}}(\mathbf{t})$ as obtained from circuit-level simulation. Note that this is only one possibility to transform a vector optimization into a scalar problem. Since $\mathbf{f}^{\,\mathrm{circ}}(\cdot)$ is nonlinear, an iterative optimization method is required that uses a simulator in a loop.

4.1.2. CIRCUIT-LEVEL CONSTRAINTS An optimization formulation according to (7) does not guarantee a technically meaningful sizing. A mathematically advantageous parameter vector $\hat{\mathbf{t}}$ might offend good design practice. Therefore, (7) has to be completed by additional constraints, which capture circuit-level design knowledge, and which are called *sizing rules* [9, 16].

Two categories suffice to classify the circuit-level sizing rules:

- Geometric/Electrical
  Geometric sizing rules directly refer to transistor geometries and are given algebraically. For example, transistor sizes must not fall below a certain minimum feature size as it is given by the production technology. Electrical rules need to be evaluated based on quick DC circuit simulations. As an example, there are often minimum operating point voltages.
- Inequality/Equality
  Inequality rules require that electrical or geometric circuit quantities exceed or remain below certain limits. Equality rules exist only for geometric quantities. Since these rules are given as explicit algebraic equations, they can be used to reduce the dimension of the parameter design space.

After the parameter space reduction on the basis of geometric equality sizing rules, the remaining parameters $\mathbf{t}'$ have to satisfy a number of inequalities, which are either explicitly given as algebraic expressions, or require simulation. After elementary algebraic transformations, we obtain a single vector inequality $\mathbf{c}^{\,\mathrm{circ}}(\mathbf{t}') \geq \mathbf{0}$ [3]. Consequently, the complete optimization formulation is

$$\hat{\mathbf{t}}' = \underset{\mathbf{t}'}{\operatorname{argmin}} \parallel \mathbf{f}^{\,\mathrm{circ}}(\mathbf{t}') - \hat{\mathbf{f}} \parallel \quad \text{s.t.} \quad \mathbf{c}^{\,\mathrm{circ}}(\mathbf{t}') \geq \mathbf{0}\,. \tag{8}$$

The circuit-level constraints proved essential to a successful automatic circuit optimization [1]. They define a technically meaningful parameter subspace, also referred to as *feasibility region* [16]:

$$\mathcal{T} = \{\mathbf{t}' \mid \mathbf{c}^{\,\mathrm{circ}}(\mathbf{t}') \geq \mathbf{0}\}\,. \tag{9}$$

For ease of notation, the superscript $'$ indicating the difference between the original and the reduced transistor parameter vector is omitted in the remainder of the article.

[3]Vector inequalities are interpreted element-wise.

4.2. System-level optimization

4.2.1. PROBLEM FORMULATION System-level optimization corresponds to circuit-level optimization with the difference that behavioral parameters $\mathbf{b}$ are adjusted instead of transistor parameters $\mathbf{t}$, and that simulations are carried out at system level:

$$\hat{\mathbf{b}} = \underset{\mathbf{b}}{\operatorname{argmin}} \parallel \mathbf{f}^{\,\mathrm{sys}}(\mathbf{b}) - \hat{\mathbf{f}} \parallel \quad \text{s.t.} \quad \mathbf{c}^{\,\mathrm{sys}}(\mathbf{b}) \geq \mathbf{0}\,. \tag{10}$$

4.2.2. SYSTEM-LEVEL CONSTRAINTS The optimization formulation given in (10) relies on a suitable representation of system-level constraints $\mathbf{c}^{\,\mathrm{sys}}(\mathbf{b}) \geq \mathbf{0}$. The associated system feasibility region

$$\mathcal{B} = \{\mathbf{b} \mid \mathbf{c}^{\,\mathrm{sys}}(\mathbf{b}) \geq 0\} \tag{11}$$

comprises all technically meaningful system parameters. Restrictions on these behavioral parameters arise from both system- and circuit-level requirements:

$$\mathbf{c}^{\,\mathrm{sys}}(\mathbf{b}) = \begin{pmatrix} \mathbf{c}^{\,\mathrm{sys}}_{\,\mathrm{sys}}(\mathbf{b}) \\ \mathbf{c}^{\,\mathrm{sys}}_{\,\mathrm{circ}}(\mathbf{b}) \end{pmatrix}. \tag{12}$$

On the one hand, there are constraints which are given directly at system level. Such system requirements include, for example, stability criteria in control loops. These requests are usually not expressly specified but reflect common system design knowledge, which can directly be formulated as system constraints:

$$\mathbf{c}^{\,\mathrm{sys}}_{\,\mathrm{sys}}(\mathbf{b}) \geq \mathbf{0}\,. \tag{13}$$

These constraints can directly be evaluated by system simulations.

On the other hand, since system parameters are circuit performances, only those system parameter values are meaningful that can be realized by the circuit topologies of the composing functional blocks. Given values $\tilde{\mathbf{b}}$ of the block performances can be achieved if and only if there exists a circuit-level parameter vector $\tilde{\mathbf{t}}$ such that

$$\tilde{\mathbf{b}} = \mathbf{b}^{\,\mathrm{circ}}(\tilde{\mathbf{t}}) \quad \wedge \quad \mathbf{c}^{\,\mathrm{circ}}(\tilde{\mathbf{t}}) \geq \mathbf{0}\,. \tag{14}$$

The task of propagating constraints from circuit to system level is to find a representation of the system constraints according to (10) that does not include circuit-level simulations any longer:

$$\mathbf{c}^{\,\mathrm{sys}}_{\,\mathrm{circ}}(\mathbf{b}) \geq \mathbf{0} \quad \Leftrightarrow \quad \mathbf{b} = \mathbf{b}^{\,\mathrm{circ}}(\mathbf{t}) \;\wedge\; \mathbf{c}^{\,\mathrm{circ}}(\mathbf{t}) \geq \mathbf{0}\,. \tag{15}$$

A new technique to calculate an approximation to $\mathbf{c}^{\,\mathrm{sys}}_{\,\mathrm{circ}}(\cdot)$ will be presented in Section 5.

4.3. Hierarchical optimization

Given an appropriate description of the system feasibility region as in (11), an optimization can be performed corresponding to the hierarchical simulation from (3) and (6).

In a first step, a single optimization is carried out entirely at system level:

$$\hat{\mathbf{b}} = \underset{\mathbf{b}}{\operatorname{argmin}} \parallel \mathbf{f}^{\,\mathrm{sys}}(\mathbf{b}) - \hat{\mathbf{f}} \parallel \quad \text{s.t.} \quad \mathbf{c}^{\,\mathrm{sys}}(\mathbf{b}) \geq \mathbf{0}\,. \tag{16}$$

A given set of system specifications $\hat{\mathbf{f}}$ leads to corresponding behavioral parameters $\hat{\mathbf{b}}$ at system level. In a second step, the entries in vector $\hat{\mathbf{b}} = (\hat{\mathbf{b}}_1^T \ldots \hat{\mathbf{b}}_N^T)^T$ *turn into the specifications* of the individual blocks. Then, multiple circuit-level optimizations, one for each block, can be carried out in parallel:

$$\underset{i\in\{1\ldots N\}}{\forall} \; \hat{\mathbf{t}}_i = \underset{\mathbf{t}_i}{\operatorname{argmin}} \parallel \mathbf{b}_i^{\mathrm{circ}}(\mathbf{t}_i) - \hat{\mathbf{b}}_i \parallel \quad \text{s.t.} \quad \mathbf{c}_i^{\mathrm{circ}}(\mathbf{t}_i) \geq \mathbf{0}\,. \tag{17}$$

This step yields the sizing $\hat{\mathbf{t}} = (\hat{\mathbf{t}}_1^T \ldots \hat{\mathbf{t}}_N^T)^T$ of the entire system at circuit level. This technique is depicted in Fig. 3.

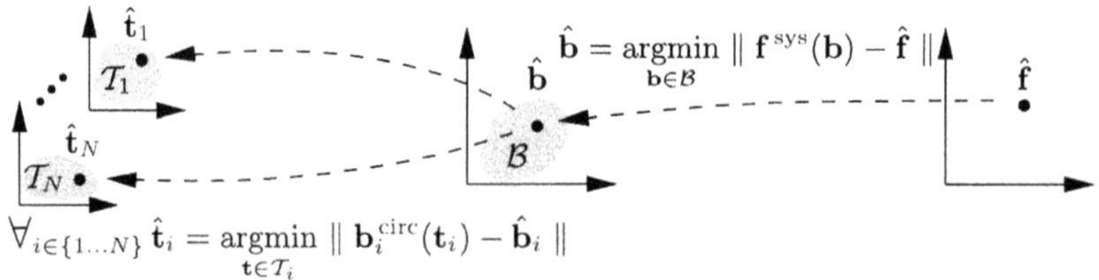

FIGURE 3. Hierarchical optimization

Note that to this approach it is of crucial importance that the system-level optimization (16) does not yield illusory specifications for the subsequent circuit-level optimizations (17). The system-level feasibility region $\mathcal{B}$, which comprises all technically meaningful block performances, ensures the solvability of the circuit-level optimization problems. This approach relies on a suitable description of $\mathcal{B}$ entirely at system level, which is needed to effectively separate the abstraction levels. In the following section, a method is introduced to gain a linear approximation to the system feasibility region.

5. System feasibility region

It was discussed in Section 4.2.2 that the system feasibility region reflects system-level and circuit-level design knowledge. System design knowledge can readily be formulated at system level in analogy to circuit design knowledge at circuit level. Therefore, in the following, we focus on the bottom-up propagation of circuit-level constraints and assume

$$\mathbf{c}^{\,\mathrm{sys}}(\mathbf{b}) = \mathbf{c}_{\mathrm{circ}}^{\,\mathrm{sys}}(\mathbf{b})\,. \tag{18}$$

Then, the system feasibility region is an image of the circuit feasibility region:

$$\underbrace{\mathbf{c}^{\,\mathrm{circ}}(\mathbf{t}) \geq \mathbf{0}}_{\substack{\text{(I)}\\ \text{circuit feasibility region}\\ \mathcal{T}}} \quad \wedge \quad \underbrace{\mathbf{b}^{\,\mathrm{circ}}(\mathbf{t}) = \mathbf{b}}_{\substack{\text{(II)}\\ \text{mapping: circuit to system}\\ \text{parameter space}\\ \mathbf{t} \mapsto \mathbf{b}}} \quad \Leftrightarrow \quad \underbrace{\mathbf{c}^{\,\mathrm{sys}}(\mathbf{b}) \geq \mathbf{0}}_{\substack{\text{(III)}\\ \text{system feasibility region}\\ \mathcal{B}}} \tag{19}$$

The goal is to find a description of $\mathcal{B}$ entirely at system level according to (19)/(III).

5.1. Linear approximation

An approximate representation of the system feasibility region can be found as follows: Based on a linearized representation of the circuit feasibility region (19)/(I) and of the mapping from the circuit to the system parameter space (19)/(II), a linearized description of the system feasibility region (19)/(III) can be calculated. These three steps will be discussed in the following subsections.

5.1.1. Linearization of circuit-level feasibility region

The circuit-level constraints

$$\mathbf{c}^{\,\mathrm{circ}}(\mathbf{t}) \geq \mathbf{0} \tag{19/I}$$

can be approximated at a technically meaningful operating point $\mathbf{t}_0$ by a linear Taylor expansion:

$$\mathbf{c}^{\,\mathrm{circ}}(\mathbf{t}) \approx \mathbf{c}^{\,\mathrm{circ}}(\mathbf{t}_0) + \left.\frac{\partial \mathbf{c}^{\,\mathrm{circ}}(\mathbf{t})}{\partial \mathbf{t}}\right|_{\mathbf{t}_0} \cdot \Delta\mathbf{t}\,. \tag{20}$$

This yields the following linear approximation to (19)/(I):

$$\underbrace{\left.\frac{\partial \mathbf{c}^{\,\mathrm{circ}}(\mathbf{t})}{\partial \mathbf{t}}\right|_{\mathbf{t}_0}}_{\mathbf{S}_{\mathbf{ct}}} \cdot \Delta\mathbf{t} \geq \underbrace{-\,\mathbf{c}^{\,\mathrm{circ}}(\mathbf{t}_0)}_{\mathbf{c}_{\mathbf{t}}}\,. \tag{21}$$

The Jacobian matrix $\mathbf{S}_{\mathbf{ct}}$ contains the sensitivities of the circuit-level constraints with respect to the transistor parameters. It can be approximated by finite differences from a number of quick DC circuit simulations. Geometrically, (21) describes a polytope in the transistor parameter space [15].

5.1.2. Linearization of mapping from circuit to system level

The mapping from circuit to system level

$$\underbrace{\mathbf{b}^{\,\mathrm{circ}}(\mathbf{t})}_{\substack{\text{circuit-level}\\ \text{performances}}} = \underbrace{\mathbf{b}}_{\substack{\text{system-level}\\ \text{parameters}}} \tag{19/II}$$

can be treated similarly to the constraints in the previous section:

$$\mathbf{b}^{\,\mathrm{circ}}(\mathbf{t}) \approx \mathbf{b}(\mathbf{t}_0) + \left.\frac{\partial \mathbf{b}^{\,\mathrm{circ}}(\mathbf{t})}{\partial \mathbf{t}}\right|_{\mathbf{t}_0} \cdot \Delta\mathbf{t}\,. \tag{22}$$

From (19)/(II) and (22) we obtain the wanted linearization

$$\underbrace{\left.\frac{\partial \mathbf{b}^{\text{circ}}(\mathbf{t})}{\partial \mathbf{t}}\right|_{\mathbf{t}_0}}_{\mathbf{S_{bt}}} \cdot \Delta \mathbf{t} = \underbrace{\mathbf{b} - \mathbf{b}(\mathbf{t}_0)}_{\Delta \mathbf{b}} \ . \tag{23}$$

In (23), the difference between the constant vectors can be seen as variations of the behavioral parameters $\Delta\mathbf{b}$ and the Jacobian matrix $\mathbf{S_{bt}}$ describes the sensitivity of the behavioral parameters $\mathbf{b}$ with respect to the transistor parameters $\mathbf{t}$.

5.1.3. Linearized system-level feasibility region

Based on the results from the previous sections, namely the linearized circuit feasibility region (21) and the linearized map from the circuit to the system parameter space (23), a linear approximation to the system-level feasibility region is sought:

$$\underbrace{\mathbf{S_{ct}} \cdot \Delta \mathbf{t} \geq \mathbf{c_t}}_{(21)} \wedge \underbrace{\mathbf{S_{bt}} \cdot \Delta \mathbf{t} = \Delta \mathbf{b}}_{(23)} \quad \Rightarrow \quad \underbrace{\mathbf{S_{cb}} \cdot \Delta \mathbf{b} \geq \mathbf{c_b}}_{(24)} \ .$$

If $\mathbf{S_{bt}}$ is nonsingular, the transistor parameter vector $\Delta\mathbf{t}$ in (23) can be obtained by matrix inversion:

$$\Delta \mathbf{t} = \mathbf{S_{bt}^{-1}} \cdot \Delta \mathbf{b} \, . \tag{24}$$

Using this result in (21), we obtain

$$\underbrace{\mathbf{S_{ct}} \cdot \mathbf{S_{bt}^{-1}}}_{\mathbf{S_{cb}}} \cdot \Delta \mathbf{b} \geq \underbrace{\mathbf{c_t}}_{\mathbf{c_b}} \ . \tag{25}$$

Usually, however, the dimension of the transistor parameter vector is larger than the dimension of its behavioral correspondence,

$$\underbrace{\dim(\Delta \mathbf{t})}_{d_\mathbf{t}} > \underbrace{\dim(\Delta \mathbf{b})}_{d_\mathbf{b}} \, , \tag{26}$$

and $\mathbf{S_{bt}}$ is not invertible. In this case, we suggest a two-step method to calculate (24) as outlined below, assuming that $\mathbf{S_{bt}}$ has full rank.

Step 1: Equation-Based Elimination

We can rewrite (23) as

$$(\mathbf{S_{bt\square}} \ \mathbf{S_{bt\diamond}}) \cdot \begin{pmatrix} \Delta \mathbf{t}_\square \\ \Delta \mathbf{t}_\diamond \end{pmatrix} = \Delta \mathbf{b} \, , \tag{27}$$

with a partitioning of $\mathbf{S_{bt}}$ into a rectangular part $\mathbf{S_{bt\square}}$ and a remainder $\mathbf{S_{bt\diamond}}$:

$$\mathbf{S_{bt}} = (\mathbf{S_{bt\square}} \ \mathbf{S_{bt\diamond}}) ; \quad \dim(\mathbf{S_{bt\square}}) = d_\mathbf{b} \times d_\mathbf{b} \, , \ \dim(\mathbf{S_{bt\diamond}}) = d_\mathbf{b} \times (d_\mathbf{t} - d_\mathbf{b}) \, . \tag{28}$$

If $\mathbf{S_{bt\square}}$ is regular, then $\Delta\mathbf{t}_\square$ can formally be obtained from

$$\Delta \mathbf{t}_\square = \mathbf{S_{bt\square}^{-1}} \cdot \Delta \mathbf{b} \ - \ \mathbf{S_{bt\square}^{-1}} \mathbf{S_{bt\diamond}} \cdot \Delta \mathbf{t}_\diamond \, . \tag{29}$$

By the same token, (21) corresponds to

$$(\mathbf{S}_{\mathbf{ct}\Box}\ \mathbf{S}_{\mathbf{ct}\Diamond}) \cdot \begin{pmatrix} \Delta\mathbf{t}_\Box \\ \Delta\mathbf{t}_\Diamond \end{pmatrix} \geq \mathbf{c}_\mathbf{t} \, . \tag{30}$$

The substitution of $\Delta\mathbf{t}_\Box$ in (30) by the expression found in (29) yields

$$\underbrace{(\mathbf{S}_{\mathbf{ct}\Box}\ \mathbf{S}_{\mathbf{ct}\Diamond}) \begin{pmatrix} \mathbf{S}_{\mathbf{bt}\Box}^{-1} & -\mathbf{S}_{\mathbf{bt}\Box}^{-1}\mathbf{S}_{\mathbf{bt}\Diamond} \\ \mathbf{0} & \mathbf{I} \end{pmatrix}}_{\mathbf{S}_{\mathbf{cbt}}} \begin{pmatrix} \Delta\mathbf{b} \\ \Delta\mathbf{t}_\Diamond \end{pmatrix} \geq \mathbf{c}_\mathbf{t} \, , \tag{31}$$

where $\mathbf{I}$ is an identity matrix of dimension $(d_\mathbf{t} - d_\mathbf{b}) \times (d_\mathbf{t} - d_\mathbf{b})$. In (31) a maximum number of transistor parameters has been replaced by behavioral parameters. A comparison to (24) reveals that, for the desired format of the system-level constraints, the remaining transistor parameters $\Delta\mathbf{t}_\Diamond$ have to be eliminated as well.

Instead of determining $\mathbf{S}_{\mathbf{bt}\Box}^{-1}$ in (31) explicitly, Gaussian elimination or any other method to solve a system of linear equations can be used to numerically calculate the expressions including the inverse matrix. Here, it becomes apparent that a suitable partitioning of $\mathbf{S}_\mathbf{bt}$ is essential to a successful calculation of $\mathbf{S}_\mathbf{cbt}$. While a partitioning aiming at a good condition of $\mathbf{S}_{\mathbf{bt}\Box}$ seems obvious at first sight, it is still unclear how the partitioning affects the properties of $\mathbf{S}_\mathbf{cbt}$ with respect to the second elimination step.

Step 2: Inequality-Based Elimination
In a second step, the remaining $(d_\mathbf{t} - d_\mathbf{b})$ transistor parameters $\Delta\mathbf{t}_\Diamond$ have to be eliminated from (31). This can be done by *Fourier-Motzkin elimination* [6, 7]. This elimination technique for inequalities bears some resemblance to Gaussian elimination for equations. It is described in detail in Section 6. The second elimination step leads to

$$\mathbf{S}_\mathbf{cb} \cdot \Delta\mathbf{b} \geq \mathbf{c}_\mathbf{b} \, , \tag{24}$$

which is the sought linearized description of the system feasibility region. As described in Section 4.3, this approximation is a cornerstone of hierarchical optimization because it establishes the link between the circuit and the system abstraction level.

Although the Fourier-Motzkin elimination has been known for a long time, it has only recently gained increased attention in the area of combinatorial optimization and compiler optimization [4, 5, 11, 12, 13]. Due to the discrete nature of the problems in these areas, available realizations use integer arithmetic to avoid numeric inaccuracies. Since the sensitivity matrices in (21), (23) and (24) consist of floating-point numbers, we are currently implementing a new version of the Fourier-Motzkin elimination suitable for this field of application.

5.1.4. RESULTS We found that the linear approximation to the system feasibility region can resemble the actual feasibility region surprisingly well. The reason is that the circuit sizing rules only allow a comparatively small circuit feasibility region and that within this region the circuit behavior is usually not far from linear

[16]. Fig. 4 shows a result for a folded cascode operational amplifier. When it is used as a building block of a larger system, its feasible performance region turns into a system feasibility region. The filled polygon represents the 2-dimensional projection of the linearized feasibility region and the dots show the results of a large number of simulation runs. A good agreement is evident.

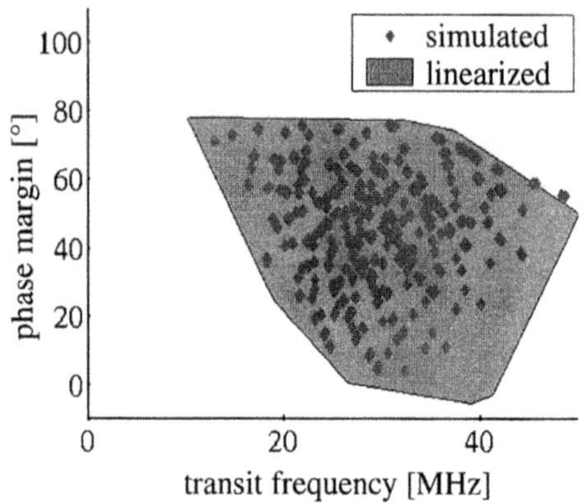

FIGURE 4. Linearized feasibility region of an operational amplifier

In industrial practice, there are often several implementation alternatives for a certain building block. Feasibility regions can also help to compare their potential. Fig. 5 shows the linearized feasibility regions of two different operational amplifiers. It is obvious that, for a high transit frequency, the folded cascode is the architecture of choice while the Miller compensated operational amplifier is slightly superior in terms of phase margin. In this way, a topology selection can be included in the optimization process.

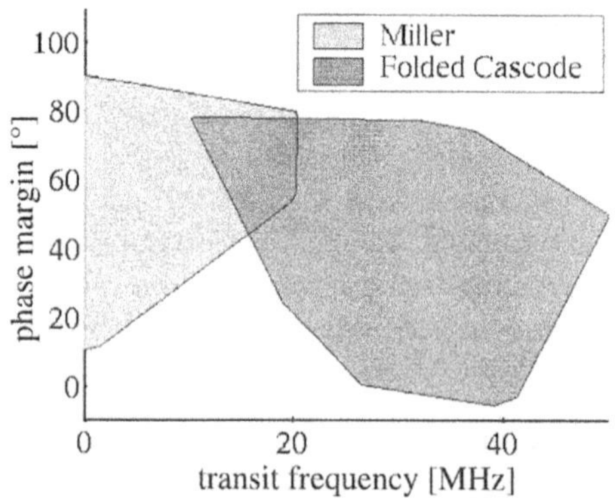

FIGURE 5. Feasibility regions of different operational amplifiers

6. Fourier-Motzkin elimination

6.1. General description

The elimination of a variable x_k, $1 \leq k \leq N$, from a system of linear inequalities

$$\mathbf{A}\mathbf{x} \geq \mathbf{b}\,, \quad \mathbf{A} = \begin{pmatrix} \mathbf{a}_1^T \\ \vdots \\ \mathbf{a}_N^T \end{pmatrix} = (a_{ij})\,, \quad \mathbf{x} = \begin{pmatrix} x_1 \\ \vdots \\ x_N \end{pmatrix}, \quad \mathbf{b} = \begin{pmatrix} b_1 \\ \vdots \\ b_N \end{pmatrix}, \quad N \geq 2\,, \tag{32}$$

corresponds to calculating a new vector inequality with $a_{ik} = 0$, $1 \leq i \leq N$. To do this, Fourier-Motzkin elimination exploits two properties of inequalities:

$$a \geq b \;\wedge\; c \geq d \;\Rightarrow a + c \geq b + d \tag{33}$$

$$a \geq b \;\wedge\; c \geq 0 \;\Rightarrow\; a \cdot c \geq b \cdot c\,. \tag{34}$$

The elimination of a variable x_k from (32) comprises two steps:

1. Sorting of inequalities
   The individual inequalities $\mathbf{a}_i^T\mathbf{x} \geq b_i$ from (32) are partitioned into three sets:

$$\begin{array}{ll} (I) & I_{>} = \{\mathbf{a}_g^T\mathbf{x} \geq b_g \mid a_{gk} > 0 \wedge g \in G\} \\ (II) & I_{<} = \{\mathbf{a}_l^T\mathbf{x} \geq b_l \mid a_{lk} < 0 \;\wedge\; l \in L\} \\ (III) & I_{=} = \{\mathbf{a}_e^T\mathbf{x} \geq b_e \mid a_{ek} = 0 \wedge e \in E\} \end{array} \tag{35}$$

   with

$$G \cap L = G \cap E = L \cap E = \emptyset \quad \wedge \quad G \cup L \cup E = \{1, 2, \ldots, N\}\,. \tag{36}$$

   Note that the inequalities in $I_{=}$ are already independent of x_k.

2. Elimination by linear combination
   The set $I_{=}$ is augmented by all pairwise linear combinations of inequalities from $I_{>}$ and $I_{<}$ according to

$$I_{=} \cup \{(a_{gk} \cdot \mathbf{a}_l^T - a_{lk} \cdot \mathbf{a}_g^T) \cdot \mathbf{x} \geq a_{gk} b_l - a_{lk} b_g \mid g \in G \;\wedge\; l \in L\}\,. \tag{37}$$

   This set of inequalities can be written in vector/matrix notation as

$$\mathbf{A}^{\backslash k} \cdot \mathbf{x} \geq \mathbf{b}^{\backslash k}\,. \tag{38}$$

   The entries in the kth column of $\mathbf{A}^{\backslash k}$ are all zeros. Hence, the new system of inequalities is independent of x_k.

These two steps are repeated for all variables to be eliminated. A numeric example in the next section illustrates this algorithm.

6.2. Example

To demonstrate the Fourier-Motzkin algorithm, an example with three transistor parameters, t_1 to t_3, and one circuit performance, $b_1 = b_1(t_1, t_2, t_3)$ is considered. Assume that the circuit feasibility region was linearized at a suitable operating point according to (21) and the circuit performance according to (23). With

Gaussian elimination, Δt_1 was replaced by Δb_1:

$$\begin{pmatrix} 5 & 0 & 7 \\ 0 & -2 & 10 \\ 5 & 0 & -3 \\ 1 & 1 & -1 \\ -7 & 0 & -2 \end{pmatrix} \cdot \begin{pmatrix} \Delta b_1 \\ \Delta t_2 \\ \Delta t_3 \end{pmatrix} \geq \begin{pmatrix} 100 \\ -20 \\ 50 \\ 10 \\ -140 \end{pmatrix} . \tag{39}$$

Wanted is the value range of Δb_1 with the remaining transistor parameters varying within the allowed bounds. Consequently, Δt_2 and Δt_3 have to be eliminated. In the first elimination step, Δt_2 is to be removed. Therefore, the inequalities are sorted according to (35):

$$\begin{array}{c} (I) \\ (II) \\ \\ (III) \\ \\ \end{array} \left(\begin{array}{rrr||r} 1 & 1 & -1 & 10 \\ \hline 0 & -2 & 10 & -20 \\ \hline 5 & 0 & 7 & 100 \\ 5 & 0 & -3 & 50 \\ -7 & 0 & -2 & -140 \end{array} \right) \begin{array}{c} (a) \\ (b) \\ (c) \\ (d) \\ (e) \end{array} \tag{40}$$

For ease of notation, the resulting system of inequalities can be written as an extended matrix with a double line separating the left-hand from the right-hand side. The horizontal lines separate the three sets of inequalities. The individual rows are marked by lowercase letters for further reference. It is only possible to set up a single new inequality according to (37). Consequently, inequalities (a) and (b) are replaced by their linear combination (a, b):

$$\Rightarrow \begin{array}{c} (I) \\ \\ (II) \\ \\ \end{array} \left(\begin{array}{rrr||r} 2 & 0 & 8 & 0 \\ 5 & 0 & 7 & 100 \\ \hline 5 & 0 & -3 & 50 \\ -7 & 0 & -2 & -140 \end{array} \right) \begin{array}{c} (a,b) \\ (c) \\ (d) \\ (e) \end{array} \tag{41}$$

Since, in this new matrix, there are two positive, two negative and no zero coefficients of Δt_3, a number of four new inequalities can be set up:

$$\Rightarrow \begin{array}{c} (I) \\ \\ (II) \\ \\ \end{array} \left(\begin{array}{rrr||r} 46 & 0 & 0 & 400 \\ 50 & 0 & 0 & 650 \\ \hline -52 & 0 & 0 & -1120 \\ -39 & 0 & 0 & -780 \end{array} \right) \begin{array}{c} (ab,d) \\ (c,d) \\ (ab,e) \\ (c,e) \end{array} \tag{42}$$

With (42), the elimination is finished and the resulting inequalities in Δb_1 are

$$\Leftrightarrow \left\{ \begin{array}{lll} \Delta b_1 & \geq & \frac{200}{23} \approx 8.7 \qquad (ab,d) \\ \Delta b_1 & \geq & 13 \qquad\qquad (c,d) \\ \Delta b_1 & \leq & \frac{280}{13} \approx 21.5 \qquad (ab,e) \\ \Delta b_1 & \leq & 20 \qquad\qquad (c,e) \end{array} \right. . \tag{43}$$

Obviously, only (c, d) and (c, e) are relevant. Therefore, the original inequalities (a) and (b) do not have any influence on the feasible circuit performance range:

$$13 \leq \Delta b_1 \leq 20 . \tag{44}$$

The Fourier-Motzkin elimination can geometrically be interpreted as a projection parallel to the coordinate axes, as depicted in Fig. 6. The original vector inequality (40) describes a polytope in the 3-dimensional space. The elimination of Δt_2 yields a triangle in the $\Delta t_3/\Delta b_1$ space. The final removal of Δt_3 leads to a line indicating the value range of Δb_1.

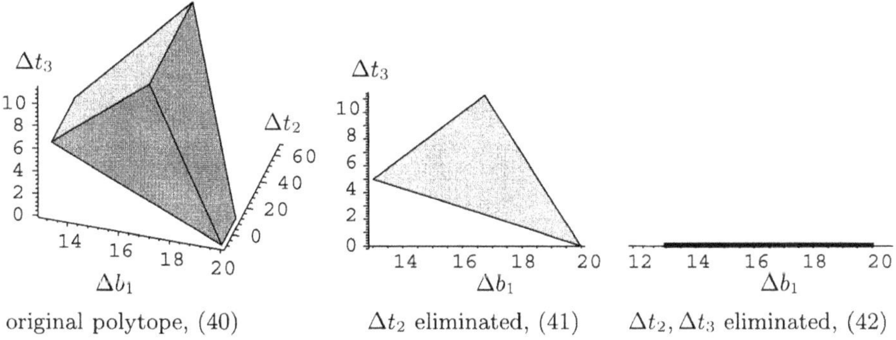

FIGURE 6. Visualization of Fourier-Motzkin elimination

6.3. Practical implementation aspects

The example above was deliberately simple. Therefore, it does not immediately reveal complexity as the severest weakness of the Fourier-Motzkin elimination. It turned out in (43) that two inequalities were redundant. In fact, especially in high-dimensional spaces, and after several elimination steps, a huge number of inequalities can exhaust computing resources easily if no special care is taken.

The number M of inequalities after an elimination step can be calculated from the powers of the three inequality sets before the elimination:

$$M = |I_>| \cdot |I_<| - |I_>| - |I_<| + |I_=| \,. \tag{45}$$

Let M_i be the number of inequalities before step i. In the worst case, the inequalities split up equally between $I_>$ and $I_<$ resulting in

$$M_{i+1} = \left(\frac{M_i}{2}\right)^2 - M_i \,. \tag{46}$$

Practical experience shows that typical application data comes close to this worst-case scenario. Considering the fact that there are usually several hundred inequalities in the original problem (39), it becomes obvious that even a complexity of second order can lead to excessive resource consumption. There are two ways to attack this problem.

First of all, the elimination order has a major impact on the required resources. Therefore, the most efficient way to cope with redundant inequalities would be their avoidance. In the example above, an advantageous elimination order was chosen, because, in the first elimination, only one new inequality could be calculated replacing two original ones. The final result included two redundant inequalities. The opposite elimination order would have led to four redundant inequalities in the result. Unfortunately, it is still an unresolved problem how to choose a globally optimal elimination order. A greedy-style heuristic is to choose the most advantageous variable for each elimination step by determining the inequality sets from (35) and calculating the associated number of equations according to (45). Always locally choosing the variable leading to a minimum inequality count, however, does not necessarily lead to a globally optimal elimination order.

Even with the best elimination order, the generation of redundant inequalities cannot be avoided altogether. Therefore, it is mandatory to detect redundancies after each elimination step. There are two criteria due to Chernikov [3], which allow a fast and easy identification of redundant inequalities:

1. *An inequality is redundant after step i, if it is based on more than $i+1$ original inequalities.*
   The examination of (41) and (42) shows that this criterion would not eliminate any inequality. After the first step, a maximum number of two inequalities were combined: (a, b). After step two, at most three original inequalities contribute to one equality: (ab, d), (ab, e).

2. *An inequality is redundant if it shares all its constituting original inequalities with another inequality in the same step.*
   In the example there is no complete overlap of supporting original inequalities. For example, (ab, d) and (ab, e) in (42) share the two original inequalities (a) and (b), but differ in the third inequality, which is (d) and (e), respectively.

In this example, the Chernikov criteria did not detect any redundancy. It turns out that their success rate increases for a growing number of elimination steps. However, practical experience showed that an additional test identifying *all* redundancies is required to keep the necessary computer resources within practical limits. We apply the Chernikov criteria as "cheap" detection methods after the termination of an elimination step to expunge obvious redundancies. After that, we use a technique based on linear programming to examine the remaining inequalities as explained in the following.

In the general case, a polyhedron is defined by

$$\mathbf{A}\mathbf{x} \geq \mathbf{b}\,. \tag{32}$$

The i th inequality $\mathbf{a}_i^T\mathbf{x} \geq b_i$ is relevant or *facet-inducing* [15], if and only if

$$b_i = \min_{\mathbf{x}} \mathbf{a}_i^T\mathbf{x} \quad \text{s.t.} \quad \mathbf{A}\mathbf{x} \geq \mathbf{b}\,. \tag{47}$$

Practically, the following test is more advantageous because it avoids a numeric test for equality:

$$\tilde{b}_i = \min_{\mathbf{x}} \mathbf{a}_i^T \mathbf{x} \quad \text{s.t.} \quad \begin{pmatrix} \mathbf{a}_1^T \\ \vdots \\ \mathbf{a}_i^T \\ \vdots \\ \mathbf{a}_N^T \end{pmatrix} \mathbf{x} \geq \begin{pmatrix} b_1 \\ \vdots \\ b_i - 1 \\ \vdots \\ b_N \end{pmatrix} \tag{48}$$

Inequality i is facet-inducing if and only if $\tilde{b}_i < b_i$.

The subtraction $b_i - 1$ on the right-hand side of the constraints means a relaxation of the i th constraint.

7. Conclusions

This paper explains a hierarchical simulation-based design technique for analog and mixed -signal electronic systems. There are two abstraction levels: system and circuit level. At system level, the functional building blocks are described by behavioral models. Since the parameters of these models are circuit-level block performances, a system optimization yields specifications for the functional blocks. At circuit level, the blocks are described by transistor models. They are sized at this level to meet the specifications from the system optimization, with one optimization run per block. It has been shown how feasibility regions can ensure technically meaningful sizings at both levels of abstraction. While the circuit-level feasibility region can be derived from elementary design knowledge, the system feasibility region has to be calculated numerically. It has been shown how Fourier-Motzkin elimination can be applied to circuit design in order to calculate a linear approximation to the system feasibility region.

Acknowledgments

The authors would like to thank Dr. Th. Theobald, Technical University of Munich, for helpful advice.

References

[1] K. Antreich, J. Eckmueller, H. Graeb, M. Pronath, F. Schenkel, R. Schwencker, and S. Zizala. WiCkeD: Analog circuit synthesis incorporating mismatch. In *IEEE Custom Integrated Circuits Conference (CICC)*, pages 511–514, May 2000.

[2] H. Chang, E. Charbon, U. Choudhury, A. Demir, E. Felt, E. Liu, E. Malavasi, A. Sangiovanni-Vincentelli, and I. Vassiliou. *A Top-Down, Constraint-Driven Design Methodology for Analog Integrated Circuits.* Kluwer Academic Publishers, 1997.

[3] S.N. Chernikov. *Lineare Ungleichungen.* Deutscher Verlag der Wissenschaften, 1971.

[4] T. Christof and A. Loebel. Porta – polyhedron representation transformation algorithm. http://www.zib.de/Optimization/Software/Porta/index.html.

[5] T. Christof and G. Reinelt. Combinatorial optimization and small polytopes. *Top (Spanish Statistical and Operations Research Society)*, 4:1–64, 1996.

[6] G. Dantzig. *Linear Programming and Extensions*. Princeton University Press, 1963.

[7] G. Dantzig and B. Eaves. Fourier-Motzkin elimination and its dual. *Journal of Combinatorial Theory (A)*, 14:288–297, 1973.

[8] N. Dhanwada, A. Nunez-Aldana, and R. Vemuri. Hierarchical constraint transformation using directed interval search for analog system synthesis. In *Design, Automation and Test in Europe (DATE)*, 1999.

[9] H. Graeb, S. Zizala, J. Eckmueller, and K. Antreich. The sizing rules method for analog integrated circuit design. In *IEEE/ACM International Conference on Computer-Aided Design (ICCAD)*, pages 343–349, 2001.

[10] R. Harjani and J. Shao. Feasibility and performance region modeling of analog and digital circuits. *Analog Integrated Circuits and Signal Processing*, 10(1):23–43, June 1996.

[11] M. Jimenez, J. Llaberia, and A. Fernandez. Loop bounds computation for multilevel tiling. In *Sixth Euromicro Workshop on Parallel and Distributed Processing*, 1998.

[12] C.W. Kessler. Parallel Fourier-Motzkin elimination. In *Euro-Par '96*, 1996.

[13] W. Pugh. A practical algorithm for exact array dependence analysis. *Communications of the ACM*, 1992.

[14] G. Van der Plas, G. Debyser, F. Leyn, K. Lampaert, J. Vandenbussche, G. Gielen, W. Sansen, P. Veselinovic, and D. Leenaerts. AMGIE–A synthesis environment for CMOS analog integrated circuits. *IEEE Transactions on Computer-Aided Design of Circuits and Systems*, 20(9):1037–1058, Sept. 2001.

[15] G.M. Ziegler. *Lectures on Polytopes*. Springer Verlag, New York, 1995.

[16] S. Zizala, J. Eckmueller, and H. Graeb. Fast calculation of analog circuits' feasibility regions by low level functional measures. In *IEEE Int. Conf. on Electronics, Circuits and Systems*, pages 85–88, Sept. 1998.

Guido Stehr
Institute for Electronic Design Automation
TU Munich
80290 Munich, Germany
email: guido.stehr@ei.tum.de

Coupling of Devices and Circuits

Modeling, Simulation and Optimization of Integrated Circuits
(Eds.) K. Antreich, R. Bulirsch, A. Gilg and P. Rentrop

International Series of Numerical Mathematics Vol. 146, 187–201

Modeling and Discretization of a Thermal-Electric Test Circuit

Andreas Bartel, Michael Günther, and Martin Schulz

Abstract. Already today, self-heating of semiconductor devices is an important aspect in electrical network design. Especially for SOI circuits, the influence of thermal aspects on the overall behavior cannot be longer neglected. Inspecting a simple but instructive example, we propose an alternative modeling ansatz to the common discrete thermal network approach. Our discussion concentrates on the interplay between different types of models, in particular coupling of differing spatial dimensions, and the associated spatial discretization of thermally and electrically active structures.

1. Introduction

The continuous reduction in the physical extension of on-broad devices increases the importance of secondary effects. On the on-chip scale, the SOI technology (silicon-on-insulator) facilitates lower switching capacities by a floating body. This is electrically achieved by an insulation layer, which is also a thermal insulation, since both effects generally occur concomitantly.

The motivation for a first order thermal modeling, and its scope is the following: as a state of the art in circuit simulation, the thermal aspect is addressed as a network, as well; that network is established by analogy, see [5], and introduces thermal nodes, representing temperatures, and corresponding thermal capacities and resistance, representing heat exchange to some environment; thus thermal RC-links are formed for each thermally active element. Commonly in chip design, these nodes are not interlinked, that is, only the local heating, storage of energy and cooling against an ambient temperature is covered. Of course, the simplified structure allows a huge number of thermal nodes (thousands of transistors), whereas the full 3D simulation of such a chip could not be handled.

The idea to improve on this is to use a medium-grained modeling: a thermal description, which enables for lumped and uni-dimensional elements. This, indeed, reflects the SOI-chip technology; there we encounter the basic gates to be lined up, while between these line-structures the interconnections are placed. Thus

2000 *Mathematics Subject Classification.* Primary 93A30; Secondary 35K05; Tertiary 47N70.
Key words and phrases. circuit simulation, modeling, thermal-electric coupling, coupled systems.

these lines are nearly insulated from each other, though a cooling to a common environment occurs, which is attenuated by the oxide layer built-in.

In the following this concept is introduced by the means of a simple but suggestive test example. We start with the presentation of that test circuit and then we set up the according electric and thermal model. Along with this, we discuss the possibilities of inter-dimensional couplings of thermal elements under the perspective of efficient algorithmic. Afterwards, this is translated to a suitable semi-discretization. Further exploitation of the coupling structure in accordingly designed simulation techniques are roughly sketched and postponed to a forthcoming paper.

2. A generic thermal-electric test circuit

As a generic example for the electric-thermal coupling, we employ the simple, but sufficiently broad test circuit, which is given in Fig. 1. It consists of an operational amplifier, two resistors, a diode and some capacitor. The resistor $R(T)$ is modeled as a structure that produces and transports heat along a uni-dimensional extension between the amplifier and the diode. Furthermore, the amplifier is modeled as heat source, whereas the diode has a temperature dependent characteristics.

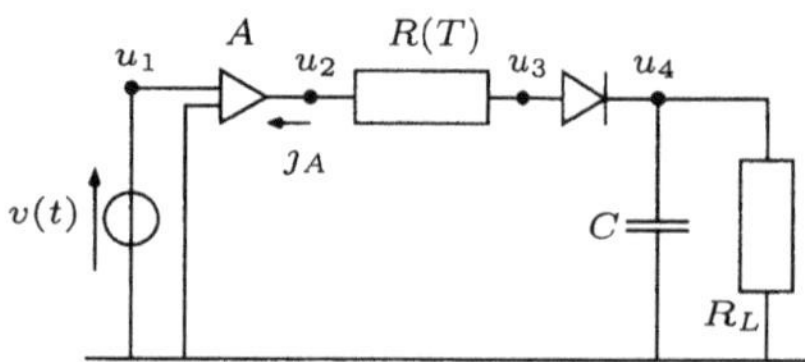

FIGURE 1. Electric description of the test circuit.

The electric behavior of that circuit is simple enough: a positive input signal charges the capacitor with constant C via the diode; later on, the capacitor is discharged via the ohmic load resistance R_L. Thermally, the amplifier heats the diode via the conducting resistor $R(T)$. Indeed, the SOI-chip features a 1D sub- or macro-structure as described above, and these can be exemplary studied in that circuit.

The main interest is how the temperature affects the system, and therefore we first focus on the mathematical setup.

3. The continuous model

Before we treat the mathematical model in detail, we summarize the overall, multiphysical setup of our test problem in Box 1 (the circuit is depicted in Fig. 1). First, the *electric network equations* are set up using classical modified nodal analysis [9]; thus a differential-algebraic description (3.1a) is obtained in terms of the node voltages u_i ($i = 1, \ldots, 4$) and the current through the amplifier $\jmath_A$.[1]

[1]The amplifier operates with amplification A.

This equation is accomplished by some consistent initial values (3.1b) and a time-dependent input signal $v(t)$, which shall satisfy $v(0) = 0$. To be precise, we have omitted the trivial equation defining the current through the independent voltage source as zero.
The intermediate coupling interface comprises all exchanged quantities such as dissipated powers of thermal active elements, branch temperatures and thermal dependent parameters.
Secondly, for the *thermal part* we have a uni-dimensional heat equation (3.1c) for the 1D-structure, supplementary boundary conditions (3.1e) and an energy balance for the attached lumped elements (3.1d). Finally, an initial condition (3.1f) concludes the system.

Box 1: **Thermal-electric problem (test circuit)**

electric network: (DAE-IVP)

$$\begin{aligned} 0 &= \jmath_A + (u_2 - u_3)/R(T) \\ 0 &= -(u_2 - u_3)/R(T) + i_{\mathrm{di}}(u_3 - u_4, T_{\mathrm{di}}) \\ C\dot{u}_4 &= i_{\mathrm{di}}(u_3 - u_4, T_{\mathrm{di}}) - u_4/R_L \\ 0 &= u_1 - v(t) \\ 0 &= u_2 - Au_1, \end{aligned} \tag{3.1a}$$

$$u_1(0) = u_2(0) = u_3(0) = u_4(0) = \jmath_A(0) = 0 \qquad \text{(IV)} \tag{3.1b}$$

coupling interface:

$$P_{\mathrm{op}} = |(v_{\mathrm{op}} - |u_1|) \cdot \jmath_A|, \quad P_{\mathrm{w}} = (u_2 - u_3)^2/R,$$
$$T_{\mathrm{op}}, \quad T_{\mathrm{di}}, \quad R = \int_0^l \widetilde{\rho}(x,T)\,\mathrm{d}x$$

heat equation: (PDAE-BIVP)

$$M'_{\mathrm{w}}\dot{T} = \frac{\partial}{\partial x}\left(\Lambda \frac{\partial T}{\partial x}\right) - \gamma S'_{\mathrm{w}} \cdot (T - T_{\mathrm{env}}) + P_{\mathrm{w}} \cdot \frac{\widetilde{\rho}}{R} \qquad \text{(1D)} \tag{3.1c}$$

$$\begin{aligned} M_0\dot{T}_{\mathrm{op}}(t) &= \Lambda T_x(0,t) - \gamma S_{\mathrm{op}} \cdot (T_{\mathrm{op}}(t) - T_{\mathrm{env}}) + P_{\mathrm{op}} \\ M_1\dot{T}_{\mathrm{di}}(t) &= -\Lambda T_x(l,t) - \gamma S_{\mathrm{di}} \cdot (T_{\mathrm{di}}(t) - T_{\mathrm{env}}) \end{aligned} \qquad \text{(0D)} \tag{3.1d}$$

$$T_{\mathrm{op}}(t) = T(0,t), \quad T_{\mathrm{di}} = T(l,t) \qquad \text{(BC)} \tag{3.1e}$$

$$T(x,0) = T_{\mathrm{env}} = 293\,\mathrm{K} \qquad \text{(IC)} \tag{3.1f}$$

4. Mathematical modeling

In the following we go into the details of the mathematical model's derivation. First, the thermal dependence of the electric network is discussed. Then the thermal aspect and the coupling electric effects are addressed.

4.1. The electric network and its thermal dependence

Since the setup of the electric network equations is fairly standard, see, e.g., [9, 7], we focus on the coupling and denote the independent spatial variable as x and time as t; furthermore temperature is generally denoted as T (plus indices). In the description (3.1a), we have already indicated that the resistance linking node two and three depends on temperature, $R(T)$, as well as the diode, $i_{\text{di}}(., T_{\text{di}})$. Thus we need to endow the basic (electric) elements with their thermal relevant property; the thermal characteristics of the electric effects (referred to as thermal-electric coupling). Generally we are going to cover two types of thermal elements: lumped elements, which have an elementwide, assigned temperature, and 1D-structures realizing a uni-dimensional heat conduction. Now, the test circuit includes the following:

- *1D resistor:* Let this element by made of a thin *structure* of length l with variable cross-section $a(x)$, i.e., a rotational symmetric geometry is assumed (not degenerated $a(x) > 0$). For the test circuit, we adopt the following cross-section type

$$a(x) = a_0 \cdot 1/\big(1 + b(l - x)x\big),$$

composed of the basic cross-section a_0 and a superposed rational characteristics of second order. The latter describes a 1D structure, which reduces towards its center point and factor b governs the influence of that varying cross-section. Choosing $b = (2/l)^2$, we obtain a structure of half the basic cross-section at the mid point. Electrically, halving the cross-section means a doubling of the resistance. Now, the local resistance ρ shall exhibit a quadratic dependence of temperature

$$\rho(T) = r_0\big(1 + \alpha(T - T_{\text{meas}}) + \beta(T - T_{\text{meas}})^2\big), \qquad \text{with } T = T(t, x), \tag{4.1}$$

where r_0 denotes the material's specific resistance per unit cube (at temperature T_{meas}). So, $[\rho] = \Omega\,\text{m}$ is the local resistance for unit length per unit cross-section. Then the local and 1D-structure relevant quantity is obtained as

$$\widetilde{\rho}(x, T) = \rho(T)/a(x).$$

Employing (4.1), the total resistance $R(T)$ is a functional of thermal distributions:

$$R(T) = \int_0^l \frac{\rho(\xi, T(t,\xi)}{a(\xi)}\,\mathrm{d}\xi = \int_0^l \widetilde{\rho}(\xi, T(t,\xi))\,\mathrm{d}\xi. \tag{4.2}$$

The corresponding current i_R through the 1D-structure is constant along $x \in [0, 1]$ and it is given by Ohm's law:

$$i_R = u_R/R(T),$$

where u_R denoting the voltage drop across the resistor. Thus the 1D-structure allows an electrically lumped description. – An possible assignment of the involved parameters is given in Box 2. ◇

Box 2:		**Electric parameters of 1D-resistor**	
material	Cu (copper)	cross-section	$a_0 = 540\,\mu\text{m}^2$
spec. resist.	$r_0 = 1.7\,\mu\Omega\text{m}$	profile	$b = (2/l)^2$
ref. temp.	$T_{\text{meas}} = 291\,\text{K}$	1st thm. coeff.	$\alpha = 1/(273\,\text{K})$
length	$l = 0.03\,\text{m}$	2nd thm. coeff.	$\beta = 1/(273\,\text{K})^2$

• *Diode:* This element is incorporated as voltage (u_{di}) controlled current source employing the function $i_{\text{di}}(.)$; furthermore that characteristics is controlled by the element's temperature T_{di}. For simplicity, the diode is modeled to be electrically and thermally lumped. Following [10], we have the exponential description

$$i_{\text{di}}(u_{\text{di}}, T_{\text{di}}) = \widehat{I}_S(T_{\text{di}})\Big(\exp((u_{\text{di}})/v_T) - 1\Big)$$
$$\text{with } \widehat{I}_S(T_{\text{di}}) = 10^{-12}\Big(\frac{T_{\text{di}}}{300\,K}\Big)^3 \exp\Big(-\frac{qE_g(300\,K)}{k_B T_{\text{di}}}\Big(1 - \frac{T_{\text{di}}}{300K}\Big)\Big),$$

where we deploy the thermal voltage v_T (reference temperature is 300 K).[2] ◇

• *Operational amplifier:* Like the diode, this element is regarded as thermally lumped, but not thermally dependent, so nothing has to be supplied, here. ◇

The remaining network parameters for the benchmark are given in Box 3, including the input voltage $v(t)$. This completes the description of the electrical part of the test circuit and the thermal-electric coupling.

Box 3:		**Electric parameters of 0D elements**	
diode		*remainder*	
energy gap	$E_g(300\,\text{K}) = 1.11\,\text{V}$	op. power	$v_{op} = 15\,\text{V}$
Boltzmann const.	$k_B = 1.381 \cdot 10^{-23}\,\text{J/K}$	amplification	$A = 20\,000$
unit charge	$q = 1.602 \cdot 10^{-19}\,\text{C}$	load resistance	$R_L = 0.3\,k\Omega$
therm. voltage	$v_T = k_B \cdot T_{\text{di}}/q$	capacitance	$C = 500\,\text{nF}$
pulsed voltage source	$v(t) = \begin{cases} 0.5\sin(\pi t/(2.5 \cdot 10^5\text{s}))\text{mV} & \text{if } t < 2.5 \cdot 10^{-5}\text{s} \\ 0\,\text{V} & \text{otherwise} \end{cases}$		

[2] Moreover, the energy gap can be formulated as thermal dependent quantity, according to [10]

$$E_g(T) = T_g(0) - (c_1 T^2)/(c_2 + T)$$

with experimentally fitted parameters for Silicon (Si): $E_g(0) = 1.16\,\text{V}$, $c_1 = 7.02 \cdot 10^{-4}\,\text{V/K}$, and $c_2 = 1108\,\text{K}$.

4.2. The thermal aspect and contributing electric effects

The core of the test circuit's thermal part is a uni-dimensional heat conduction for the resistor to which the attached thermally lumped elements contribute. In the coupling, we recognize Joule's law for self-heating, but neglect both Thomson and Peltier effect.

- *Heat conduction (1D):* That backbone of our thermal description is the diffusive heat transport

$$M'_{\mathrm{w}}\dot{T} = \frac{\partial}{\partial x}\left(\Lambda(x)\frac{\partial T}{\partial x}\right) + \text{ sources} \tag{4.3}$$

in one space dimension. Here the thermal mass (per unit length) M'_{w} is the following capacity

$$M'_{\mathrm{w}}(x) = a(x) d_{\mathrm{w}} c_{\mathrm{w}},$$

using the above cross-section a, density d_{w}, and specific heat capacity c_{w}. Furthermore, we assign the initial temperature to the environmental temperature T_{env}

$$\text{(IC)} \quad T(0,x) = T_{\mathrm{env}} = 293\,K.$$

The involved parameters are found in Box 4. Boundary conditions are postponed, and will serve for the attachment of lumped thermal elements.

Box 4:		**Thermal Parameters of the 1D element**	
for Cu			
density	$d_{\mathrm{w}} = 8.98 \cdot 10^3\,\mathrm{kg/m^3}$	heat conductivity	$\lambda_{\mathrm{w}} = 390\,\mathrm{W/(m\,K)}$
spec. heat	$c_{\mathrm{w}} = 385\,\mathrm{J/(kg\,K)}$	transition coeff.	$\gamma = 1.0\,\mathrm{W/(m^2\,K)}$

We now discuss and complete the right-hand side of (4.3). Its first part is derived from the experimental Fourier law and describes the diffusive heat leveling mastered by the heat conductivity: $\Lambda(x) = \lambda(x) \cdot a(x)$, the local 1D-conductivity.
The additional source term include heat generation due to the local resistivity and the drain according to the adaption to the ambient temperature T_{env} by the virtue of Newton's cooling. We have to add models for both effects:
Local self-heating is due Joule's law (other effects are neglected); thus the electric network submits the total power:

$$P_{\mathrm{w}} = |u_R \cdot \jmath_A| = u_R^2/R$$

(coupling interface, Box 1). This quantity needs to be spatially distributed. Therefore we define $R(x_1)$ to be the resistance from $x = 0$ up to $x = x_1$ (i.e., the upper integral bound in (4.2) is replaced by the location x_1). As remarked above, we have a spatially constant electric current i_R for the 1D element. So, we can derive a local potential $u(x)$ to be

$$u(x) = u_+ - R(x) \cdot i_R,$$

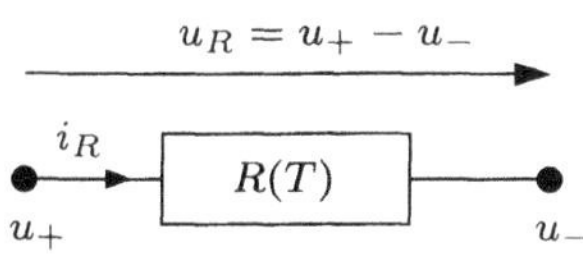

using Ohm's law; by (4.2) the local voltage drop (per unit length) is then

$$\frac{\partial u}{\partial x}(x) = -\widetilde{\rho}(x, T) \cdot i_R.$$

Thus we have the local power dissipation per unit length to be

$$-\frac{\partial u}{\partial x}(x) \cdot i_R = \widetilde{\rho}(x_1, T) \cdot i_R^2 = \widetilde{\rho}(x_1, T) \cdot (u_R)^2 / R^2 \tag{4.4}$$

(by Joule's law, since the direction of i_R is defined parallel to $-\frac{\partial u}{\partial x}$). Note that the quantity is always positive, since the resistor is a passive element. This straightforward approach is further investigated in the perspective of finite volumes below.

Newton's cooling shall render the radiation of heat to the environment. Consider an averaged body temperature T, and assume the cooling on its surface S to be proportional to the temperature difference (Newton). For volume V, we have

$$(V dc)\,\dot{T} = -\gamma S(T - T_{\text{env}}). \tag{4.5}$$

This translates to the 1D-structure by the transition of volumes to cross-sections $a(x)$ and surfaces to perimeters[3] $S'(x)$: so, the effect of cooling is described by

$$(a(x)\, d_{\text{w}}\, c_{\text{w}})\, \dot{T}(x) \;= M'_{\text{w}} \dot{T}(x) = -\gamma S'(x)(T - T_{\text{env}}).$$

This completes the description of (4.3) and we have derived the parabolic PDE (3.1c) in the overall. ◇

• *Diode:* For the above indicated thermal dependence, we used the temperature T_{di} – having small devices in mind, a thermally lumped version is reasonable. Moreover its thermal inactivity, the neglect of the diode's heat dissipation, is justified by the assumption of short switching times. For a thermally complete modeling we denote its thermal mass M_{di}, and surface S_{di}.
Note that the representative thermal mass of a lumped thermal element is generally the product of the corresponding volume, density and heat capacity: $M = V\, d\, c$. ◇

• *Operational amplifier:* Similar as for the diode, we have equipped the temperature T_{op} and possibly a thermal mass M_{op} and surface S_{op}. In addition, the operational amplifier shall produce heat: according to Joule's law the dissipated power is proportional to $u_{\text{op}} \cdot i_{\text{op}}$, so we assume for the test circuit and the linear part of the amplifiers characteristics:

$$P_{\text{op}} = \kappa |(v_{\text{op}} - |u_2|) \cdot \jmath_A|,$$

where v_{op} denotes the power supply of the amplifier (as positive value), and κ a transition constant being 1, generally. This term is part of the coupling interface of the problem formulation Box 1. ◇

[3]Note that, for our rotational symmetric structure, we simply have $S'(x) = 2\sqrt{\pi a(x)}$.

So far, we have settled the elementwise thermal behavior, and it remains the thermal interaction between these elements. This is addressed, now.

4.3. Thermal interaction – boundary conditions

The 1D partial differential equation (3.1c) requires boundary conditions for well-posedness: On the left $x \leq 0$, the operational amplifier generates heat according to its power dissipation, which heats not only the amplifier, but also the 1D structure; on right $x \geq l$, the diode is attached and gets heated via the resistor. – Basically, for the connection of a 0D element $\mathcal{X}$ and an 1D element $\mathcal{Y}$ we may distinguish three cases by the neglect of certain interactions for the first two cases:

(I)	$\mathcal{X}$ influences $\mathcal{Y}$:	(neglect of flux F for $\mathcal{X}$)
(II)	$\mathcal{Y}$ influences $\mathcal{X}$:	(neglect of flux F for $\mathcal{Y}$)
(III)	$\mathcal{X}, \mathcal{Y}$ influences each other:	(full description).

Mathematically, let the coupling of $\mathcal{X}, \mathcal{Y}$ be at $x = 0$, then these cases have the following formulation:

• **ad I:** Here, the inter-dimensional coupling does not influence $\mathcal{X}$ (but $\mathcal{Y}$); therefore $\mathcal{X}$ is thermally active. So, we have two subcases: (a) $\mathcal{X}$ has thermal mass $M_{\mathcal{X}}$. Thus the element can store energy, and its change of energy is given by the corresponding sources. That is, we have an *evolution equation* for $\mathcal{X}$, which is independent of $\mathcal{Y}$ and a temperature identification for the boundary of $\mathcal{Y}$, a *Dirichlet* condition:

$$M_{\mathcal{X}} \dot{T}_{\mathcal{X}} = P_{\mathcal{X}} - \gamma S_{\mathcal{X}} \cdot (T_{\mathcal{X}} - T_{\text{env}}), \qquad T_{\mathcal{Y}}(0) = T_{\mathcal{X}}. \tag{4.6}$$

(b) Without thermal mass, $\mathcal{X}$ cannot store any energy, hence the dissipated power $P_{\mathcal{X}}$ is directly submitted to $\mathcal{Y}$ as flux F and the temperature of $\mathcal{X}$ is obtained by identification:

$$F = -\Lambda \frac{\partial}{\partial x} T_{\mathcal{Y}} = P_{\mathcal{X}}, \qquad T_{\mathcal{X}} = T_{\mathcal{Y}}(0). \tag{4.7}$$

This is a *von Neumann* condition for $\mathcal{Y}$ and having supplied an identification, we need no additional evolution equation for $\mathcal{X}$. Note, the sign of the power term $P_{\mathcal{X}}$ is due to the direction of the outward normal of the 1D element.

Examples: In case (a) the modeling hypothesis is the following: the heat flux exchanged with $\mathcal{Y}$ is negligible for the thermal evolution of $\mathcal{X}$ (e.g., cooling dominates). So, we can think of a relatively large amplifier attached to a relatively small conduction path, where the main power dissipation occurs at the amplifier. In other words, it dominates the overall heat production and serves as reservoir for the attached element.
For (b), the element is still thermally active, but so small that the produced heat is not stored at the element but directly introduced to the conduction path.

Resumee: In case (a) we neglect certain transfers of energy. Therefore the energy balance is violated and the error has to be in the order of magnitude of a prescribed modeling error. ◇

• **ad II:** Because we neglect the exchanged thermal flux for $\mathcal{Y}$, we have immediately the homogeneous *von-Neumann* condition for $\mathcal{Y}$. If $\mathcal{X}$ had mass $M_\mathcal{X} \neq 0$, then an evolution equation for $\mathcal{X}$ has to be supplied with a given flux function $F(T_\mathcal{X}, T_\mathcal{Y}(0))$:

$$M_\mathcal{X}\dot{T}_\mathcal{X} = P_\mathcal{X} - \gamma S_\mathcal{X}(T_\mathcal{X} - T_{\text{env}}) - F(T_\mathcal{X}, T_\mathcal{Y}(0)), \qquad \frac{\partial}{\partial x}T_\mathcal{Y}(0) = 0.$$

Thus we generally obtain $T_\mathcal{X} \neq T_\mathcal{Y}$, which is rather bizarre in the context of attached elements. Therefore let $M_\mathcal{X} = 0$, then temperature $T_\mathcal{X}$ is obtained by identification and we have

$$T_\mathcal{X} = T_\mathcal{Y}(0), \qquad \frac{\partial}{\partial x}T_\mathcal{Y}(0) = 0. \tag{4.8}$$

Example: Here we can consider the diode: being thermally modeled as inactive ($P_\mathcal{X} = 0$) and having negligible mass, its temperature is just aligned to the temperature of $\mathcal{Y}$. ◇

• **ad III:** Here all occurring fluxes shall be exchanged, that is, the flux leaving $\mathcal{X}$ equals to the flux entering $\mathcal{Y}$. Therefore, we need $M_\mathcal{X} > 0$ (since otherwise a thermal flux cannot be recognized by $\mathcal{X}$). This flux is $F = -\Lambda\frac{\partial}{\partial x}T_\mathcal{Y}$, with the uni-dimensional conductivity Λ proportional to the common interface of $\mathcal{X}$ and $\mathcal{Y}$. Thus $\mathcal{X}$ is described by an evolution equation and as boundary condition the temperatures are identified:

$$M_\mathcal{X}\dot{T}_\mathcal{X} = \Lambda\frac{\partial}{\partial x}T_\mathcal{Y} - \gamma S_\mathcal{X}(T_\mathcal{X} - T_{\text{env}}) + P_\mathcal{X}, \qquad T_\mathcal{Y}(0) = T_\mathcal{X}. \tag{4.9}$$

Example: Generally any element is in the scope of this modeling.

Resumee: This case provides the full thermal model and by its definition (thermal) energy is conserved. ◇

We conclude this section with some remarks. i) The reason for such a classification can be found in the numerical handling: whenever we can neglect certain influences, we can solve smaller sub-problems sequentially. ii) The mentioned energy conservation refers to the interthermal aspect. In a multiphysical energy conservation one needs to take into account the dissipated power of all elements. iii) The simplified case I is consistent with (4.9): simply let in (4.9) either flux $F = -\Lambda\frac{\partial}{\partial x}T_\mathcal{Y}$ or mass $M_\mathcal{X}$ vanish for case (a) or (b), respectively. iv) Last for the test circuit, we have stated in Box 1 the full model of case III in equations 3.1c-3.1e.

Next we investigate the numerical treatment of the thermal aspect and thereby discuss the behavior of the various formulations.

5. Semi-discretization

In the scope of circuit simulation, we naturally have a time integration based algorithm. To apply such methods to the coupled problem (3.1), we need to discretize space, first (methods of lines). Therefore we introduce a grid[4] I_h: $X_i = i \cdot h$, $i = 0, \ldots, N$, with grid size $h > 0$ such that $X_N = N \cdot h = l$. Moreover, let $T_i(t) = T(t, X_i)$.

Employing the finite volume (FV) approach to deduce a discrete version for (3.1c), we have to set up an energy balance and deal rather with cells than discretization points. So the ith cell is the interval $[X_i - h/2, X_i + h/2]$, while boundary cells are of half the size: $[X_0, X_0 + h/2]$ and $[X_N - h/2, X_N]$ (see Fig. 2). Integrating the heat conduction (3.1c) over a single cell, and applying the divergence theorem, we obtain that the diffusion is equal to the total flux over the cell's boundaries. That is, the change of energy is inflow minus outflow. Hence we obtain for the flux leaving the ith cell and entering to the $(i+1)$st cell the approximation

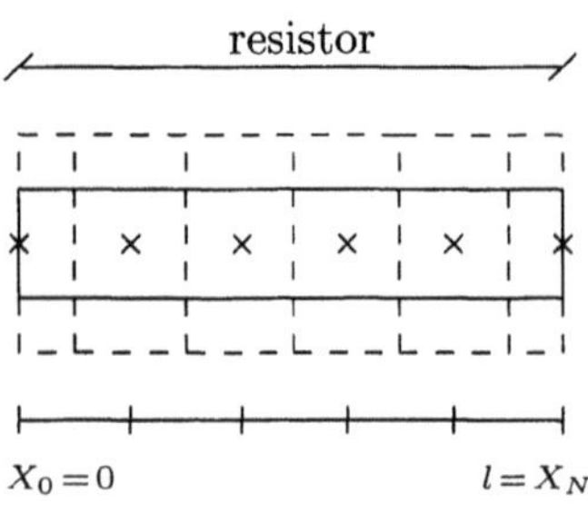

FIGURE 2. Cells for the 1D structure.

$$-\Lambda \left.\frac{\partial T}{\partial x}\right|_{X_{i+1/2}} = -\Lambda \frac{T_{i+1} - T_i}{h} + \mathcal{O}(h^2), \qquad \text{and define} \qquad f_{i+\frac{1}{2}} := -\Lambda \frac{T_{i+1} - T_i}{h};$$

hence we deduce for the ith cell the energy conservative form

$$M'_{\mathrm{w},i}\, h \dot{T}_i = \left(f_{i-\frac{1}{2}} - f_{i+\frac{1}{2}}\right) + P'_{\mathrm{w},i} \cdot h - \gamma\, S'_{\mathrm{w},i}\, h \cdot (T_i - T_{\mathrm{env}}), \tag{5.1}$$

using the short hands

$$M'_{\mathrm{w},i} = a(X_i) \cdot d_{\mathrm{w}} \cdot c_{\mathrm{w}}, \quad S'_{\mathrm{w},i} = S'_{\mathrm{w}}(X_i), \quad \text{and} \quad P'_{\mathrm{w},i} = \frac{(u_R)^2}{R} \cdot \frac{\widetilde{\rho}(X_i, T_i)}{R}.$$

The insertion of the fluxes yields

$$M'_{\mathrm{w},i} \dot{T}_i = \Lambda \frac{T_{i+1} - 2T_i + T_{i-1}}{h^2} + P'_{\mathrm{w},i} - \gamma S'_{\mathrm{w},i}(T_i - T_{\mathrm{env}}), \tag{5.2}$$

for $i = 1, \ldots, N-1$. These are exactly the equations one obtains for the classical finite differences. Furthermore, the source term for the self-heating of the resistor is consistent with the finite volume approach.[5]

[4] Of course, equidistancy is for the mere simplicity of notation.

[5] The total dissipated power on the segment $[x_0 - h/2, x_0 + h/2]$ is

$$P_{\mathrm{w},i} = \left(u(X_i - \tfrac{h}{2}) - u(X_i + \tfrac{h}{2})\right) \cdot i_R^2 = \int_{X_i - h/2}^{X_i + h/2} \widetilde{\rho}(x, T)\, \mathrm{d}x \cdot i_R^2$$
$$= \widetilde{\rho}(X_i) \cdot i_R^2\, h + \mathcal{O}(h^2) = \frac{(u_2 - u_3)^2 \widetilde{\rho}(X_i, T)}{R^2} + \mathcal{O}(h^2)$$

(midpoint rule) since i_R is constant. Similar for the boundary cells.

Now, we turn towards the *handling of BC* via finite volumes. Following the cases of the previous section, we have for a connection at $x = 0$ the following:

• **ad I:** (a) The discrete version of the evolution equation (4.6) reads

$$M_{\mathcal{X}} \dot{T}_0 = P_{\mathcal{X}} - \gamma S_{\mathcal{X}} \cdot (T_0 - T_{\text{env}}) \tag{5.3}$$

(having already plugged in the identification $T_0 = T_{\mathcal{Y}}(0) = T_{\mathcal{X}}$). (b) The flux condition for vanishing thermal mass (4.7) is

$$f_0 = P_{\mathcal{X}}, \tag{5.4}$$

which directly enables the setup of the discrete partial differential equation for the boundary via (5.1) ($T_{\mathcal{X}} = T_0$)

$$M'_{\text{w},0} \frac{h}{2} \dot{T}_0 = P_{\mathcal{X}} + \Lambda \frac{T_1 - T_0}{h} + P'_{\text{w},0} \frac{h}{2} - \gamma S'_{\text{w},0} \frac{h}{2} \cdot (T_0 - T_{\text{env}}). \tag{5.5}$$

• **ad II:** The homogeneous von-Neumann condition (4.8) is a special case of I(b) Thus, setting $f_0 = 0$ yields

$$M'_{\text{w},0} \frac{h}{2} \dot{T}_0 = \Lambda \frac{T_1 - T_0}{h} + P'_{\text{w},0} \frac{h}{2} - \gamma S'_{\text{w},0} \frac{h}{2} \cdot (T_0 - T_{\text{env}}) \tag{5.6}$$

(for the boundary cell), and we have $T_{\mathcal{X}} = T_0$, again. ◇

• **ad III:** For a complete thermal modeling, we assume a thermal mass for the lumped element $\mathcal{X}$. In the FV-context the attached element $\mathcal{X}$ enlarges the boundary cells, Fig. 3. Thus for the continuous model (3.1c-3.1d) with (4.9), we obtain the discrete model for the boundary temperature by summing (5.3) and (5.6), here the interface f_0 does not computationally exist:

$$\begin{aligned}\left(M_{\mathcal{X}} + \frac{h}{2} M'_{\text{w},0}\right) \dot{T}_0 = P_{\mathcal{X}} + \Lambda \frac{T_1 - T_0}{h} + \frac{h}{2} P'_{\text{w},0} \\ - \gamma \left(S_{\mathcal{X}} + S'_{\text{w},0} \frac{h}{2}\right) (T_0 - T_{\text{env}}).\end{aligned} \tag{5.7}$$

◇

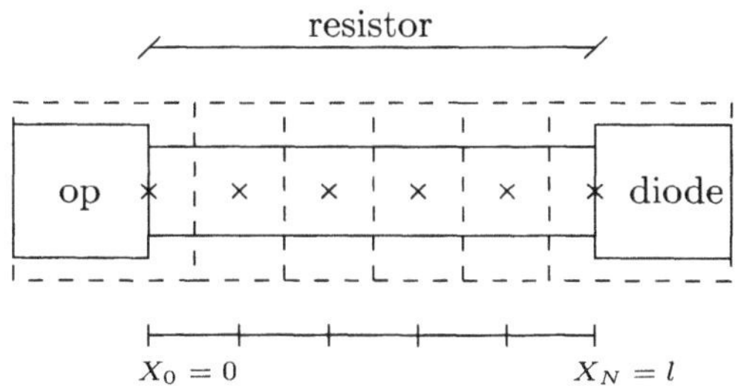

FIGURE 3. Cell discretization with enlarged cells.

Following the examples given in the previous section, there are various possibilities given to perform the semi-discretization. In any case, a set of ordinary differential equations is deduced.
The enlarged boundary cell in case III involves only fluxes in 'regular' regions, i.e., not directly at the material boundaries. Thus the flux is defined very rigorously compared to a corresponding finite difference approach (FD), which needs further approximations for the derivatives at the boundary. The actual difference between FD and FV is merely the term $M_{\mathrm{w},0} \cdot h/2$, thus FV is slightly more accurate and, furthermore, the presence of this term constitutes the smooth transition for vanishing thermal masses. Also for the case of von-Neumann boundary, FV is slightly more accurate, since the involved flux is centered at the cell boundary, but in the FD approach, a one-side formulation is generally applied, which basically differ by an approximation order. However, they can be made to coincide using the ghost-cell concept, which allows to set up the discrete PDE for the boundary node (FD), too.
Therefore, the overall set of the semi-discretized equations in Box 6 is stated by applying the third case: the system of semi-discretized heat equations (5.8d, 5.8e, 5.8f) is coupled to the network equations (5.8a). – The assignment for involved extension parameters are found in Box 5.

Quadrature for 1D resistive structure. It remains to discretize the integral equation for the total wire resistance (4.2): utilizing the trapezoidal rule on the grid I_h, we obtain equation (5.8c), which completes Box 6. Thus system (5.8) results in a DAE.

Next, we have to investigate the differential index of the coupled system.

6. Index of semi-discrete equations

In the following, we show that the overall DAE (5.8) is of differential index-1 (see [8] for the index definition). Note that the FV-discretized thermal equations (5.8d-5.8e) are already ODEs for any thermal variable. Therefore all variables $\mathbf{T} = (T_i)$ $(i = 0, \ldots, N)$ and u_4 are of index-0; and the remaining $\mathbf{z} = (\jmath, u_1, u_2, u_3)$ are variables of at least index-1. Computing the partial derivatives of the algebraic

Box 5:		**Extension parameters for 0D elements**	
amplifier	cubic	*diode*	cubic
material	Al (aluminum)	material	Si (Silicon)
size	$e_{\mathrm{op}} = 0.5\mathrm{mm}$	size	$e_{\mathrm{di}} = 0.167\mathrm{mm}$
heat cap.	$c_{\mathrm{Fe}} = 449\,\mathrm{J/(kg\,K)}$	heat cap.	$c_{\mathrm{Si}} = 700\,\mathrm{J/(kg\,K)}$
		density	$d_{\mathrm{Si}} = 2.33 \cdot 10^3\,\mathrm{kg/m^3}$

Box 6: **Semi-discretized model**

$$
\begin{aligned}
0 &= \jmath_A + (u_2 - u_3)/R(T) \qquad (5.8a)\\
0 &= -(u_2 - u_3)/R(T) + i_{\mathrm{di}}(u_3 - u_4, T_N)\\
C\dot{u}_4 &= i_{\mathrm{di}}(u_3 - u_4, T_N) - u_4/R_L\\
0 &= u_1 - v(t)\\
0 &= u_2 - Au_1,
\end{aligned}
$$

$$P_{\mathrm{op}} = |(v_{\mathrm{op}} - |u_1|)\cdot \jmath_A|, \quad P_{\mathrm{w}} = (u_2 - u_3)^2/R \qquad (5.8b)$$

$$R(T) = \left(\frac{1}{2}\rho(0,T_0) + \sum_{i=1}^{N-1}\rho(i,T_i) + \frac{1}{2}\rho(l,T_N)\right)\cdot h \qquad (5.8c)$$

$$
\begin{aligned}
M'_{\mathrm{w},i}h\dot{T}_i &= \Lambda\frac{T_{i+1} - 2T_i + T_{i-1}}{h} + P_{\mathrm{w}}\,\frac{\widetilde{\rho}(X_i,T_i)}{R}\,h \qquad (5.8d)\\
&\quad - \gamma S'_{\mathrm{w},i}h\,(T_i - T_{\mathrm{env}}) \qquad (i = 1,\dots,N-1)\\
(M'_{\mathrm{w},0}\cdot\frac{h}{2} + M_{\mathrm{op}})\dot{T}_0 &= \Lambda\,\frac{T_1 - T_0}{h} + P_{\mathrm{w}}\,\frac{\widetilde{\rho}(0,T_0)}{R}\,\frac{h}{2} \qquad (5.8e)\\
&\quad - \gamma(S'_{\mathrm{w},0}\,\frac{h}{2} + S_{\mathrm{op}})\cdot(T_0 - T_{\mathrm{env}}) + P_{\mathrm{op}}\\
(M'_{\mathrm{w},N}\cdot\frac{h}{2} + M_{\mathrm{di}})\dot{T}_N &= \Lambda\,\frac{T_{N-1} - T_N}{h} + P_{\mathrm{w}}\,\frac{\widetilde{\rho}(X_N,T_N)}{R}\,\frac{h}{2} \qquad (5.8f)\\
&\quad - \gamma(S'_{\mathrm{w},N}\,\frac{h}{2} + S_{\mathrm{di}})\cdot(T_N - T_{\mathrm{env}})
\end{aligned}
$$

network equations $\mathbf{G}(u_4,\mathbf{T},\mathbf{z}) = 0$,

$$\mathbf{G}(u_4,\mathbf{T},\mathbf{z}) = \begin{pmatrix} \jmath_A + (u_2 - u_3)/R(\mathbf{T}) \\ -(u_2 - u_3)/R(\mathbf{T}) + i_{\mathrm{di}}(u_3 - u_4, T_N) \\ u_1 - v(t) \\ u_2 - Au_1 \end{pmatrix} \qquad (6.1)$$

with respect to $\mathbf{z}$, we obtain

$$\frac{\partial\mathbf{G}}{\partial\mathbf{z}} = \begin{pmatrix} 1 & 0 & 1/R(\mathbf{T}) & -1/R(\mathbf{T}) \\ 0 & 0 & -1/R(\mathbf{T}) & 1/R(\mathbf{T}) + \partial i_{\mathrm{di}}/\partial u_3 \\ 0 & 1 & 0 & 0 \\ 0 & -A & 1 & 0 \end{pmatrix}.$$

Since

$$1/R(\mathbf{T}) + \partial i_{\mathrm{di}}/\partial u_3 = 1/R(\mathbf{T}) + \widehat{I}_S(T_N)\exp(q\cdot(u_3 - u_4)/(k_B\,T_N))\frac{q}{k_B T_N} > 0$$

($\widehat{I}_S(T_N)/T_N > 0$), the above Jacobian is regular. Hence we have index-1. And thus, the network index, which is for the charge-oriented network equations at least one, is not increased by to the thermal coupling for the semi-discretized equations (or by the diode's characterisics). Moreover, an index increase cannot happen in general, if we provide an ODE-discretization for total thermal aspect.

7. Conclusions and outlook

The self-heating of semiconductor devices strongly influences the overall behavior of electric circuits in SOI-technology. To describe these thermal effects at sufficient accuracy and reasonable computational cost, one-dimensional models have to be incorporated into the electrical network design for elements both electrically and thermally active or common thermal structures. In this paper we have applied this modeling approach for an instructive test circuit, and discussed in full detail the interplay and linkage of different types of modeling and associated spatial associated spatial discretization. This approach leads to an abstract modeling ansatz [4] and can be matched with macro-structures of chips [2].

Numerical time integration of the approximated DAE system (5.8) remains as the last step in the simulation chain: modeling – spatial discretization – numerical time integration. As for most multiphysical systems with largely differing time constants, a co-simulation is first choice within an industrial environment.

For the coupled thermal-electric problem, a large multirate potential is evident ($\tau_{\text{circuit}} = 10^{-6}\,\text{s}$ and $\tau_{\text{heat}} = 1\,\text{s}$). And indeed a multirate co-simulation is quite feasible for that test system, see [3]. This is enabled by an 'energy conserving' procedure in the dynamic iteration based on an averaging technique [6]. These first results are quite encouraging and lead to questions on general stability for dynamic iteration schemes on the continuous (3.1) and semi-discretized system: its seems, energy-conserving-procedure works as regularization, such that convergence of the dynamic iteration can be guaranteed according to [1] with contraction parameter being always zero. However further mathematical investigations will have to prove this rigorously.

References

[1] M. Arnold, and M. Günther, *Preconditioned dynamic iteration for coupled differential algebraic systems.* BIT **41**:1 (2000), pp. 1–25.

[2] A. Bartel. *First order thermal PDAE models in electric circuit design.* Submitted to Mathmod4 Proceedings.

[3] A. Bartel and M. Günther, *Multirate co-simulation of first order thermal models in electric circuit design.* Submitted to SCEE 2002 proceedings.

[4] A. Bartel and M. Günther, *From SOI to abstract electric-thermal-1D multiscale modeling for first order thermal effects.* Submitted to MCMDS.

[5] F.E. Cellier, *Continuous System Modeling*, Springer-Verlag, New York (1991).

[6] Ch. Deml and P. Türkes, *Fast Simulation Technique for Power Electronic Circuits with Widely Different Time Constants.* IEEE Transactions on Industry Applications. **35**:3 (1999), 657–662.

[7] M. Günther and U. Feldmann, *CAD-based electric-circuit modeling in industry,* Surv. Math. Ind. **8** (1999), pp. 97–157.

[8] E. Hairer and G. Wanner, *Solving Ordinary differential equations, Vol. II,* Springer-Verlag, Berlin (1996).

[9] C.W. Ho, A.E. Ruehli, and P.A. Brennan, *The modified nodal approach to network analysis.* IEEE Trans. Circuits and Systems, CAS **22** (1975), 505–509.

[10] G. Massobrio and P. Antognetti, *Semiconductor Device Modeling with SPICE.* 2nd. ed., McGraw-Hill, New York (1993).

Acknowledgment

We are grateful for many fruitful discussions with P. Rentrop (TU Munich) and U. Feldmann (Infineon Technologies).
This work is part of the project "Numerische Simulation von elektrischen Netzwerken mit Wärmeeffekten" (No. 03BEM3KA), which is supported by the German federal ministry for education and research (BMBF) within the program "Neue mathematische Verfahren in Industrie und Dienstleisungen".

Institut für Wissenschaftliches Rechnen und
Mathematische Modellbildung
Univerität Karlsruhe (TH)
Engesserstr. 6
D-76128 Karlsruhe, Germany
email: {bartel,guenther,schulz}@iwrmm.math.uni-karlsruhe.de

Modeling, Simulation and Optimization of Integrated Circuits
(Eds.) K. Antreich, R. Bulirsch, A. Gilg and P. Rentrop

International Series of Numerical Mathematics Vol. 146, 203–218

A Refined PDAE Network Model for a CMOS Ring Oscillator

Michael Günther

Abstract. Parasitic and second order effects such as transmission line effects can be linked to the common DAE network equations of circuit design by generalized PDE network models. This modeling approach allows for constructing spatial discretization schemes that are appropriate for the respective PDE system and transform the mixed initial-boundary value problem of PDEs and DAEs (PDAE system) into an approximate DAE (ADAE) system. In many simulation packages, however, one is stuck to companion network models that can be regarded as non-adaptive semi discretization schemes applied already at the modeling level. Inspecting the PDAE model and its analytical properties, these approximate DAE system can be classified and possible short-comings can be determined. If necessary, methods to overcome these short-comings are to be proposed, either on the level of spatial discretization (new models) or in the framework of efficient time integration (exploiting multirate potential). In this paper, all steps of this analysis chain are carefully discussed for a benchmark system: a CMOS ring oscillator with transmission line effects.

1. Introduction

In network simulation packages, real circuit elements and interconnections are commonly replaced by companion models of ideal and compact network elements. Therefore these models are uniquely determined by fixing electrical parameters like capacitances or inductances. This yields a unique modeling, which allows for including parasitic and second order effects into the differential-algebraic (DAE) network approach. Examples are transistor models which approximate the physical behavior of semiconductor devices by companion circuits of different modeling levels, or transmission line models, which consist of RLC links and controlled sources. Mathematically, this approach corresponds to a spatial discretization of

2000 *Mathematics Subject Classification.* Primary 65L05; Secondary 35L99.
Key words and phrases. Refined Network Models, Partial Differential-Algebraic Equations (PDAEs), Appropriate Spatial Discretization, Approximate Differential-Algebraic Equations (ADAEs), Multirate Potential, Dynamic Iteration, Gauss-Seidel Iteration, Distributed Time Integration.

the governing partial differential equations (PDEs) already at the modeling level. One short-coming of this approach is the frequent use of arbitrary coupled sources that may destroy the structure of the network equations and thus leads to high-index systems [8, 13]. In other words: the type of semi discretization realized by a companion model may not be appropriate for the given PDE system or coupling structure.

An alternative is the use of generalized network models. Refined models for interconnects and semiconductor devices are given by the characteristic PDE models. Here numerical methods can be tailored exactly to the resulting mathematical models – the spatial discretization is not yet made at modeling level. Mathematically spoken, this approach leads to a coupled system of DAEs and PDEs, with the boundary conditions for the PDEs linked to the DAEs at the boundary nodes. Such systems are called partial differential-algebraic equations, for short PDAE systems.

Having this PDAE model at hand, different spatial discretization schemes can be classified by comparing the properties of the underlying PDAE model and the respective approximate DAE (ADAE) system obtained after semi discretization: does the approximate DAE system properly reflect the behavior of the original PDAE system? Or does one detect an artificial smoothing effect? Or even coarsening? It is already known that the last but one question has to be answered in the affirmative. Semi discretization may act like a regularization [1, 3]: the ADAE system is less sensitive with respect to input data than the PDAE model, and may yield physically incorrect solutions. In this contribution we will affirm the last question, too.

We will discuss modeling, semi discretization and numerical simulation of a CMOS ring oscillator under consideration of transmission line effects. For this system, we will derive a PDAE model in Section 2, which is based on a generalized network model for transmission line systems, and discuss its analytical properties. In Section 3, these results will help to understand the short-comings of the approximate DAE systems obtained by an RLGC companion network for the interconnect system that is commonly used in circuit simulation packages: an artificial sensitivity with respect to data not present in the underlying PDAE model, and a spurious oscillation of high frequency that superposes the physical oscillation of the ring oscillator. These effects demand very small time steps and thus increase computational costs significantly. One may overcome these flaws by applying an appropriate semi discretization; however, if semi discretization is too strongly intertwined with modeling and thus must not be changed in a given network design environment, an alternative is given by exploiting the multirate behavior by dynamic iteration. This approach and its realization by distributed time integration will be discussed in Sections 4 and 5.

2. PDAE model – the generalized network approach

The CMOS ring oscillator in Fig. 1 with three CMOS inverters connected by transmission lines will serve as an example for electrical networks which are coupled by lossy transmission lines. The task of a ring oscillator is to generate autonomously an oscillating signal, which may be used for driving other parts of a circuit, but in many cases serves only for measuring the maximal clock rates which can be achieved with a given technology. The basic principle is to connect an odd number of inverter stages (in our case: three) in a loop. Since transmission lines effects such as signal delay, reflection, attenuation, dispersion and crosstalk may yield malfunctions in high speed digital circuits, they have to be included in the model.

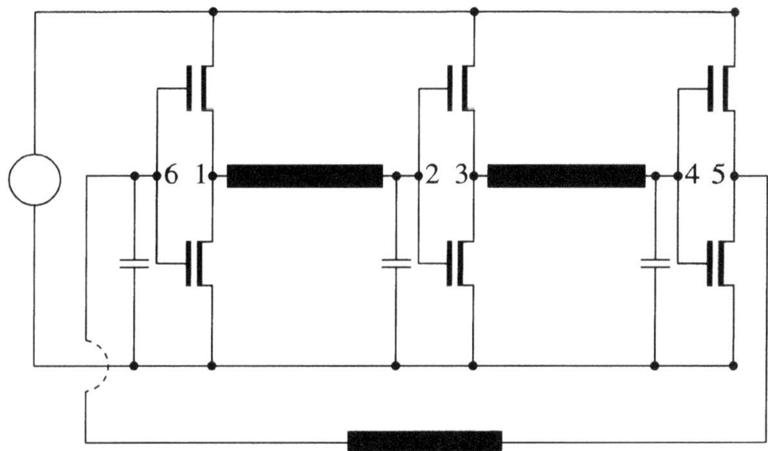

FIGURE 1. System of CMOS ring oscillator consisting of three inverter stages connected by transmission lines.

To derive a mathematical model, we use a generalized network approach: electrical circuits are described by DAE models, whereas transmission lines are governed by a PDE model. Both models are linked via boundary node voltages and currents.

2.1. PDE network model for transmission lines

Assuming quasi stationary behavior transverse to the wave propagation, the signal propagation in the block of d uniform, lossy transmission lines shown in Fig. 2 can be characterized by the telegrapher's equation

$$\begin{aligned} 0 &= V_z(z,t) + LJ_t(z,t) + RJ(z,t), \\ 0 &= J_z(z,t) + CV_t(z,t) + GV(z,t), \end{aligned} \tag{2.1a}$$

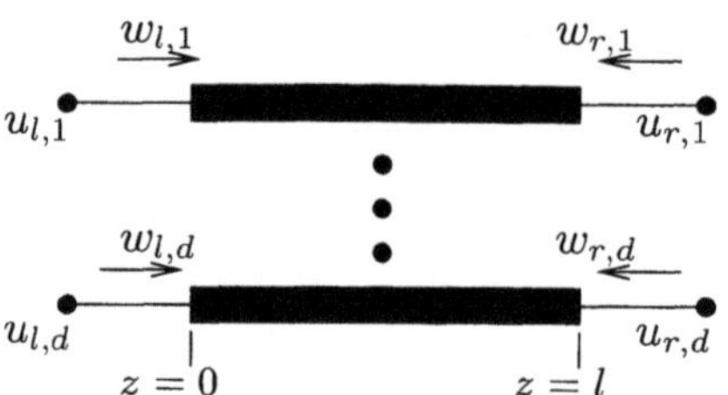

FIGURE 2. PDE-network model for a block of d uniform lossy transmission lines.

where R, L, G and $C \in \mathbb{R}^{d\times d}$ are the positive-definite symmetric resistance, inductance, conductance and capacitance matrices per unit length. $V(z,t)$ is a d-dimensional vector of line voltages with respect to ground, and $J(z,t)$ is a d-dimensional vector of line currents.

This first order hyperbolic system of partial differential equations is supplemented by a set of initial values

$$V(z,0) = V^0(z), \quad J(z,0) = J^0(z) \tag{2.1b}$$

for all $z \in I := [0,l]$.

After introducing $2d$ virtual current sources $w_l := (w_{l,1},\dots,w_{l,d})^\top$ and $w_r := (w_{r,1},\dots,w_{r,d})^\top$ at the boundaries, the characteristic equation in admittance form for the block of d transmission lines reads

$$w := \begin{pmatrix} w_l \\ w_r \end{pmatrix} = \begin{pmatrix} J(0,t) \\ -J(l,t) \end{pmatrix}, \tag{2.1c}$$

where the line currents J at the right-hand side are defined by the telegrapher's equation (2.1a). The PDE model is closed by the boundary condition

$$\begin{pmatrix} V(0,t) \\ V(l,t) \end{pmatrix} = \begin{pmatrix} u_l \\ u_r \end{pmatrix} \tag{2.1d}$$

for (yet unknown) boundary node potentials $(u_l, u_r)^\top$. Thus system (2.1) defines the characteristic equation for a generalized network element, a PDE model for a single block of d uniform transmission lines.

2.2. DAE network model for electrical circuits

The electrical part of the CMOS ring oscillator consists of three inverter stages (Fig. 3, left) composed of one p-channel MOS transistor T_2 and one n-channel MOS transistor T_1, a constant operator voltage U_{op} and a linear load capacitance C_0. Both transistors are described by companion models due to Shichman and Hodges (see Fig. 3, right, with two different parameter sets for p- and n-channel type) that link nonlinear capacitors, resistors and controlled current sources. Since only topology and no spatial dimension is considered, charge-oriented Nodal Analysis,

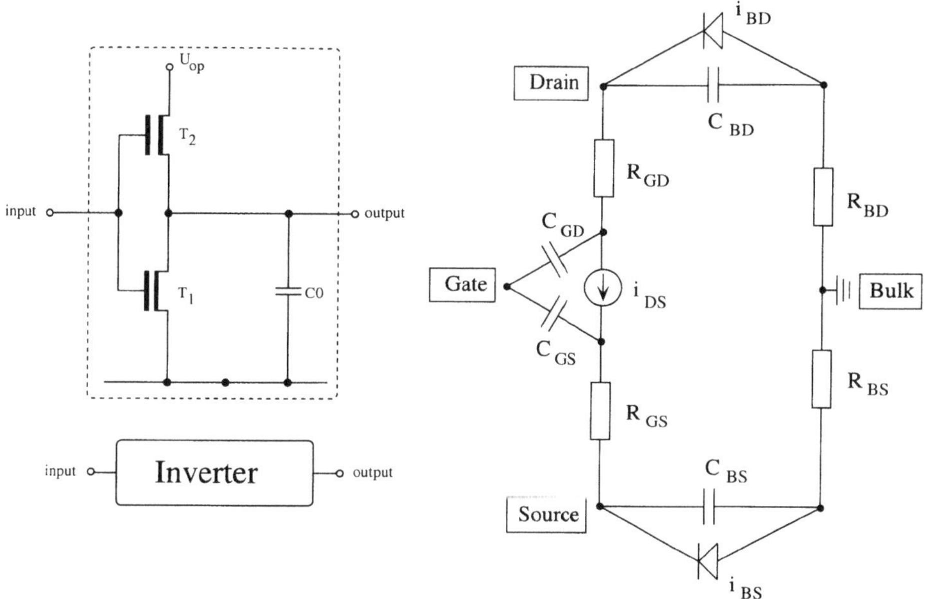

FIGURE 3. Network model for a CMOS inverter (left) and MOS transistor companion model due to Shichman and Hodges (right).

i.e., the current balance for each node but ground, yields a system of differential-algebraic equations in the node potentials u:

$$0 = A_C \frac{d}{dt}\left[q_C(A_C^\top u(t))\right] + A_R r(A_R^\top u(t), t) + A_I \imath(A_C^\top u(t), t) \tag{2.2a}$$

with consistent initial values

$$u(0) = u_0. \tag{2.2b}$$

The element-related partial incidence matrices A_C, A_R and A_I describe the branch-current relations for capacitive and resistive branches and branches for current sources.

Remark 2.1. If the Jacobian matrices of q_C and r are positive-definite, all charge storing elements and resistors are strictly passive [13]. In this case, the initial-value problem (2.2) has index 0 (or 1), if $\ker A_C = \{0\}$ ($\neq \{0\}$): with Q_C projecting onto $\ker A_C^\top$ and $P_C := I - Q_C$, the network variables u can be split into the differential part $P_C u$ and the algebraic components $Q_C u$. Note that all index concepts (differential, perturbation, tractability, etc.) coincide for this case [13].

2.3. Coupling interface

Let the circuit comprise a number of n blocks of uniform transmission lines, where each block is described by the characteristic equation (2.1) for d_i transmission lines of length l_i with line voltages V^i, line currents J^i and positive-definite symmetric resistance, inductance, conductance and capacitance matrices R_i, L_i, G_i and $C_i \in \mathbb{R}^{d_i \times d_i}$. The boundary currents are given by

$$w^i = \begin{pmatrix} w_l^i \\ w_r^i \end{pmatrix} \quad \text{with} \quad (w_l^i)^\top := (w_{l,1}^i, \dots, w_{l,d_i}^i)^\top, \; (w_r^i)^\top := (w_{r,1}^i, \dots, w_{r,d_i}^i)^\top,$$

and u_l^i and u_r^i denote the boundary node potentials at the left-hand and right-hand side, respectively. Let A_{w_i} denote the partial incidence matrix that assembles the coupling current w^i at the respective boundary nodes of the electrical network. Then a current contribution

$$+\sum_{i=1}^{n} A_{w_i} w^i = +A_w w \tag{2.3a}$$

with

$$A_w := (A_{w_1}, \dots, A_{w_n}), \quad w := ((w^1)^\top, \dots, (w^n)^\top)^\top$$

has to be added to the current balance (2.2a). On the other hand, the partial incidence matrix A_{w_i} links the boundary condition (2.1d) of the i-th interconnect system to the boundary node potentials of the electrical circuit:

$$\begin{pmatrix} V^i(0,t) \\ V^i(l_i,t) \end{pmatrix} = \begin{pmatrix} u_l^i \\ u_r^i \end{pmatrix} = A_{w_i}^\top u. \tag{2.3b}$$

Overall, equations (2.1–2.3) define an initial-boundary value problem of differential-algebraic and hyperbolic equations coupled via source terms (2.3a) and a linear contact condition (2.3b).

Remark 2.2. In this approach we assume that all n blocks are electromagnetically decoupled. If crosstalk of two or more blocks may not be neglected, they have to be combined to one block and modeled altogether.

2.4. Analytical properties

Before discussing discretization and numerical simulation, we have to address the analytical properties of this model.

EXISTENCE AND UNIQUENESS. Well-posedness is guaranteed, if three conditions hold [7]:

a) The Jacobian matrices of q_C and r are positive-definite, i.e., all charge storing elements and resistors are strictly passive.
b) All boundary nodes that link the electrical network with the transmission lines are connected to ground via capacitive paths, i.e., $\ker P_C^\top A_w = \{0\}$.
c) All initial values and time-dependent sources are smooth enough:

$$V^{i,0}, J^{i,0} \in H^1(0,l_i)^{d_i} \quad (i = 1, \dots, n), \qquad \imath(0,\cdot) \in H^1(0,t_{\text{end}})^{n_\imath}.$$

Then the partial differential-algebraic equation (2.1–2.3) has a unique solution (V,J,u), $V^\top := ((V^1)^\top,\dots,(V^n)^\top)^\top$, $J^\top := ((J^1)^\top,\dots,(J^n)^\top)^\top$. Both V, J and V_t, J_t are bounded on the finite time interval $[0, t_{\text{end}}]$ with

$$V^i, J^i \in H^1(0,l_i)^{d_i}, \quad V_t^i, J_t^i \in L^2(0,l_i)^{d_i}, \qquad i=1,\dots,n.$$

For the network variables one gets

$$u \in H^1(0,t_{\text{end}})^{n_u}.$$

A PRIORI ESTIMATES. Defining appropriate norms

$$\|w(\cdot,t)\|_{\mathcal{V}} := \sqrt{\int_0^l \|w(x,t)\|_2^2\,dx}, \quad \|\imath\|_{L^2(0,t)} := \sqrt{\int_0^t \|\imath(\tau)\|_2^2\,d\tau}$$

for the $1d$ and $0d$ components, one gets in the differential components

$$\rho(t) := \|V(\cdot,t)\|_{\mathcal{V}}^2 + \|J(\cdot,t)\|_{\mathcal{V}}^2 + \|P_C u(t)\|_2^2,$$

the estimate

$$\rho(t) \le \text{const}\cdot\left(\rho(0) + \|\imath\|_{L^2(0,t)}^2\right), \tag{2.4}$$

and finally for the algebraic components

$$\|Q_C u(t)\|_2^2 \le \text{const}\left(\rho(t) + \|\imath(t)\|_2^2\right). \tag{2.5}$$

The constants depend on the model parameters, but not on initial and boundary data. Therefore the PDAE initial-boundary value problem (2.1–2.3) is well posed. The solution depends continuously on the data. In the DAE sense it is well conditioned, too, as only data enter the solution, but not its derivatives.

PERTURBATION ANALYSIS. To investigate the sensitivity of the PDAE system (2.1–2.3) with respect to perturbations, we insert slightly perturbed solutions $(\widehat{V},\widehat{J},\widehat{u})$ into the system, defining a residual $\delta(z,t)$ and $\mu(t)$ in the left-hand side of (2.1a) and (2.2a). Similar to the a priori estimates, we get estimates for the sensitivity of the solution, i.e., for the difference

$$(\mathbf{\Delta}V, \mathbf{\Delta}J, \mathbf{\Delta}u) := (V,J,u) - (\widehat{V},\widehat{J},\widehat{u})$$

between both solutions in terms of the residuals δ and μ and the initial difference:

$$\begin{aligned}\widetilde{\rho}(t) &:= \|\mathbf{\Delta}V(\cdot,t)\|_{\mathcal{V}}^2 + \|\mathbf{\Delta}J(\cdot,t)\|_{\mathcal{V}}^2 + \|\mathbf{\Delta}P_C(t)\|_2^2\\ &\le \text{const}\cdot\left(\widetilde{\rho}(0) + \|\delta\|_{L^2([0,t],\mathcal{V})}^2 + \|\mu\|_{L^2(0,t)}^2\right),\\ \|\mathbf{\Delta}Q_C u(t)\|_2^2 &\le \text{const}\left(\widetilde{\rho}(t) + \|\delta(\cdot,t)\|_{\mathcal{V}}^2 + \|\mu(t)\|_2^2\right)\end{aligned}$$

with the norm

$$\|\delta\|_{L^2([0,t],\mathcal{V})} := \sqrt{\int_0^t \|\delta(\cdot,\tau)\|_{\mathcal{V}}^2\,d\tau}.$$

Thus the norm of the difference of reference and perturbed solution can be estimated by the norm of the perturbations itself, and does not depend on derivatives of the perturbations, neither with respect to space nor time. Hence the PDAE system (2.1–2.3) has perturbation index 1. For a precise definition of the perturbation index of hyperbolic-type PDAE systems we refer to [10, 14].

3. Semi discretization by companion networks

Many circuit simulation packages treat transmission line effects within the framework of standard circuit theory: transmission lines are modeled by equivalent circuits of lumped elements, generally coupled RLGC elements and controlled sources. A standard circuit companion model for a transmission line with capacitance C, inductance L, resistance R and conductance G per unit length is shown in Fig. 4. It models a section of length Δx.

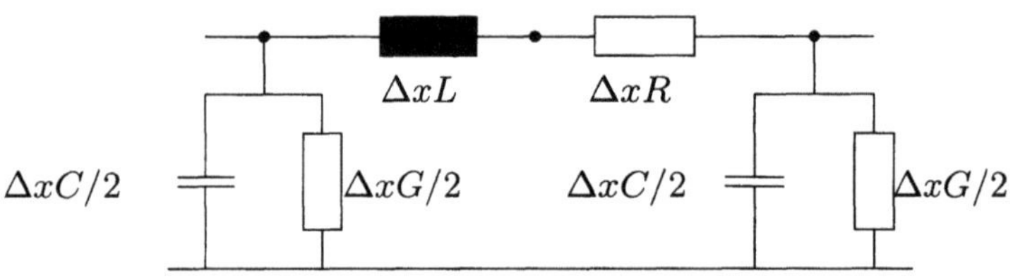

FIGURE 4. RLGC-model for a transmission line of length Δx.

The overall system is thus described by a coupled system of approximate differential-algebraic equations, for short, RLGC-ADAEs. Such an RLGC-ADAE model for the CMOS ring oscillator with transmission lines consists of three parts:

- *Network model for the circuit part:*

$$0 = A_C \frac{d}{dt}\left[q_C(A_C^\top u)\right] + A_R r(A_R^\top u, t) + A_I \imath(A_C^\top u, t) + A_w w \tag{3.1a}$$

- *RLGC companion model for the whole interconnect system, i.e., all n blocks of transmission lines (in our case: $n = 3$):*

$$0 = \begin{pmatrix} C_{tl} & 0 \\ 0 & L_{tl} \end{pmatrix} \dot{v} + Fv + \begin{pmatrix} \bar{A}_w \\ 0 \end{pmatrix} w, \text{ with} \tag{3.1b}$$

  $C_{tl} := \operatorname{diag}(\Delta x_1 C_1, \ldots, \Delta x_n C_n)$ and $L_{tl} := \operatorname{diag}(\Delta x_1 L_1, \ldots, \Delta x_n L_n)$; the vector $v^\top := (v_1^\top, v_2^\top)^\top$ of unknowns consists of all node potentials v_1 and branch currents v_2 through inductors of the companion model. The constant matrix F depends on the discretization parameters Δx_i used for the different blocks. The incidence matrix $\bar{A}_w$ assembles the boundary current w at the respective boundary nodes $\bar{A}_w^\top v_1$ of the companion model. See [7] for a detailed description of this model and related parameters.

- *Coupling system:*

$$A_w^\top u + \bar{A}_w^\top v_1 = 0. \tag{3.1c}$$

First of all, this approximate DAE system shows an artificial sensitivity not given in the underlying PDAE model: since all boundary nodes are connected to ground via a capacitive path, the coupling system can be written as

$$A_w^\top P_C u + \bar{A}_w^\top v_1 = 0,$$

an algebraic relation involving only differential components! Hence the index is at least two, and one can show that the index is indeed two.

A second flaw of this semi discretization approach becomes visible if we plot the numerical solution to the ring oscillator. The zoom in Fig. 5 shows a spurious oscillation of high frequency that superposes the physical oscillation of the ring oscillator for the mid line voltage (second transmission line) between 8 and 12 psec.

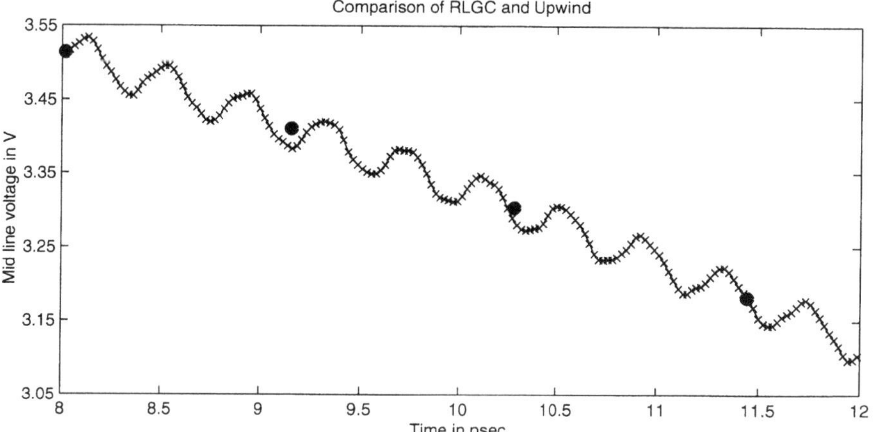

FIGURE 5. Numerical simulation of mid line voltage (second transmission line) for two different types of semi discretization:
$\times$: RLGC companion circuit, i.e., 2nd order central differences;
$\bullet$: 2nd oder upwind scheme

This spurious oscillation of high frequency in the line voltages can be easily explained. From a mathematical point of view, the use of RLGC companion circuits to model the telegraphers' equations for the transmission lines can be characterized as a non-adaptive semi discretization with respect to space. To obtain approximation order two, this approach uses central differences on a staggered grid. Of course, this ansatz is not appropriate to describe the information flow in hyperbolic systems properly. Consequently, spurious oscillations are invoked with a frequency that increases for decreasing discretization parameter $\Delta x_i := \Delta x$, $i = 1, 2, 3$.

Is there a way out? A look at the properties of the underlying PDAE model shows that semi discretization should neither regularize nor deregularize the system; in addition, it should be consistent with the hyperbolic information flow of the transmission lines. An example for such an approach is given by an upwind scheme of order two: now the telegrapher's equation together with the coupling equation leads to an initial-value problem of linear ordinary differential equations:

$$\dot{v} = \left(\frac{1}{\Delta x} D_2 + B_2\right) v + \left(\frac{1}{\Delta x} D_1 + B_1\right) A_w^\top P_C u \quad \text{and} \quad w = w(A_w^\top P_C u, v),$$

with constant matrices D_2, B_2, D_1 and B_1; the coupling current w on the right-hand side of the network model is explicitly given as a linear function of $A_w^\top P_C u$ and v. Obviously, the index of the approximate system is now one. Additionally, no spurious oscillations occur any longer: Fig. 5 shows that now the physical oscillation is nicely resolved, but approximately one hundred times larger integration steps can be used.

In many circuit simulation packages, however, this alternative is not offered. Modeling and semi discretization are highly intertwined – changing the type of semi discretization would demand the time and cost consuming inclusion of new models. Hence we are stuck to RLGC models in most simulation packages, and have to cope with this problem: spurious oscillations of high frequency, which stem from an inappropriate discretization of a PDE model, demand very small time steps and thus increase computational costs significantly.

One idea is to use the numerical damping of implicit integration schemes and construct some sort of numerical low pass filter: tracking physical oscillations, but damping out very rapidly artificial oscillations of high frequency caused by numerical noise. See [9] for more details.

In this paper we will follow another idea: exploiting the multirate potential of the approximate DAE system by applying dynamic iteration based on distributed

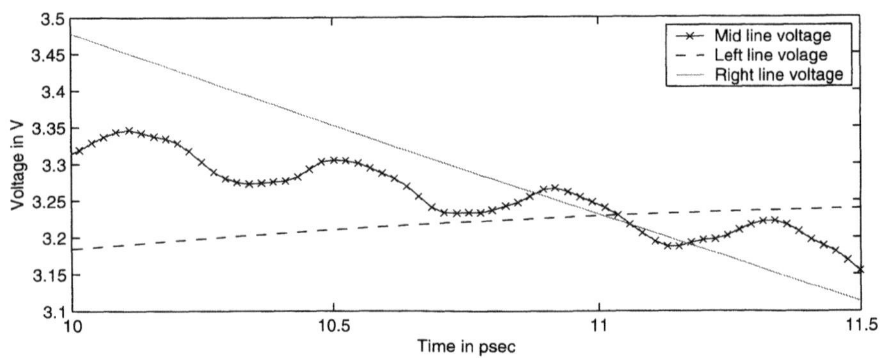

FIGURE 6. Multirate potential at RLGC-ADAEs

time integration. But is there any multirate potential? To motivate this, we consider Fig. 6 that zooms again at the line voltages for the second transmission line. The line voltages at both ends are smooth and vary only at a low time scale. These quantities correspond to the boundary node voltages of the circuit part, and a similar activity level can be found for all components of the circuit model. However, the mid line voltage shows a very fast varying oscillation superposing the physical solution. Consequently, the coupled system shows a multirate behavior that can be exploited by dynamic iteration schemes.

4. Dynamic iteration

To apply dynamic iteration to the RLGC-ADAE system (3.1), we first split the whole time interval $[0, t_{\text{end}}]$ into time windows $[T_n, T_{n+1}]$ of size $H_n := T_{n+1} - T_n$ $(0 = 0 < T_1 < \ldots < T_{N+1} = T_{\text{end}}, n = 0, \ldots, N)$. We consider a window $[T_n, T_{n+1}] \subset [0, T_{\text{end}}]$ and suppose that the numerical solution has already been computed for $t \in [0, T_n]$. In $[T_n, T_{n+1}]$ a finite number k_n of dynamic iteration steps is applied to get the numerical solution

$$\begin{aligned} \tilde{u}|_{(T_n,T_{n+1}]} := \tilde{u}_n^{(k_n)}|_{(T_n,T_{n+1}]}\,, \quad \tilde{v}|_{(T_n,T_{n+1}]} := \tilde{v}_n^{(k_n)}|_{(T_n,T_{n+1}]}\,, \\ \tilde{w}|_{(T_n,T_{n+1}]} := \tilde{w}_n^{(k_n)}|_{(T_n,T_{n+1}]}\,. \end{aligned} \tag{4.1}$$

The iterates $\tilde{u}_n^{(k)}$, $\tilde{v}_n^{(k)}$, and $\tilde{w}_n^{(k)}$ with $k \geq 1$ are defined recursively by a Gauss–Seidel method that is adapted to the specific structure of (3.1). Each iteration step starts with the solution of an initial value problem for the first subsystem to get $\tilde{u}_n^{(k)}$:

$$\begin{aligned} 0 &= A_C \tfrac{d}{dt}\left[q_C(A_C^\top \tilde{u}_n^{(k)})\right] + A_R r(A_R^\top \tilde{u}_n^{(k)}, t) + \\ &\quad + A_I \imath(A_C^\top \tilde{u}_n^{(k)}, t) + A_w \tilde{w}_n^{(k-1)} \\ P_C \tilde{u}_n^{(k)}(T_n) &= P_C \tilde{u}(T_n) \end{aligned} \quad . \tag{4.2a}$$

Afterwards the iterates $(\tilde{v}_n^{(k)}, \tilde{w}_n^{(k)})$ are obtained as solution of an initial value problem for the second subsystem together with the coupling equation:

$$\begin{aligned} 0 &= \begin{pmatrix} C_{tl} & 0 \\ 0 & L_{tl} \end{pmatrix} \dot{\tilde{v}}_n^{(k)} + F\tilde{v}_n^{(k)} + \begin{pmatrix} \bar{A}_w \\ 0 \end{pmatrix} \tilde{w}_n^{(k)} \\ 0 &= A_w^\top \tilde{u}_n^{(k)} + \begin{pmatrix} \bar{A}_w \\ 0 \end{pmatrix}^\top \tilde{v}_n^{(k)} \\ \tilde{v}_n^{(k)}(T_n) &= \tilde{v}^{(k)}(T_n) \end{aligned} \quad , \tag{4.2b}$$

where the computed components $\tilde{u}^{(k)}$ are used already. The iteration starts with initial guesses $(\tilde{u}_n^{(0)}, \tilde{v}_n^{(0)}, \tilde{w}_n^{(0)})$ that may be computed extrapolating $\tilde{u}, \tilde{v}$ and $\tilde{w}$ continuously from $(T_{n-1}, T_n]$ to $[T_n, T_{n+1}]$, (set $(T_{-1}, 0] := \{0\}$ for a unified notation).

4.1. Stability and convergence for systems in normal form

To show stability and convergence of this method, we can apply results derived in [2] for systems in normal form

$$\begin{aligned} \dot{y}_1 &= f_1(y, z_1) & \qquad \dot{y}_2 &= f_2(y, z_2) \\ 0 &= h_1(y, z_1, w) & \qquad 0 &= h_2(y, z_2, w) \end{aligned} \qquad (4.3)$$
$$0 = g(y, z)$$

that fulfill the following index-1 conditions: all matrices

$$\begin{pmatrix} \frac{\partial h}{\partial z} & \frac{\partial h}{\partial w} \\ \frac{\partial g}{\partial z} & 0 \end{pmatrix}, \quad \text{and} \quad \begin{pmatrix} \frac{\partial h_i}{\partial z_i} & \frac{\partial h}{\partial w} \\ \frac{\partial g}{\partial z_i} & 0 \end{pmatrix}, \frac{\partial h_i}{\partial z_i} \quad (i = 1, 2) \qquad (4.4)$$

are regular in a neighborhood of the solution. Then the Gauss-Seidel method is convergent of order $\mathcal{O}(H^p)$, with p denoting the order of the extrapolation method used to derive the initial guess and $H := \max_{i=1,\ldots,N} H_i$, if and only if the following contractivity conditions hold:

- for a stable error recursion in the algebraic coupling component w within one window: we need to have $\alpha < 1$, where

$$\alpha := \max_{t \in [0,T]} \| R_2^{-1}(t) R_1(t) \| \qquad (4.5)$$

  and

$$R_i(t) := \left[\frac{\partial g}{\partial z_i} \left(\frac{\partial h_i}{\partial z_i} \right)^{-1} \frac{\partial h_i}{\partial w} \right] (y(t), z(t), w(t)) \,, \quad (i = 1, 2) \,. \qquad (4.6)$$

- for a stable error recursion in the algebraic coupling component w also from one window to the next one:

$$L \alpha^{k_n} < 1$$

  is necessary and sufficient, with L denoting the Lipschitz constant of the extrapolation operator.

4.2. Transformation of the RLGC-ADAE into normal form

To apply these results, we have to transform the approximate DAE (3.1) into the normal form (4.3): defining $z_{11} := P_C u$, $z_{12} := Q_C u$, $y_2 := v$ and using $A_w^\top Q_C = 0$, system (3.1) can be rewritten into (4.3) with

$$\begin{aligned} f_1 &= z_{13}, \\ h_1 &= \begin{pmatrix} y_1 - \left(A_C \cdot q_C(A_C^\top z_{11}) + Q_C^\top Q_C z_{11}\right), \\ Q_C^\top \left(A_R \cdot r(A_R^\top [z_{11} + z_{12}], t) + A_I \cdot \imath(A_C^\top z_{11}, t)\right) \\ z_{13} + P_C^\top \left(A_R \cdot r(A_R^\top [z_{11} + z_{12}], t) + A_I \cdot \imath(A_C^\top z_{11}, t)\right) + A_w w \end{pmatrix}, \\ f_2 &= z_2, \\ h_2 &= \begin{pmatrix} \frac{\Delta x}{2} C_{tl} & 0 \\ 0 & \Delta x L_{tl} \end{pmatrix} z_2 + f^{\Delta x}(y_2) + \begin{pmatrix} \bar{A}_w \\ 0 \end{pmatrix} w, \end{aligned}$$

introducing dummy variables y_1, z_{13} and z_2 (with $z_1^\top := (z_{11}^\top, z_{12}^\top, z_{13}^\top)^\top$). Differentiating the coupling condition

$$0 = A_w^\top z_{11} + \begin{pmatrix} \bar{A}_w \\ 0 \end{pmatrix}^\top y_2$$

once with respect to time leads to

$$0 = g(y,z) := A_w^\top \left(A_C q_C'(A_C^\top z_{11}) A_C^\top + Q_C^\top Q_C \right)^{-1} z_{13} + \begin{pmatrix} \bar{A}_w \\ 0 \end{pmatrix}^\top z_2 \, .$$

It is easy to verify that all index-1 conditions required in (4.4) are now fulfilled. Computing

$$R_1(t) = A_w^\top (A_C q_C'(A_C^\top z_{11}(t)) A_C^\top + Q_C^\top Q_C)^{-1} A_w, \quad R_2(t) = \frac{1}{\Delta x} \bar{A}_w^\top C_{tl}^{-1} \bar{A}_w$$

the contractivity constant reads

$$\alpha := \max_{t \in [0,T]} \| R_2^{-1}(t) R_1(t) \| = \mathcal{O}(\Delta x) \, .$$

For sufficiently small discretization parameters Δx, the stability constant α is smaller than 1: for an example see [7]. Since the effective capacitance of the RLGC model is of order $\Delta x \cdot C_{tl}$, the following rule of thumb applies: take the system with larger effective capacitance, i.e., the ring oscillator, as subsystem to be integrated.

Remark 4.1. For low order extrapolation schemes one gets Lipschitz constants $L \approx 1$, e.g., for constant or linear extrapolation $L = 1$. In this case, the second contractivity conditions is always fulfilled (given $\alpha < 1$), even if no iterations are performed ($k_n \equiv 1$).

5. Distributed time integration

Does this approach of dynamical iteration speed up integration time significantly, also under the restriction of using existing commercial circuit simulation packages? To answer this question, we regard distributed time integration methods based on the dynamic iteration method (4.2) with fixed window size $H = \tau^+ \geq \tau$, fixed number of iteration steps $k_n = \bar{k}$ and constant extrapolated data as initial guesses:

$$\tilde{w}_n^{(0)}(T) = \tilde{w}(T_n)$$

($T \in [T_n, T_{n+1}]$). Since almost all circuit simulation packages offer only one integration scheme in the simulation kernel, we have to apply the same method to both circuit models (circuit part and RLGC companion model) and must not aim at exploiting the possibility of multimethod approaches. Thus the initial-value problems for both subproblems (4.2a) and (4.2b) are integrated by the Rosenbrock-Wanner scheme RODAS [11] of fourth order for index-1 DAE systems (another suitable choice within simulation packages would be the use of BDF schemes). The fixed step size τ for the RLGC model is chosen such that the spurious oscillations of the mid line voltages shown in Fig. 6 are resolved properly: $\tau := 5 \cdot 10^{-2}$ psec, i.e., one oscillation is sampled by approximately 10 steps. Larger step sizes τ^+ (the window step size) are used for the circuit model. Regarding the interface restrictions in commercial network analysis packages, the necessary data at the inner grid points $T_n + \tau, T_n + 2\tau, T_n + \tau^+ - \tau$ is provided by (linear) interpolation.

The numerical results have been obtained without preconditioning [2] and, following our rule of thumb, with the electrical network as first subsystem. RODAS applied to the overall system with the basic step size $\tau := 5 \cdot 10^{-2}$ psec serves as a reference method. All numerical results have been compared with a highly accurate solution. Since the error in the mid line voltages dominates the integration error we show in Figures 7–8 the (absolute) error in the mid line voltage. For the reference method (dotted line), this error is of size 10^{-2}. In most practical applications errors up to some per cents are acceptable in circuit simulation. Besides that, more stringent error tolerances would require to capture the spurious oscillations in more detail, which does not make sense.

The distributed time integration based on the Gauss-Seidel dynamic iteration method leads to good results already for $\bar{k} = 1$ and step size $\tau^+ = 20\tau$ (Figure 7), but the error increases dramatically if $\tau^+ = 100\tau$ (Figure 8). In this case, the error can be decreased substantially if one additional iteration is performed ($\bar{k} = 2$), since the error of the dynamic iteration dominates the discretization error of RODAS. However, there is no further error reduction if $\bar{k} > 2$. This is caused by the fact that only linear interpolation can be used to define the coupling variables in the iterates. Higher order formulas, for example, dense output, are not

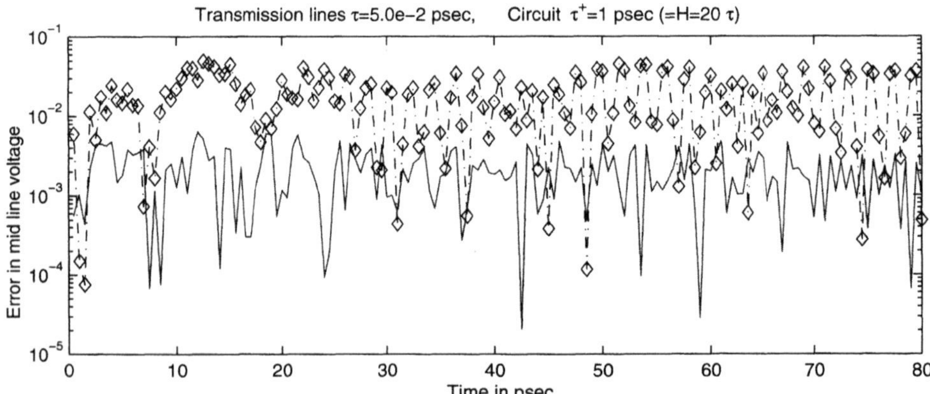

FIGURE 7. Benchmark “CMOS ring oscillator with transmission lines”: error of the distributed time integration (dashed line, “◇”) with RODAS for both subsystems, linear interpolation, $k_n = 1$, and step sizes $\tau = 5 \cdot 10^{-2}$ psec and $\tau^+ = 20\tau$; error of the reference method, RODAS with step size $\tau = 5 \cdot 10^{-2}$ psec applied to the overall system (3.1) (dotted line).

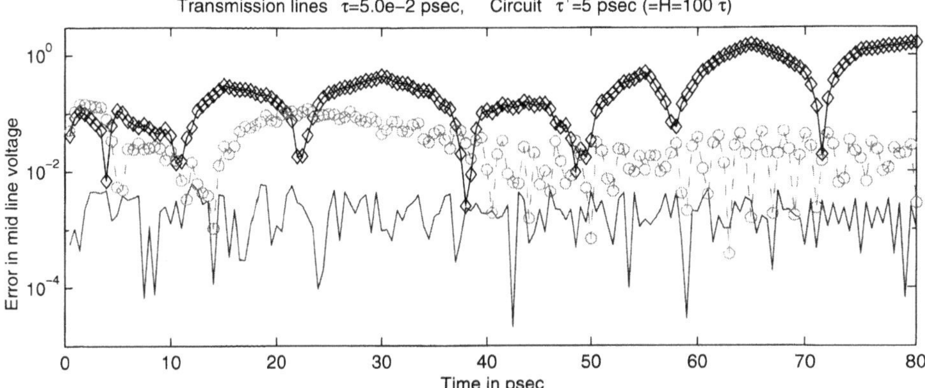

FIGURE 8. Benchmark "CMOS ring oscillator with transmission lines": error of the distributed time integration scheme with RODAS for both subsystems, linear interpolation and step sizes $\tau = 5 \cdot 10^{-2}$ psec and $\tau^+ = 100\tau$ (solid lines, "◇": $k_n = 1$, "o": $k_n = 2$) and error of the reference method (dotted line).

available within circuit simulation packages. Nevertheless, the overall effort for the simulation of coupled DAE models in circuit simulation may be reduced substantially by distributed time integration based on dynamic iteration.

6. Summary

In refined network analysis, the use of companion models for spatial discretization of general network elements may lead to approximate DAE systems that do not reflect properly the underlying PDAE model and, consequently, increase computation costs. Restricted to the use of existing commercial simulation packages, these problems cannot be avoided by appropriate discretization schemes according to the PDAE model properties. It is one demanding task for applied numerical analysis to cope with such numerical problems under software-engineering restrictions. As we have seen by inspecting the CMOS ring oscillator with transmission line effects as one example, distributed time integration based on dynamic iteration offers one approach to lower computation costs that had been increased previously by an inappropriate (but fixed) modeling. Although only constant extrapolation (for initial guesses) and linear interpolation (for coupling variables) are used, one gains simulation results of sufficient accuracy.

References

[1] M. Arnold.: *A note on the uniform perturbation index.* Rostock. Math. Kolloq. 52 (1998), 33–46.

[2] M. Arnold, and M. Günther: *Preconditioned dynamic iteration for coupled differential-algebraic systems*, BIT 41:1 (2001), 1–25.

[3] S.L. Campbell, and W. Marszalek: *ODE/DAE integrators and MOL problems.* Z. f. angew. Math. u. Mech. 76 Suppl. 1 (1996), 251–254.

[4] S.L. Campbell, and W. Marszalek: *The index of an infinite-dimensional implicit system.* Mathematical and Computer Modelling of Dynamical Systems 5 (1999), 18–42.

[5] M. Günther: *A joint DAE/PDE model for interconnected electrical networks.* Mathematical and Computer Modeling of Dynamical Systems 6 (2000), 114–128.

[6] M. Günther: *A PDAE model for interconnected linear RLC networks.* Mathematical and Computer Modelling of Dynamical Systems 7 (2001), 189–203.

[7] M. Günther: *Partielle differential-algebraische Systeme in der numerischen Zeitbereichsanalyse elektrischer Schaltungen*, VDI Verlag, Düsseldorf, 2001.

[8] M. Günther, and U. Feldmann: *CAD based electric circuit modeling in industry. I: Mathematical structure and index of network equations. II: Impact of circuit configurations and parameters.* Surv. Math. Ind. 8 (1999), 97–157.

[9] M. Günther, U. Feldmann, and P. Rentrop: *CHORAL – a one step method as numerical low pass filter in electrical network analysis.* In: Scientific Computing in Electrical Engineering. Proceedings of the 3rd International Workshop, 20-23 August 2000, Warnemünde, Germany. Eds.: Rienen, U. van; Günther, M.; Hecht, D. Springer-Verlag, Berlin, 2001, 199–215.

[10] M. Günther, and Y. Wagner: *Index Concepts for Linear Mixed Systems of Differential-algebraic and Hyperbolic-Type Equations.* SIAM J. Sci. Comp. 22:5 (2000), 1610–1629.

[11] E. Hairer and G. Wanner: *Solving Ordinary Differential Equations II. Stiff and Differential-Algebraic Problems.* Springer-Verlag, Berlin Heidelberg New York, 2nd ed., 1996.

[12] W. Lucht, K. Strehmel and C. Eichler-Liebenow: *Indexes and special discretization methods for linear partial differential algebraic equations.* BIT 39 (1999), 484–512.

[13] C. Tischendorf: *Topological index calculation of differential-algebraic equations in circuit simulation.* Surv. Math. Ind. 8, 187–199 (1999)

[14] Y. Wagner: *A further index concept for PDAEs of hyperbolic type.* Mathematics and Computers in Simulation 53 (2000), 287–291.

Michael Günther
Universität Karslruhe (TH)
Lehrstuhl für Wissenschaftliches Rechnen und
Mathematische Modellbildung
Engesserstr. 6
D-76128 Karlsruhe, Germany
email: `guenther@iwrmm.math.uni-karlsruhe.de`

Modeling, Simulation and Optimization of Integrated Circuits
(Eds.) K. Antreich, R. Bulirsch, A. Gilg and P. Rentrop

International Series of Numerical Mathematics Vol. 146, 219–227

Parasitics Reduction for Analog Circuit Simulation

Martin Kahlert

To my father

Abstract. The modeling of interconnect lines using lumped elements results in large nets that in general can only be handled after a reduction of these parasitic elements. This paper presents an alternative to the Padé via Lanczos methods like SymPVL which allows for a direct error estimation by working in the time domain instead of approximating the transfer function. The proposed method uses a direct eigenvalue calculation which is performed with the Jacobi Davidson algorithm.

1. Introduction

High frequency signals forbid the modeling of interconnect lines as ideally conducting lines. This would lead to wrong results due to the following physical effects:

- Signals travel along the lines with a velocity lower than the speed of light. This produces time delays from one end of the line to the other one and causes wrong simulation results for high speed driven devices.
- Lines which are located nearby others are coupled capacitively and inductively. Thus fast signals at one line produce crosstalk signals on nearby lines. When these crosstalk signals are large enough they lead to malfunction of the analog device.
- The damping of the lines is frequency dependent. A correct analog simulation has to describe the distortion of the signal shapes.

There are two main methods to describe coupled conducting lines in a correct manner:

One can use the telegrapher's equations for all n lines including the $n(n-1)/2$ capacitive and inductive coupling terms. This method is the most physically correct one but there is no known efficient way of solving this system together with

---

Key words and phrases. Parasitics reduction, Eigenvalue decomposition.

This work was completed with the support of the German BMBF project 'Parasitics' and Infineon AG.

thousands or even millions of mosfet equations and the electrical boundary conditions at the lines' ends. Another drawback of this method is that today's analog circuit simulators are tuned to solve differential algebraic systems emerging from Kirchhoff's laws and are not well suited for additionally handling partial differential equations.
So the electrical engineers came up with a different solution: the discretization of the lines into so-called lumped elements. This can be thought of as a space discretization of the telegrapher's equations using, e.g., the centered difference scheme (this is in general unstable as we will see in a later part of this document).

The discretization process is done using the geometrical information from the layout and either library information for typical line arrangements or three-dimensional simulations of small subparts of the layout which yield coupling capacities and inductivities.

Unfortunately the resulting net list contains so many parasitic elements that the circuit simulator runs out of memory or simulates linear devices during most of its runtime. Thus there is an emerging need for reducing the necessary effort to simulate conducting lines.

2. Today's main reduction methods

The most commonly used reduction algorithms are bases on the Padé via Lanczos process (the most successful one seems to be SymPVL by Roland Freund). There are a lot of papers dealing with this so another deeper description is considered unnecessary here. The main idea of these methods is the approximation of the so-called transfer function with a considerably lower number of degrees of freedom. The transfer function Z connects the input potentials to the output currents of any linear n-port element in the frequency domain:

$$I_B(s) = Z(s)U_B(s),$$

where $I_B(s)$ and $U_B(s)$ are the vectors of the Laplace transformed boundary currents and potentials. Thus $Z(s)$ is a matrix-valued frequency dependent function. The idea is to approximate $Z(s)$ with a much lower number of parameters.

This method has a drawback: A good fit of a Laplace transformed function does not necessarily mean a good fit after the back transformation into the time domain. The first attempts to use this method resulted in unstable approximations and produced circuit descriptions, which are not passive any more. You have to deal with these issues separately and make sure that you get correct descriptions.

The second group of algorithms used for parasitics reduction are based on the direct calculation of generalized eigenvalues as outlined in the next section.

3. The reduction idea

For simplicity the idea is shown for RC-circuits only (i.e., no parasitic inductivities). The application of nodal analysis for every linear circuit sub-component yields

$$\begin{pmatrix} C_I & C_C \\ C_C^T & C_B \end{pmatrix} \begin{pmatrix} \dot{x}_I \\ \dot{u}_B \end{pmatrix} + \begin{pmatrix} G_I & G_C \\ G_C^T & G_B \end{pmatrix} \begin{pmatrix} x_I \\ u_B \end{pmatrix} = \begin{pmatrix} 0 \\ j \end{pmatrix}.$$

Here x_I denotes the vector of unknown inner node potentials (i.e., nodes which are connected to resistors or capacities only) and u_B the vector of known boundary node potentials (nodes connected to other elements like mosfets). The vector j contains the unknown currents through the boundary nodes. The symmetric positive-definite conductivity matrix G and the symmetric positive semi-definite capacity matrix C are partitioned accordingly.

We solve the generalized symmetric eigenvalue problem

$$C_I W = G_I W \Lambda, \qquad W^T G_I W = I$$

with a diagonal matrix Λ and represent the vector of unknowns using the eigenvector basis: $x_I(t) =: Wd(t)$.
Without loss of generality we can assume, that eigenvalues and corresponding eigenvectors are sorted with respect to decreasing absolute value: $\Lambda = \operatorname{diag}(\Lambda_1, \Lambda_2)$ and $W = (W_1, W_2)$. This leads to the diagonal differential algebraic (some eigenvalues might be zero!) system

$$\begin{aligned} \Lambda_1 \dot{d}_1 + d_1 &= -W_1^T (C_C \dot{u}_B + G_C u_B) \\ \Lambda_2 \dot{d}_2 + d_2 &= -W_2^T (C_C \dot{u}_B + G_C u_B). \end{aligned}$$

The reduction idea comes from

$$f(t) = d_i(t) + \lambda_i \dot{d}_t(t) \mathrel{\dot{\approx}} d_i(t + \lambda_i) \quad \Longrightarrow \quad d_i(t) \mathrel{\dot{\approx}} f(t - \lambda_i).$$

Thus the eigenvalues can be interpreted as mode delays. Of course this is only possible in the case of RC circuits, where G_I is symmetric positive-definite and C_I is positive semi-definite and the eigenvalues are non-negative real numbers.
By neglecting the small eigenvalues Λ_2 we obtain

$$\widetilde{d_2} := -W_2^T (C_C \dot{u}_B + G_C u_B).$$

Unfortunately this simple idea has some drawbacks:

- The approximation shows large, unphysical voltage peeks. This can be seen by looking at the extreme limits: Assume that we neglect all eigenvalues. Then our solution contains derivatives of some right-hand side boundary potentials without any corresponding damping term on the left-hand side.
  The solution of this comes from a congruent transformation of the matrix: Replace the capacity matrix C by

$$\widetilde{C} := \begin{pmatrix} G_I W_1 W_1^T C_I W_1 W_1^T G_I & G_I W_1 W_1^T C_C \\ C_C^T W_1 W_1^T G_I & C_B \end{pmatrix}.$$

- Charge and energy conservation cannot be proved (and it is usually not given).
- The approximated crosstalk is some orders of magnitudes too small. Here is an easy attempt to make that clear: The crosstalk appears nearly instantly and it thus roughly corresponds to the reduced eigenvalues which are neglected. So the method throws away too much of the crosstalk.

A much better algorithm results from the idea described above but with a different start point: We do not reduce the original set of equations but the equations for the difference y_i to the "non delayed" (i.e., no capacities at all) values $x_I = -G_I^{-1}G_C u_B$:

$$y_I := x_I + G_I^{-1}G_C u_B.$$

Another interpretation of this is the splitting of the exact solution of the differential algebraic system into the general solution of the homogeneous system and a special solution of the inhomogeneous system. The approximation only acts on the general solution of the homogeneous system. As a side note, this fortunately solves the conservation problem, too.

The proposed reduction method is described by the following algorithm:

- Solve the generalized symmetric eigenvalue problem

$$C_I W = G_I W \Lambda, \qquad W^T G_I W = I.$$

In the current implementation this is done using a variant of the Jacobi Davidson algorithm. Only the largest eigenvalues Λ_1 and corresponding eigenvectors W_1 are needed. We set $\widetilde{y}(t) := W_1 d(t)$.

- Solve the system

$$\Lambda_1 \dot{d} + d = -W_1^T [C_C - C_I G_I^{-1} G_C] \dot{u}_B, \quad d(0) = 0.$$

- The boundary currents are given by

$$\widetilde{J}(t) := [C_C^T - G_C^T G_I^{-1} C_I] W_1 \dot{d}(t) + C_P \dot{u}_B(t) + G_P u_B(t),$$

where the abbreviations

$$C_P := C_B - (C_C^T G_I^{-1} G_C + G_C^T G_I^{-1} C_C) + G_C^T G_I^{-1} C_I G_I^{-1} G_C$$

and

$$G_P := G_B - G_C^T G_I^{-1} G_C$$

have been used.

During the actual simulation u_B is approximated as a piecewise linear function (i.e., linear in every time step of the analog simulator) and the solution of the differential equation can be given exactly. This eliminates the need for a different time step control mechanism for the parasitic nets.
Note that the matrices used in the algorithm can be precomputed and are of moderate size.

4. Features of the proposed algorithm

There is one main difference between the proposed algorithm and the moment matching methods working with Padé like algorithms: If at least one non-vanishing eigenvalue is neglected, exactly two moments of the transfer function are matched. In the Lanczos methods this number increases with the accuracy reached.

In contrast to the Padé like methods, all parameters bear a physical meaning (at least in the RC case): The computed eigenvalues are directly interpretable as signal delays and the eigenvectors are the corresponding spacial modes.

In the RC case (only resistors and capacities, no inductivities extracted) one can prove, that the algorithm generates stable and passive approximations. There is no further post processing step necessary to ensure that.

Based on the magnitude of the first neglected eigenvalue one can derive an error estimation of the produced solution in the time domain. Of course this estimation can be derived for the linear sub-net only. Estimations for the complete nonlinear circuit are very hard to obtain and will usually be of low practical relevance since they tend to be too pessimistic.

The proposed algorithm can be extended to the RLC case (see Kern's thesis [5]) but this might be no good idea: One looses the physical interpretation and there is no good idea how to determine whether a complex eigenpair is to be neglected or not. The RLC case has stability problems, which will be outlined in the next section.

The main reduction effort is spent during the eigenvalue calculation. This work can be parallelized trivially by reducing different linear sub-nets on different machines and by distributing the calculation effort during the block Jacobi Davidson algorithm.

The algorithm produces very high quality approximations, both for delay and crosstalk.

5. Lumped elements and stability

This section does not belong to the problem of parasitics reduction directly but rises the question whether the path physical conducting lines $\longrightarrow$ telegrapher's equation $\longrightarrow$ discretization into lumped elements $\longrightarrow$ reduction of lumped elements $\longrightarrow$ simulation of the reduced system has any sense at all in the general RLC case.

The extraction of lumped elements for a one line system into RLC elements can be described with the following figure (the so-called π-discretization):

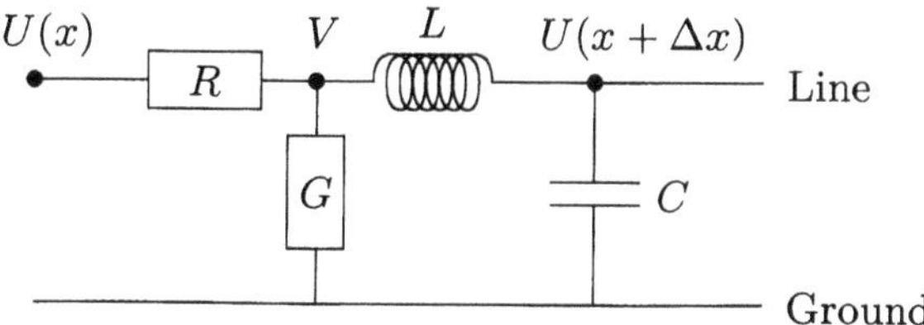

Kirchhoff's laws lead to the following equations:

$$\begin{aligned} 0 &= \frac{V - U(x)}{R} + I(x) \quad \Longrightarrow \quad V = U(x) - RI(x) \\ L\dot{I} &= V - U(x + \Delta x) \\ 0 &= -I(x) + I(x + \Delta x) + C\dot{U}(x + \Delta x). \end{aligned}$$

If we assume that the piece of line shown in the figure has the length Δx, we can calculate the resistance, capacity and inductivity value as

$$R = \widetilde{R} \cdot \Delta x, \quad C = \widetilde{C} \cdot \Delta x \quad \text{and} \quad L = \widetilde{L} \cdot \Delta x.$$

By the elimination of V from the first equation, we obtain the system

$$0 = U(x) - U(x + \Delta x) - RI(x) - L\dot{I}(x) \tag{1}$$

$$0 = I(x + \Delta x) - I(x) + C\dot{U}(x + \Delta x). \tag{2}$$

Approximating U and I with their first-order Taylor polynomial with respect to the spacial coordinate x, one obtains after the division by Δx

$$\begin{aligned} 0 &= \frac{L}{\Delta x} I_t(x) + \frac{R}{\Delta x} I(x) + U_x(x) \\ 0 &= \frac{C}{\Delta x} U_t(x) + \frac{C}{\Delta x} U_{xt} \Delta x + I_x(x), \end{aligned}$$

which leads us in the limit $\Delta x \to 0$ to the telegrapher's system

$$\begin{aligned} 0 &= \widetilde{L} I_t(x) + \widetilde{R} I(x) + U_x(x) \\ 0 &= \widetilde{C} U_t(x) + I_x(x). \end{aligned}$$

In order to investigate the stability of the discretization, we introduce the spacial coordinates $x_n := sn/N$, $n = 0, \ldots, N$, where $s = N\Delta x$ is the conductor's length and N denotes the number of π-elements used for the discretization. With the abbreviations $U_n := U(x_n)$ and $I_n := I(x_n)$ the two equations (1) and (2) can be written as

$$\begin{aligned} U_{n+1} &= U_n - RI_n - L\dot{I}_n \\ I_{n+1} &= I_n - C\dot{U}_{n+1}. \end{aligned}$$

We insert the ansatz

$$\begin{pmatrix} U_n \\ I_n \end{pmatrix} = \begin{pmatrix} a_n \\ b_n \end{pmatrix} \cdot e^{i\omega t}$$

into these equations and obtain

$$\begin{aligned} a_{n+1} &= a_n - Rb_n - i\omega L b_n \\ b_{n+1} &= b_n - i\omega C a_{n+1} \\ &= b_n - i\omega C \left[a_n - Rb_n - i\omega L b_n \right] \\ &= b_n - i\omega C a_n + i\omega RC b_n - \omega^2 LC b_n, \end{aligned}$$

or using matrix notation

$$\begin{pmatrix} a_{n+1} \\ b_{n+1} \end{pmatrix} = \underbrace{\begin{pmatrix} 1 & -(R+i\omega L) \\ -i\omega C & 1 - LC\omega^2 + i\omega RC \end{pmatrix}}_{=:A} \begin{pmatrix} a_n \\ b_n \end{pmatrix}.$$

The characteristic polynomial of the transfer matrix A is

$$\chi_A(\lambda) = \lambda^2 - \left[2 - LC\omega^2 + i\omega RC\right]\lambda + 1$$

and with the aid of the abbreviations

$$h := LC\omega^2 - i\omega RC =: \alpha - \beta i$$

the matrix A's eigenvalues are given by

$$\lambda_{1,2} = \frac{1}{2}\left(2 - h \pm \sqrt{h^2 - 4h}\right).$$

In case the line is *not* damped (i.e., $R = 0$) h is a real number and a simple calculation shows: For $h = 0$ there is a double eigenvalue $+1$. For $h \in]0,4[$ we get two conjugate complex eigenvalues with size 1 and for $h = 4$ there is a double eigenvalue -1. For $h > 4$ one eigenvalue goes to zero and the other one decreases beginning at -1. Thus the recursion for (a_n, b_n) is ("spacial"-)stable iff $h \in]0,4[$, which is equivalent to

$$0 < \omega < \frac{2}{\sqrt{LC}} \qquad \Longleftrightarrow \qquad 0 < \omega\Delta x < \frac{2}{\sqrt{\tilde{L}\tilde{C}}};$$

So it is possible to derive an upper limit for the π element's length out of the specific material constants. This is to be expected since higher frequencies ω demand a higher spacial resolution due to lower wave lengths.

The instability is moderate, because the amplification of high frequencies is limited: The corresponding input amplitudes come out of the conductor's output side increased by a factor lower than $e^{s\sqrt{\tilde{R}\tilde{C}/8}}$.

Now let us look at the damped case. Here

$$\lambda_{1,2} \approx 1 \pm \sqrt{\frac{\beta}{2}}\frac{1-i}{2}$$

holds for low frequencies ω and thus the largest eigenvalue is

$$|\lambda| \approx 1 + \sqrt{\frac{\beta}{2}} = 1 + \Delta x \cdot \sqrt{\frac{\tilde{R}\tilde{C}}{2}} > 1.$$

In the general case according to Vieta's theorem the zeros of χ_A simultaneously fulfill the equations

$$\lambda_1 \cdot \lambda_2 = 1 \qquad \text{and} \qquad \lambda_1 + \lambda_2 = 2 - h.$$

Due to the first equation we can write with a non-negative real value q

$$\lambda_1 = qe^{i\varphi} \qquad \text{and} \qquad \lambda_2 = q^{-1}e^{-i\varphi}.$$

The recursion can only be stable if no eigenvalue of A is larger than 1. So in this case we must have $q = 1$, but that yields to a contradiction because the value

$$\lambda_1 + \lambda_2 = 2\cos\varphi$$

would be a real number then. Thus the discretization of damped lines (using the simple π scheme) is instable for all possible frequencies ω.

Much better variants use instead of a discretization into lumped elements a discretization into lossless lines connected with resistors to get the damping right. These methods simulate the lossless parts with the help of the analytical solution of d'Alembert. Another method for stable simulations of parasitic nets can be found in [3].

The following simulation results show the effects which are introduced by the unstable discretization process. The figure on the left was obtained using 200 π-elements, the right one is based on d'Alembert's solution.

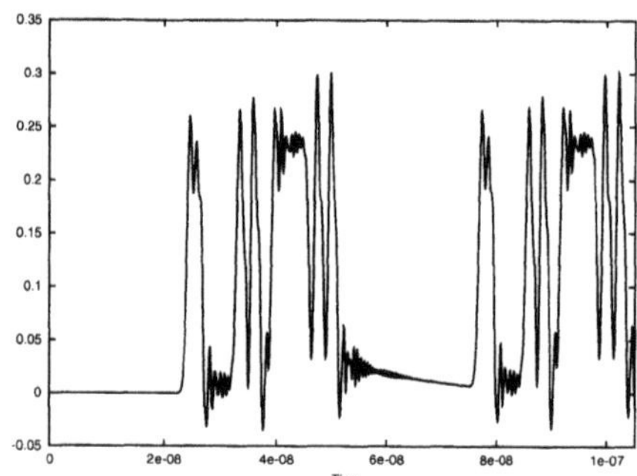

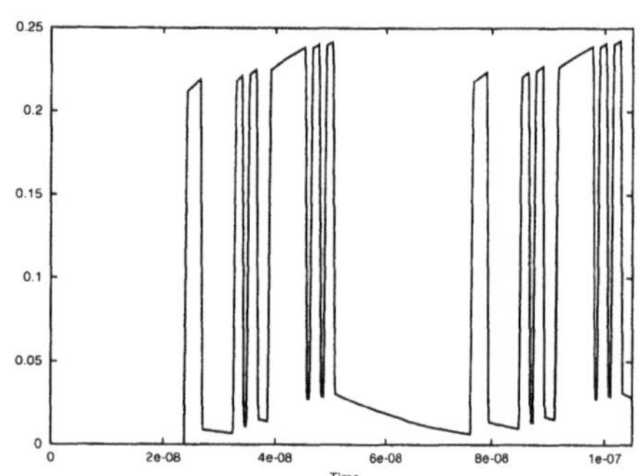

References

[1] R. Freund, P. Feldmann, *The SyMPVL Algorithm and Its Applications to Interconnect Simulation.* Numerical Analysis Manuscript (1997).

[2] R. Freund, *Passive Reduced-Order Models for Interconnect Simulation and their Computation via Krylov-Subspace Algorithms.* Numerical Analysis Manuscript (1998).

[3] M. Günther, *Partielle differential-algebraische Systeme in der numerischen Zeitbereichsanalyse elektrischer Schaltungen.* VDI Verlag GmbH, Fortschritt-Berichte VDI (2001).

[4] M. Kahlert, *Reduktion parasitärer Schaltungselemente unter Verwendung der Spektralzerlegung.* Thesis of doctorate at the Technical University of Munich, Hieronymus Buchreproduktions GmbH, München, ISBN 3-89791-244-9, 2002.

[5] K. Kerns, *Accurate and Stable Reduction of RLC Networks Using Split Congruence Transformations.* Thesis of doctorate at the University of Washington, UMI microform 9716864, 1996.

[6] K. Kerns, *Stable and Efficient Reduction of Large, Multiport RC Networks by Pole Analysis via Congruence Transformations.* IEEE Transactions on Computer-Aided Design of integrated Circuits and Systems **16** (1997) number 7.

Acknowledgment

Many thanks to Infineon Technologies AG for the support of this work.

Martin Kahlert
Sonnenstr. 16
D-85229 Markt Indersdorf, Germany
email: martin.kahlert@gmx.de

Modeling, Simulation and Optimization of Integrated Circuits
(Eds.) K. Antreich, R. Bulirsch, A. Gilg and P. Rentrop

International Series of Numerical Mathematics Vol. 146, 229–247

Modeling Circuit Systems Coupled with Distributed Semiconductor Equations

Caren Tischendorf

Abstract. For a refined network analysis, we are interested in circuit simulation including distributed models of semiconductors. We construct a mathematical model for nonlinear electric networks containing semiconductors described by the drift-diffusion equations. The focus lies on the coupling of the network DAEs and the semiconductor PDEs.
Furthermore, we study the behavior of the coupled systems with respect to time-dependent perturbations of the right-hand side using an index concept for abstract DAEs. We present a network topological criterion that guarantees index-1 systems.

1. Introduction

The miniaturization trend in the development of integrated circuits has reached a level where certain physical effects as, for example, parasitic capacities, temperature dependency and quantum effects cannot any longer be neglected. This implies a refined modeling of the devices which yields to comprehensive replacement circuits, usually called as compact models, for the circuit simulation.

The development of these compact models and the correct tuning of the big number of model parameters is nowadays so expensive that a direct coupling of circuit and device simulation becomes desirable. Naturally, a coupling of both simulations (see, e.g., [18, 27, 7, 5]) is very time consuming. However, in case of large circuits with a few critical semiconductor devices (used, e.g., in RF applications), we expect that the simulation costs of such a coupled system become sufficiently low with regard to the short development cycles.

Aiming a development of fast and reliable simulations of the coupled systems, we start with constructing and analyzing the model equations. Regarding that the sensitivity of numerical solutions of DAEs with respect to perturbations depends

---

2000 *Mathematics Subject Classification.* Primary 94C05; Secondary 65J05.
Key words and phrases. circuit, device, simulation, partial differential algebraic equations, PDAE, coupled simulation, abstract differential algebraic systems.
Supported by the DFG Research Center "Mathematics for key technologies" (FZT 86) in Berlin and by the German Federal Ministry of Education and Research (03TIM3B3).

strongly on the DAE index, we are interested in an adequate criterion for the coupled system. Therefore, we investigate the coupled system as abstract differential algebraic system (ADAS) and its index (see [14]). The ADAS index concept orientates on the tractability index for DAEs. It characterizes the sensitivity of the solution with respect to time-dependent perturbations of the right-hand side and it represents an extension of the time index introduced in [16] for time-dependent and nonlinear PDAEs. Furthermore, it is closely related to the modal index [4]. We neglect space-dependent perturbations of the right-hand side here because, in circuit simulation, the main interest is devoted to the simulation of the transient behavior of the circuit.

Considering the results concerning the coupling of different circuit models [3], we can not expect that the index of the coupled system does not exceed the index of the sub-systems. Therefore, we focused our investigations to the question under which conditions the coupled system has index one, which means that the system is well posed considering time-dependent perturbations of the right-hand side. In [2], it has been shown that the index is one for network models including the stationary drift-diffusion equations if the network itself has index 1 and all semiconductors are connected by a capacitive path. Here, we consider network models including the instationary drift-diffusion equations and we are interested in necessary as well as sufficient criteria for index-1 systems.

The present paper is organized as follows. Section 2 introduces the network models described by a differential algebraic system. In Section 3, we describe the drift-diffusion model for semiconductor elements. Section 4 is devoted to the modeling of the coupling of both systems. Finally, in Section 5, we consider the coupled system as abstract differential algebraic system and investigate the index as a measure for the behavior of the system with respect to time-dependent perturbations of the right-hand side.

2. Network modeling

The numerical simulation of electric networks is closely connected to the network modeling. Circuit models have to meet two contradicting demands. On the one hand, they have to describe the physical behavior of the circuit as correct as possible. On the other hand, the models should be as simple as possible in order to reduce the computing time to a minimum.

A well-established approach meeting both demands to a certain extent is the description of the network by the Modified Nodal Analysis (MNA). It bases on Kirchhoff's laws and the voltage-current relations of the network elements.

Introducing j as the vector of all branch currents, v as the vector of all branch voltages and e as the vector of all node potentials (voltage with respect to the mass node), we may express Kirchhoff's current law (KCL) as

$$Ai = 0, \tag{1}$$

and Kirchhoff's voltage law (KCL) as

$$v = A^{\mathrm{T}} e. \tag{2}$$

The matrix A represents the incidence matrix describing the node-to-branch relations. It is constant and has the entries 1, -1 and 0 only. The voltage-current relation of basic network elements like resistances, capacitances and inductances may be described by

$$I_R = g(v_R), \quad i_C = \frac{\mathrm{d}q_C}{\mathrm{d}t}(v_C), \quad v_L = \frac{\mathrm{d}\phi_L}{\mathrm{d}t}(i_L)$$

respectively. For the description of more complex devices like transistors, the use of so-called compact models has been proven to be very efficient. The compact models represent replacement circuits of the device using basic network elements only. In Figure 1 you see a simple compact model for a metal-oxide-semiconductor

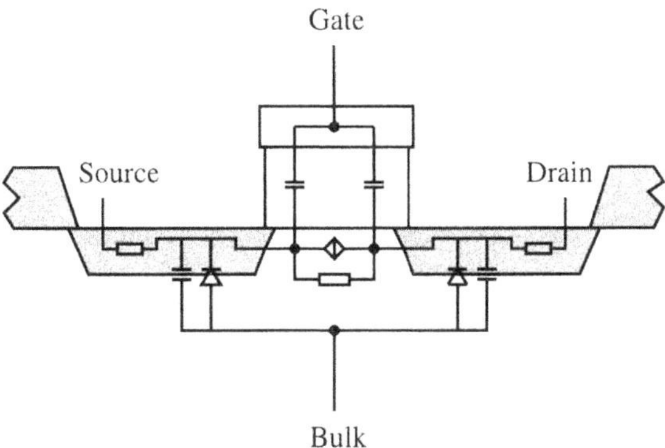

FIGURE 1. Compact model for a MOSFET

field effect transistor (MOSFET).

Finally, the MNA leads in its charge oriented form to an equation system of the form

$$\begin{aligned} A_C \frac{\mathrm{d}q(A_C^{\mathrm{T}} e)}{\mathrm{d}t} + A_R g(A_R^{\mathrm{T}} e) + A_L j_L + A_V j_V &= -A_I i_s, & (3)\\ \frac{\mathrm{d}\phi(j_L)}{\mathrm{d}t} - A_L^{\mathrm{T}} e &= 0, & (4)\\ A_V^{\mathrm{T}} e &= v_s. & (5) \end{aligned}$$

Here, q and ϕ describe the charges and fluxes, respectively. A_C describes the incidences of all capacitive branches. Correspondingly, R, L, V and I stand for resistances, inductances, voltage sources and current sources. The input functions i_S and v_S describe the current and voltage sources, respectively. For brevity, we consider only independent sources. This implies that i_S and v_S depend on time only. For a more detailed discussion of the MNA, we refer to [6].

The further miniaturization of the network elements leads to more and more complex compact models, which may contain more than 500 parameters today. The development of the compact models and the extraction of their parameters becomes so expensive that it is worth to consider the inclusion of original physical device models directly in the network simulation. The next section is devoted to the description of such models.

3. Semiconductor device modeling

Considering the literature, one finds an enormous amount of books and papers dealing with semiconductor device modeling. We want to mention here [21, 8, 9, 17, 25]. These contributions provide a comprehensive overview of the topic and focus onto the mathematical background.

Semiconductor device models describe the electron transport in the bulk of the semiconductor. In consideration of the degree of simplification one distinguishes between quantum level transport, semi-classical transport and balance equations. The first one yields the Schrödinger equation and the second one leads to the Boltzmann equation. Simplifying the Boltzmann equation further by the method of moments, we get the so-called energy balance equations (considering four moments) or the drift-diffusion equations (if we consider only two moments).

From the practical point of view, the interest in semiconductor device modeling is to replace as much laboratory testing as possible by numerical simulation in order to minimize the costs. This implies that the involved mathematical models cannot be too complicated. For most semiconductor technologies, the drift-diffusion equations seem to represent a reasonable compromise between computational efficiency and an accurate description of the underlying physics.

However with the increased miniaturization of semiconductor devices, one comes closer and closer to the limits of validity of the drift-diffusion equations. The reason for this is, on one hand, that in ever smaller devices the free carriers can not longer be modeled as a continuum. On the other hand, the drift-diffusion equations are derived through a limiting process where the mean free path of a particle tends to zero. Through miniaturization this mean free path becomes larger and larger in comparison to the size of the device. In addition, quantum mechanical effects play a more and more important role in novel device structures.

But in spite of that the drift-diffusion equations remain an important tool since the microscopic effects not described by them appear only locally. Thus, the most likely approach will be to use more sophisticated models only locally, and to use the drift-diffusion equations in the parts of the device where they are sufficient to describe the physics (usually in the bulk of the semiconductor).

Therefore, we are concentrating here onto the drift-diffusion equations considered as an important model description for the device part in coupled network and device simulation.

3.1. The drift-diffusion model for semiconductor devices

The model equations are given by the Poisson equation

$$\operatorname{div}(\varepsilon \operatorname{grad} V) = q(n - p - N) \tag{6}$$

for the electrostatic potential V and the continuity equations

$$-\partial_t n + \frac{1}{q}\operatorname{div} J_n = R(n, p, J_n, J_p, \operatorname{grad} V), \tag{7}$$

$$\partial_t p + \frac{1}{q}\operatorname{div} J_p = -R(n, p, J_n, J_p, \operatorname{grad} V) \tag{8}$$

for the electron and hole concentration, n and p, respectively. The parameter q denotes the elementary charge. Furthermore, ε represents the material-dependent dielectric constant. J_n and J_p describe the densities of the electron and hole current that is assumed to be a composition of a drift and a diffusion current. Consequently,

$$J_n = q(D_n \operatorname{grad} n - \mu_n n \operatorname{grad} V), \tag{9}$$

$$J_p = q(-D_p \operatorname{grad} p - \mu_p p \operatorname{grad} V). \tag{10}$$

Here, $qD_n \operatorname{grad} n$ describes the diffusion part of the electron current caused by the effort to achieve the chemical equilibrium. The drift part $-q\mu_n n \operatorname{grad} V$ is caused by the electric field by the effort to achieve the electrical equilibrium. The mobilities μ_n and μ_p as well as the diffusivities D_n and D_p are bounded, strictly positive functions. They may depend on position x (due to dependency on doping) and on the electrical field $\operatorname{grad} V$. For a more detailed discussion of the model equations see, e.g., [21, 22].

The system (6)–(10) represents a system of five coupled partial differential equations. The Poisson equation (6) is of elliptic type. Regarding the current density equations (9) and (10), the continuity equations (7) and (8) are of parabolic type.

Note that we assume a constant temperature. It is justified for applications with low performance devices. In case of high performance devices, one has to consider the temperature T as a variable. The drift-diffusion equations have to be completed by an energy balance equation (see, e.g., [26, 13, 1]). It is a future task to combine such energy models with the network equations. In this paper, we shall concentrate on the network-device coupling using the classical instationary drift-diffusion equations.

3.2. The boundary conditions

There are essentially three different types of materials a semiconductor is bounded by. The contacts between the network and the semiconductor are usually layers of metal. The second kind of bounding materials are insulators (e.g., oxide). Finally, they may be bounded by other semiconductors. Such semiconductor-semiconductor interfaces are called heterojunction. Here, we consider metal contacts only, since we restrict to the one-dimensional case for the first study. For a discussion of other boundary types, we refer to [20, 23].

Metal contacts are usually modeled as ohmic or Schottky contacts. Both types of contacts imply Dirichlet boundary conditions for the electrostatic potential V. Furthermore, we get Dirichlet boundary conditions for the electron and hole concentrations, n and p, in case of Ohmic contacts. For Schottky contacts, we have Neumann boundary conditions for the electron and hole current densities, J_n and J_p. For brevity and the sake of simplicity, we restrict to ohmic contacts.

Ohmic contacts are characterized by a high doping of the semiconductor. This implies a large band bending and a very thin barrier at the metal-semiconductor interface. In this case, tunneling of electrons is the dominant transport mechanism. It leads to high current densities at low voltage drops and, consequently, to a low resistance of the contact.

Since tunneling is not included in the drift-diffusion equations describing electron transport in the semiconductor volume, one should place the actual boundary for the simulation domain at the edge x_T, that means at the end of the tunneling region. At high doping concentrations, the tunneling length comprises the total depletion region, and the boundary is placed at the depletion layer edge. Consequently, we have charge neutrality at the actual boundary that means

$$n - p - N = 0.$$

Furthermore,

$$V = V_{\mathrm{ap}} + V_{\mathrm{bi}}, \tag{11}$$

where V_{ap} is the applied voltage and V_{bi} is the so-called built-in potential of the semiconductor. The built-in potential depends on semiconductor material, on doping concentration, and on temperature. Since we consider the temperature as constant, V_{bi} is a given function of the position variable x only. The applied potential V_{ap} depends on the node potentials e at the contacts.

For very high doping (ideal ohmic contact), the resistance tends to zero which implies

$$np = n_{\mathrm{i}}^2$$

with the intrinsic concentration n_i depending on material and temperature. This leads to Dirichlet boundary conditions for the electron and hole concentrations

$$n = \frac{1}{2}(\sqrt{N^2 + 4n_i^2} + N) \qquad \text{on } \Gamma, \tag{12}$$

$$p = \frac{1}{2}(\sqrt{N^2 + 4n_i^2} - N) \qquad \text{on } \Gamma. \tag{13}$$

Note, that the doping density N and the intrinsic density n_{i} depend on the position x only.

4. Coupling of the network and the device model equations

From the engineering point of view, the coupling of network and device simulation is not a new task (see, e.g., [5, 19, 27, 15]).

In contrast, the mathematical analysis of coupled network and device model equations represents a very young research field. First results have already been obtained in [10] and [9]. In [10], one semiconductor connected to a resistance has been considered. In [9] the connection of one semiconductor with a simple circuit has been studied. Simple means here that the currents entering the semiconductor may be expressed by a function of the applied voltages. In both cases, the network is treated as a special boundary condition for the semiconductor. This approach fails if more than one device belongs to the network. This is, in particular, the case for integrated circuits. However, the approach may be extended by a modification of the considered operator equation (see [23]).

More recently, networks containing uniform lossy transmission lines have been investigated in [12] and [11]. The resulting equation systems represent also coupled systems of differential algebraic equations and partial differential equations. In contrast to the case here, the PDEs are of hyperbolic type.

To the authors knowledge, an existence analysis for integrated networks containing semiconductor device models has only been developed in [2]. There, the stationary drift-diffusion model for diodes have been considered. Here, we are dealing for the first time with the instationary case.

For brevity, we shall formulate the coupled system for a network that contains exactly one semiconductor. However, one can simply verify that the following arguments remain true if one considers a network with several semiconductors.

Considering the network equations (3)–(5), the task is now to include semiconductor devices that are described by the drift-diffusion model. Consequently, the currents of these semiconductor devices have to be added to the KCL equation (3). We denote the vector of all currents leaving semiconductor devices by $\hat{j}_S$. Let b_S be the number of all these branch currents and let n_N be the number of all nodes of the network. Then, we introduce the matrix $\hat{A}_S \in L(\mathbb{R}^{b_S}, \mathbb{R}^{n_N-1})$ with the entries

$$\hat{a}_{ik} := \begin{cases} 1 & \text{if the current } j_{S_k} \text{ enters node } i, \\ 0 & \text{else.} \end{cases} \tag{14}$$

This way, the matrix $\hat{A}_S$ describes the incidence of currents of semiconductor devices. However, it differs from the other incidence matrices A_E (E corresponds to one of the basic network elements) by the fact that each column contains only entries 1 but no -1. We arrive at

$$A_C \frac{\mathrm{d}q}{\mathrm{d}t} + A_R g(A_R^{\mathrm{T}} e) + A_L j_L + A_V j_V + \hat{A}_S \hat{j}_S = -A_I i_s \tag{15}$$

instead of (3). Let Γ_k be the k-th contact. Then, the current flowing through terminal k is given by

$$j_{S_k} = \int_{\Gamma_k} J_{tot} \cdot \nu \, \mathrm{d}\sigma.$$

J_{tot} represents the total current density and ν is the outward unity normal vector at a point of Γ_k. In the semiconductor we meet three types of currents: the current of electrons, the current of holes, and the displacement current caused by the electrostatic potential. Consequently, the total current density is given by

$$J_{tot} = J_n + J_p - \varepsilon \operatorname{grad} \partial_t V.$$

This implies

$$\hat{j}_{S_k} = \int_{\Gamma_k} (J_n + J_p - \varepsilon \operatorname{grad} \partial_t V) \cdot \nu \, d\sigma. \tag{16}$$

Note that the displacement current disappears if one considers the stationary case only. But it can not be neglected in the non-stationary case in order to guarantee charge conservation (see next subsection).

The semiconductor current represents only one part of the coupling. The other part is given by the boundary condition (11) for the electrostatic potential V

$$V(x,t) = V_{\mathrm{ap}}(t) + V_{\mathrm{bi}}(x) \qquad \text{on } \Gamma_k. \tag{17}$$

$V_{\mathrm{ap}}(t)$ is the applied potential at time t. It equals $e_j(t)$ if terminal k (corresponding to Γ_k) enters node number j. Regarding equation (14), we find that

$$V_{\mathrm{ap}}(t) = \begin{cases} 0 & \text{if terminal } k \text{ enters the mass node,} \\ \hat{a}_k^{\mathrm{T}} e(t) & \text{else} \end{cases}$$

for $\hat{a}_k^{\mathrm{T}} := (\hat{a}_{1k}, \ldots, \hat{a}_{n-1,k})$. Introducing the vector $c(x)$ with the entries

$$c_k(x) = \begin{cases} 1 & \text{if } x \text{ belongs to } \Gamma_k, \\ 0 & \text{else,} \end{cases}$$

we may write equation (17) as

$$V(x,t) = c(x) \cdot \hat{A}_S^{\mathrm{T}} e(t) + W(x) \qquad \text{on } \Gamma_k. \tag{18}$$

4.1. Charge conservation and homogenization

Before we turn to the investigation of the coupled system as an abstract differential algebraic system, we want to reformulate the system as follows. First, we use the model property of charge conservation which allows the replacement of $\hat{A}_S$ by an incidence matrix A_S with the same properties as A_C, etc. Secondly, we homogenize the boundary conditions for the potential V in order to obtain function spaces that do not depend on the nodal potentials of the network.

The presented device model is charge conserving. In fact, we have

$$\operatorname{div}(J_n + J_p) = q(\partial_t n - \partial_t p)$$

if we add the continuity equations (7) and (8). Additionally,

$$\operatorname{div}(\varepsilon \operatorname{grad} \tfrac{\partial}{\partial t} V) = q(\partial_t n - \partial_t p),$$

if we differentiate the Poisson equation (6) with respect to time. This implies

$$\operatorname{div} J_{tot} = \operatorname{div}(J_n + J_p - \varepsilon \operatorname{grad} \partial_t V) = 0$$

and, by Gauss law,

$$\frac{\mathrm{d}Q}{\mathrm{d}t} = \oint_{\Gamma} J_{tot} \cdot \nu \, \mathrm{d}\sigma = \int_{\Omega} \operatorname{div} J_{tot} \, \mathrm{d}x = 0,$$

where Ω denotes the whole domain of the semiconductor. Furthermore, we get

$$\begin{aligned} \sum_k \hat{\jmath}_{S_k} &= \sum_{\substack{k \\ \Gamma_k \subseteq \Gamma_O \cup \Gamma_S}} \int_{\Gamma_k} (J_n + J_p - \varepsilon \operatorname{grad} \partial_t V) \cdot \nu \, \mathrm{d}\sigma \\ &\quad - \varepsilon \sum_{\substack{k \\ \Gamma_k \subseteq \Gamma_{MI}}} \int_{\Gamma_k} \operatorname{grad} \partial_t V \cdot \nu \, \mathrm{d}\sigma = \oint_{\Gamma} J_{tot} \cdot \nu \, \mathrm{d}\sigma = 0. \end{aligned}$$

That means, that the current flowing through one terminal of the semiconductor may be described by the negative sum of the currents flowing through the other terminals. We choose one terminal (usual the bulk terminal) and call it the reference terminal. We delete the current of the reference terminal from $\hat{\jmath}_S$ and denote the resulting vector by j_S. This implies

$$\hat{A}_S \hat{\jmath}_S = A_S j_S$$

if the entries of the matrix A_S are defined as

$$a_{ik} := \begin{cases} 1 & \text{if the current } j_{S_k} \text{ enters node } i, \\ -1 & \text{if the reference terminal is connected to node } i, \\ 0 & \text{else.} \end{cases}$$

This way A_S has the same form as the other incidence matrices A_C, etc. Including the row for the mass node, each column of A_S has exactly one 1 and one -1. Finally, we arrive at

$$A_C C A_C^{\mathrm{T}} \frac{\mathrm{d}}{\mathrm{d}t} e + A_R G A_R^{\mathrm{T}} e + A_L j_L + A_V j_V + A_S j_S + A_I i_s = 0 \tag{19}$$

instead of equation (15). Obviously, equation (16) is equivalent to

$$j_{S_k} = \int_{\Gamma_k} (J_n + J_p - \varepsilon \operatorname{grad} \partial_t V) \cdot \nu \, \mathrm{d}\sigma. \tag{20}$$

Note that, in case of several semiconductors, one has to choose one reference terminal for each semiconductor and to follow the procedure for each semiconductor.

Considering existence results and Galerkin approaches for partial differential equations, Dirichlet boundary conditions are usually treated by a suitable choice of a function space as solution space. In our case, the Dirichlet boundary conditions (18) for the Poisson equation depend on the node potentials that are described only implicitly by the network equations. But it seems not comfortable to use a function space that is given implicitly only. Therefore, we shall homogenize the conditions (18) as follows.

From now we restrict to the one-dimensional case. Then, the semiconductor has only two boundary contacts. Without loss of generality, we choose $x = 0$ as the left and $x = l$ as the right boundary, where l is the length of the semiconductor. Let f be a smooth function such that

$$f(0) = 1, \quad f(1) = 0, \quad f'(0) = 0. \tag{21}$$

If the reference terminal enters node e_j, then we have

$$c \cdot \hat{A}_S^{\mathrm{T}} e = e_j + f(x) A_S^{\mathrm{T}} e \qquad \text{on } \Gamma.$$

Introducing

$$\tilde{V}(x,t) := V(x,t) - e_j(t) - f(x) \cdot A_S^{\mathrm{T}} e(t) - V_{\mathrm{bi}}(x), \tag{22}$$

we get homogeneous boundary conditions for $\tilde{V}$. Introducing

$$\tilde{R}(n, p, J_n, J_V, \operatorname{grad} \tilde{V}) := R(n, p, J_n, J_V, \operatorname{grad} V)$$

and summarizing all equations we arrive at the coupled system

$$A_C \frac{\mathrm{d}q(A_C^{\mathrm{T}} e)}{\mathrm{d}t} + A_R g(A_R^{\mathrm{T}} e) + A_L j_L + A_V j_V + A_S j_S = -A_I i_s, \tag{23}$$

$$\frac{\mathrm{d}\phi(j_L)}{\mathrm{d}t} - A_L^{\mathrm{T}} e = 0, \tag{24}$$

$$A_V^{\mathrm{T}} e = v_s, \tag{25}$$

$$\operatorname{div}(\varepsilon \operatorname{grad} \tilde{V}) = q(n - p - N) - \operatorname{div}(\varepsilon \operatorname{grad}(f A_S^{\mathrm{T}} e + V_{\mathrm{bi}})), \tag{26}$$

$$\partial_t n - \tfrac{1}{q} \operatorname{div} J_n = -\tilde{R}(n, p, J_n, J_p, \operatorname{grad} \tilde{V}), \tag{27}$$

$$\partial_t p + \tfrac{1}{q} \operatorname{div} J_p = -\tilde{R}(n, p, J_n, J_p, \operatorname{grad} \tilde{V}), \tag{28}$$

$$J_n = q(D_n \operatorname{grad} n - \mu_n n \operatorname{grad}(\tilde{V} + f A_S^{\mathrm{T}} e + V_{\mathrm{bi}})), \tag{29}$$

$$J_p = q(-D_p \operatorname{grad} p - \mu_p p \operatorname{grad}(\tilde{V} + f A_S^{\mathrm{T}} e + V_{\mathrm{bi}})), \tag{30}$$

$$j_S = [(J_n + J_p) - \varepsilon \operatorname{grad}(\partial_t \tilde{V})]_{x=0} \tag{31}$$

with the boundary conditions

$$\tilde{V} = 0, \quad n = \tfrac{1}{2}(\sqrt{N^2 + 4n_i^2} + N), \quad p = \tfrac{1}{2}(\sqrt{N^2 + 4n_i^2} - N). \tag{32}$$

5. The ADAS index of the coupled system

In order to treat coupled systems of partial differential equations and differential-algebraic equations in a systematic way, abstract differential algebraic systems (ADASs) of the form

$$\mathcal{A}(t) \frac{\mathrm{d}}{\mathrm{d}t}(\mathcal{D}(t) u(t)) + \mathcal{B}(t) u(t) = q(t) \qquad \text{for } t \in [t_0, T] \tag{33}$$

have been investigated in [14]. This equation is to be understood as an operator equation with operators $\mathcal{A}(t)$, $\mathcal{D}(t)$ and $\mathcal{B}(t)$ acting in real Hilbert spaces. More precisely, let X, Y, Z and $\tilde{Z}$ be Hilbert spaces and

$$\mathcal{A}(t) : Z \to Y, \quad \mathcal{D}(t) : X \to Z, \quad \mathcal{B}(t) : X \to Y.$$

As we will see later, we may formulate the coupled problem (23)–(32) as abstract DAEs with operators $\mathcal{A}$, $\mathcal{D}$ and $\mathcal{B}$ that do not explicitly depend on the time t. Therefore, we omit the argument t for these operators in the following. Having the coupled problem in mind, we assume $\mathcal{A}$ to be linear.

From the finite-dimensional case we know that the sensitivity of solutions of DAEs with respect to perturbations depends on its index. Since we are interested in the solution behavior of the coupled system, we want to follow the concept in [14] extending the tractability index to abstract DAEs. Therefore, we assume that the Fréchet derivatives $\mathcal{B}_0$ and $\mathcal{D}_0$ of the operators $\mathcal{B}$ and $\mathcal{D}$ exist. Furthermore, we introduce $\mathcal{G}_0 := \mathcal{A}\mathcal{D}_0$.

Then, the abstract differential algebraic system

$$\mathcal{A}\frac{\mathrm{d}}{\mathrm{d}t}(\mathcal{D}u(t)) + \mathcal{B}u(t) = q(t) \qquad \text{for } t \in [t_0, T] \tag{34}$$

is called to have the index μ if

(i) $\dim(\operatorname{im}\mathcal{W}_i) > 0$ for all $i = 0, \ldots, \mu - 1$, $\dim(\operatorname{im}\mathcal{W}_\mu) = 0$ and
(ii) $\mathcal{G}_\mu$ is injective,

where $\mathcal{W}_i$ and $\mathcal{G}_i$ are recursively defined as follows. $\mathcal{W}_i$ is a (bounded) projector with

$$\ker\mathcal{W}_i = \overline{\operatorname{im}\mathcal{G}_i}.$$

Furthermore,

$$\mathcal{G}_{i+1} = \mathcal{G}_i + \mathcal{B}_i\mathcal{Q}_i, \qquad \mathcal{B}_{i+1} = \mathcal{B}_i\mathcal{P}_i,$$

where $\mathcal{P}_i = \mathcal{I} - \mathcal{Q}_i$ and $\mathcal{Q}_i$ is a (bounded) projector with $\operatorname{im}\mathcal{Q}_i = \overline{\ker\mathcal{G}_i}$.

We want to note here, that the so defined index for an abstract differential algebraic system shall characterize the behavior of the system with respect to time-dependent perturbations of the right-hand side. It should not be confused with the Fredholm index of operators.

Omitting the tilde notation for the potential V of the coupled system (23)–(31) and regarding the boundary conditions (32), we are interested in solutions

$$u(t) = (e(t), j_L(t), j_V(t), j_S(t), V(\cdot,t), n(\cdot,t), p(\cdot,t), J_n(\cdot,t), J_p(\cdot,t))$$

that belong to

$$\begin{aligned} X \;:=\; & \mathbb{R}^{n_N-1} \times \mathbb{R}^{n_L} \times \mathbb{R}^{n_V} \times \mathbb{R} \times (H^2(\Omega) \cap H_0^1(\Omega)) \times H^1(\Omega) \times H^1(\Omega) \times \\ & H(\operatorname{div};\Omega) \times H(\operatorname{div};\Omega). \end{aligned}$$

for the coupled system (23)–(31). Here, n_N, n_L and n_V denote the number of nodes, the number of inductances and the number of voltage sources of the network, respectively. Recall that $\Omega = (0, l)$. We need the higher regularity for V since we have to evaluate the gradient of V at the boundary for the determination of the semiconductor current j_S. Furthermore, we choose

$$\begin{aligned} Y &:= \mathbb{R}^{n_N-1} \times \mathbb{R}^{n_L} \times \mathbb{R}^{n_V} \times (L_2(\Omega))^5 \times \mathbb{R}, \\ Z &:= \mathbb{R}^{n_C} \times \mathbb{R}^{n_L} \times \mathbb{R}^{n_V} \times \mathbb{R} \times (H^1(\Omega))^2, \end{aligned}$$

where n_C denotes the number of capacities in the network. Note that then we have to choose $(L_2(\Omega))^{5n_S}$ and $(H^1(\Omega))^{2n_S}$ instead of $(L_2(\Omega))^5$ and $(H^1(\Omega))^2$, respectively, if the network contains not only one but n_S semiconductors. Correspondingly, the space X has also to be extended.

We may write the coupled system (23)–(31) as ADAS

$$\mathcal{A}\frac{\mathrm{d}}{\mathrm{d}t}(\mathcal{D}u(t)) + \mathcal{B}u(t) = q(t) \qquad \text{for } t \in [t_0, T] \tag{35}$$

with

$$\mathcal{A} = \begin{pmatrix} A_C & 0 & 0 & 0 & 0 \\ 0 & I & 0 & 0 & 0 \\ 0 & 0 & 0 & 0 & 0 \\ 0 & 0 & 0 & 0 & 0 \\ 0 & 0 & 0 & I & 0 \\ 0 & 0 & 0 & 0 & I \\ 0 & 0 & 0 & 0 & 0 \\ 0 & 0 & 0 & 0 & 0 \\ 0 & 0 & I & 0 & 0 \end{pmatrix}, \quad \mathcal{D}(u) = \begin{pmatrix} q(A_C^{\mathrm{T}} e) \\ \phi(j_L) \\ -\mathfrak{r}\,\varepsilon \operatorname{grad} V \\ n \\ p \end{pmatrix}, \quad q = \begin{pmatrix} -A_I i_s \\ 0 \\ v_s \\ 0 \\ 0 \\ 0 \\ 0 \\ 0 \\ 0 \end{pmatrix},$$

and

$$\mathcal{B}(u) = \begin{pmatrix} A_R g(A_R^{\mathrm{T}} e) + A_L j_L + A_V j_V + A_S j_S \\ -A_L^{\mathrm{T}} e \\ A_V^{\mathrm{T}} e \\ \operatorname{div}(\varepsilon \operatorname{grad} \tilde{V}) - q(n-p-N) + \operatorname{div}(\varepsilon \operatorname{grad}(f A_S^{\mathrm{T}} e + V_{\mathrm{bi}})) \\ -\frac{1}{q} \operatorname{div} J_n + \tilde{R}(n,p,J_n,J_p,\operatorname{grad} V) \\ \frac{1}{q} \operatorname{div} J_p + \tilde{R}(n,p,J_n,J_p,\operatorname{grad} V) \\ J_n - q(D_n \operatorname{grad} n - \mu_n n \operatorname{grad}(V + f A_S^{\mathrm{T}} e + V_{\mathrm{bi}})) \\ J_p + q(-D_p \operatorname{grad} p - \mu_p p \operatorname{grad}(V + f A_S^{\mathrm{T}} e + V_{\mathrm{bi}})) \\ j_S - \mathfrak{r}\, J_n - \mathfrak{r} J_p \end{pmatrix}$$

where $\mathfrak{r}$ is the restriction operator onto the left boundary, i.e.,

$$\mathfrak{r}J(\cdot, t) := J(0, t).$$

From [24] we know that the network DAEs have an index ≤ 1 if the network neither contains LI-cutsets nor CV-loops with at least one voltage source. LI-cutsets are cutsets including inductances and current sources only. CV-loops are loops containing capacitances and voltage sources only. As we will see in the next theorem, LI-cutsets are also relevant for the coupled systems. However CV-loops have to be replaced by CVS-loops, i.e., loops consisting of capacitances, voltage sources and semiconductors.

Remark 5.1. The network does not contain LI-cutsets if and only if the matrix

$$(A_C, A_R, A_V, A_S) \quad \text{has full row rank.} \tag{36}$$

The network does not contain CVS-loops with at least one voltage source or one semiconductor if and only if the matrix

$$(Q_C^T A_V, Q_C^T A_S) \quad \text{has full column rank} \tag{37}$$

for any projector Q_C onto $\ker A_C^{\mathrm{T}}$.

Furthermore, we will need the capacitance, resistance and inductance matrices

$$C(v) := \frac{\mathrm{d}q(v)}{\mathrm{d}v}, \quad G(v) := \frac{\mathrm{d}g(v)}{\mathrm{d}v}, \quad \textit{and} \quad L(j) := \frac{\mathrm{d}\phi(j)}{\mathrm{d}j},$$

respectively.

Theorem 5.2. *Let $C(v)$, $G(v)$ and $L(j)$ be positive definite for all voltages v and currents j. Then, the ADAS (35) has index 1 if and only if the network contains neither LI-cutsets nor CVS-loops with at least one voltage source or one semiconductor.*

Proof: We start with the Fréchet derivatives

$$\mathcal{D}_0(u) = \begin{pmatrix} C(A_C^{\mathrm{T}}e)A_C^{\mathrm{T}}e & 0 & 0 & 0 & 0 & 0 & 0 & 0 & 0 \\ 0 & L(j_L)j_L & 0 & 0 & 0 & 0 & 0 & 0 & 0 \\ 0 & 0 & 0 & 0 & -\tau\,\varepsilon\mathrm{grad} & 0 & 0 & 0 & 0 \\ 0 & 0 & 0 & 0 & 0 & I & 0 & 0 & 0 \\ 0 & 0 & 0 & 0 & 0 & 0 & I & 0 & 0 \end{pmatrix}$$

and

$$\mathcal{B}_0(u) = \begin{pmatrix} A_R G(A_R^{\mathrm{T}}e)A_R^{\mathrm{T}} & A_L & A_V & A_S & 0 & 0 & 0 & 0 & 0 \\ -A_L^{\mathrm{T}} & 0 & 0 & 0 & 0 & 0 & 0 & 0 & 0 \\ A_V^{\mathrm{T}} & 0 & 0 & 0 & 0 & 0 & 0 & 0 & 0 \\ \varepsilon\Delta(f)A_S^{\mathrm{T}} & 0 & 0 & 0 & \varepsilon\Delta & -q & q & 0 & 0 \\ 0 & 0 & 0 & 0 & \tilde{R}_V(u)\mathrm{grad} & \tilde{R}_n(u) & \tilde{R}_p(u) & \tilde{R}_{J_n}(u) - \frac{1}{q}\,\mathrm{div} & \tilde{R}_{J_p}(u) \\ 0 & 0 & 0 & 0 & \tilde{R}_V(u)\mathrm{grad} & \tilde{R}_n(u) & \tilde{R}_p(u) & \tilde{R}_{J_n}(u) & \tilde{R}_{J_p}(u) + \frac{1}{q}\,\mathrm{div} \\ q\mu_n n\mathrm{grad}\,(f)A_S^{\mathrm{T}} & 0 & 0 & 0 & q\mu_n n\mathrm{grad} & \mathrm{j}_n(u) & 0 & I & 0 \\ q\mu_p p\mathrm{grad}\,(f)A_S^{\mathrm{T}} & 0 & 0 & 0 & q\mu_p p\mathrm{grad} & 0 & \mathrm{j}_p(u) & 0 & I \\ 0 & 0 & 0 & I & 0 & 0 & 0 & -\tau & -\tau \end{pmatrix}$$

For brevity, we have introduced

$$\begin{aligned} \mathrm{j}_n(u) &:= q\mu_n\mathrm{grad}\,(V + fA_S^{\mathrm{T}}e + V_{\mathrm{bi}}) - qD_n\mathrm{grad}\,, \\ \mathrm{j}_p(u) &:= q\mu_p\mathrm{grad}\,(V + fA_S^{\mathrm{T}}e + V_{\mathrm{bi}}) + qD_p\mathrm{grad}\,. \end{aligned}$$

Then, the image space of

$$\mathcal{G}_0(u) = \begin{pmatrix} A_C C(A_C^{\mathrm{T}}e)A_C^{\mathrm{T}} & 0 & 0 & 0 & 0 & 0 & 0 & 0 & 0 \\ 0 & L(j_L) & 0 & 0 & 0 & 0 & 0 & 0 & 0 \\ 0 & 0 & 0 & 0 & 0 & 0 & 0 & 0 & 0 \\ 0 & 0 & 0 & 0 & 0 & 0 & 0 & 0 & 0 \\ 0 & 0 & 0 & 0 & 0 & I & 0 & 0 & 0 \\ 0 & 0 & 0 & 0 & 0 & 0 & I & 0 & 0 \\ 0 & 0 & 0 & 0 & 0 & 0 & 0 & 0 & 0 \\ 0 & 0 & 0 & 0 & 0 & 0 & 0 & 0 & 0 \\ 0 & 0 & 0 & 0 & -\tau\,\varepsilon\mathrm{grad} & 0 & 0 & 0 & 0 \end{pmatrix}$$

is given by

$$\mathrm{im}\, A_C^{\mathrm{T}} \times \mathbb{R}^{n_L} \times 0 \times 0 \times H^1(\Omega) \times H^1(\Omega) \times 0 \times 0 \times \mathbb{R}.$$

Obviously,

$$\mathcal{W}_0 = \begin{pmatrix} Q_C^{\mathrm{T}} & & & & & & & & \\ & 0 & & & & & & & \\ & & I & & & & & & \\ & & & I & & & & & \\ & & & & 0 & & & & \\ & & & & & 0 & & & \\ & & & & & & I & & \\ & & & & & & & I & \\ & & & & & & & & 0 \end{pmatrix}$$

is a projection operator along $\overline{\operatorname{im} \mathcal{G}_0(u)}$ for any projector Q_C onto the nullspace of A_C^{T}. Furthermore, $\mathcal{G}_0(u)$ has the nontrivial nullspace

$$\ker A_C^{\mathrm{T}} \times 0 \times \mathbb{R}^{nV} \times \mathbb{R}^{nS} \times \mathcal{N}_{\mathfrak{t}\,\mathrm{grad}} \times 0 \times 0 \times H^1(\Omega) \times H^1(\Omega)$$

where

$$\mathcal{N}_{\mathfrak{t}\,\mathrm{grad}} = \{v \in H^2(\Omega) \cap H_0^1(\Omega) : \ \mathfrak{t}\,\mathrm{grad}\, v = 0\}.$$

We choose

$$\begin{array}{rccc} \mathcal{Q}_{\mathfrak{t}\,\mathrm{grad}} : & H^2(\Omega) \cap H_0^1(\Omega) & \to & H^2(\Omega) \cap H_0^1(\Omega) \\ & u & \mapsto & v \end{array}$$

with

$$v(x) = u(x) - \frac{x(l-x)}{l}\mathfrak{t}\,\mathrm{grad}\, u \qquad \forall x \in \Omega$$

as projection operator onto $\mathcal{N}_{\mathfrak{t}\,\mathrm{grad}}$. Then,

$$\mathcal{Q}_0 = \begin{pmatrix} Q_C & 0 & 0 & 0 & 0 & 0 & 0 & 0 & 0 \\ 0 & 0 & 0 & 0 & 0 & 0 & 0 & 0 & 0 \\ 0 & 0 & I & 0 & 0 & 0 & 0 & 0 & 0 \\ 0 & 0 & 0 & I & 0 & 0 & 0 & 0 & 0 \\ 0 & 0 & 0 & 0 & \mathcal{Q}_{\mathfrak{t}\,\mathrm{grad}} & 0 & 0 & 0 & 0 \\ 0 & 0 & 0 & 0 & 0 & 0 & 0 & 0 & 0 \\ 0 & 0 & 0 & 0 & 0 & 0 & 0 & 0 & 0 \\ 0 & 0 & 0 & 0 & 0 & 0 & 0 & I & 0 \\ 0 & 0 & 0 & 0 & 0 & 0 & 0 & 0 & I \end{pmatrix}$$

is a projection operator onto $\ker \mathcal{G}_0$. It yields

$$\mathcal{G}_1(u) = \left(\begin{array}{ccccccccc} A_C C(A_C^{\mathrm{T}} e)A_C^{\mathrm{T}} + A_R G(A_R^{\mathrm{T}} e)A_R^{\mathrm{T}} Q_C & 0 & A_V & A_S & 0 & 0 & 0 & 0 & 0 \\ -A_L^{\mathrm{T}} Q_C & L(j_L) & 0 & 0 & 0 & 0 & 0 & 0 & 0 \\ A_V^{\mathrm{T}} Q_C & 0 & 0 & 0 & 0 & 0 & 0 & 0 & 0 \\ \varepsilon \Delta(f) A_S^{\mathrm{T}} Q_C & 0 & 0 & 0 & \varepsilon \Delta \mathcal{Q}_{\mathfrak{t}\,\mathrm{grad}} & 0 & 0 & 0 & 0 \\ 0 & 0 & 0 & 0 & \tilde{R}_V(u)\mathrm{grad}\,\mathcal{Q}_{\mathfrak{t}\,\mathrm{grad}} & I & 0 & \tilde{R}_{J_n}(u) - \frac{1}{q}\,\mathrm{div} & \tilde{R}_{J_p}(u) \\ 0 & 0 & 0 & 0 & \tilde{R}_V(u)\mathrm{grad}\,\mathcal{Q}_{\mathfrak{t}\,\mathrm{grad}} & 0 & I & \tilde{R}_{J_n}(u) & \tilde{R}_{J_p}(u) + \frac{1}{q}\,\mathrm{div} \\ q\mu_n n\,\mathrm{grad}\,(f) A_S^{\mathrm{T}} Q_C & 0 & 0 & 0 & q\mu_n n\,\mathrm{grad}\,\mathcal{Q}_{\mathfrak{t}\,\mathrm{grad}} & 0 & 0 & I & 0 \\ q\mu_p p\,\mathrm{grad}\,(f) A_S^{\mathrm{T}} Q_C & 0 & 0 & 0 & q\mu_p p\,\mathrm{grad}\,\mathcal{Q}_{\mathfrak{t}\,\mathrm{grad}} & 0 & 0 & 0 & I \\ 0 & 0 & 0 & I & -\mathfrak{t}\,\varepsilon\mathrm{grad} & 0 & 0 & -\mathfrak{t} & -\mathfrak{t} \end{array}\right)$$

(ii) We show that $\mathcal{G}_1(u)$ is not injective if the network contains LI-cutsets or CVS-loops. If the network contains an LI-cutset, then we find a nontrivial $w \in \mathbb{R}^{n_N-1}$ with

$$A_C^{\mathrm{T}} w = 0,\ A_R^{\mathrm{T}} w = 0,\ A_V^{\mathrm{T}} w = 0,\ A_S^{\mathrm{T}} w = 0.$$

It implies $w = Q_C w$ and, consequently,

$$v = (w, L^{-1}(j_L) A_L^{\mathrm{T}} w, 0, 0, 0, 0, 0, 0, 0)$$

belongs to the nullspace of $\mathcal{G}_1(u)$. Conversely, if the network contains a CVS-loop, then we find a nontrivial $(w_1, w_2, w_3) \in \mathbb{R}^{n_C} \times \mathbb{R}^{n_V} \times \mathbb{R}^{n_S}$ with

$$A_C w_1 + A_V w_2 + A_S w_3 = 0.$$

Since $C(u)$ is positive definite, we find a $w_4 \in \operatorname{im} P_C$ such that

$$A_C w_1 = A_C C(A_C^{\mathrm{T}} e) A_C^{\mathrm{T}} w_4.$$

Choosing $w_5 \in H^2(\Omega) \cap H_0^1(\Omega)$ with

$$w_5(x) = \frac{x(l-x)}{\varepsilon\, l} w_3$$

the vector function

$$v = (w_4, 0, w_2, w_3, w_5, 0, 0, 0, 0)$$

belongs to the nullspace of $\mathcal{G}_1(u)$. The existence of at least one voltage source or one semiconductor in the CVS-loop ensures that the constructed v is nontrivial because at least w_2 or w_3 is nontrivial.

(ii) Now we show that $\mathcal{G}_1(u)$ is injective if the network contains neither LI-cutsets nor CVS-loops with at least one voltage source or one semiconductor. We assume that

$$\mathcal{G}_1(u) w = 0. \tag{38}$$

Integration of the 4th line

$$\varepsilon \Delta f A_S^{\mathrm{T}} Q_C w_1 + \varepsilon \Delta Q_{\mathfrak{r}\,\mathrm{grad}}\, w_5 = 0$$

yields to

$$\operatorname{grad} f A_S^{\mathrm{T}} Q_C w_1 + \operatorname{grad} Q_{\mathfrak{r}\,\mathrm{grad}}\, w_5 = \mathfrak{r} \operatorname{grad} f A_S^{\mathrm{T}} Q_C w_1 = 0,$$

since $\mathfrak{r} \operatorname{grad} f = 0$ for the chosen function f (see (21)). Integrating once again, we obtain

$$f A_S^{\mathrm{T}} Q_C w_1 + Q_{\mathfrak{r}\,\mathrm{grad}}\, w_5 = \mathfrak{r}\, f A_S^{\mathrm{T}} Q_C w_1 + \mathfrak{r}\, Q_{\mathfrak{r}\,\mathrm{grad}}\, w_5 = A_S^{\mathrm{T}} Q_C w_1, \tag{39}$$

because $\mathfrak{r}\, f = 1$ (cf. (21)) and $Q_{\mathfrak{r}\,\mathrm{grad}}\, w_5$ belongs to $H_0^1(\Omega)$. Evaluating this equation at the point $x = l$, we arrive at

$$0 = A_S^{\mathrm{T}} Q_C w_1.$$

From the 3rd line of (38), we know that $A_V^{\mathrm{T}} Q_C w_1 = 0$. The 1st line of (38) implies

$$Q_C^{\mathrm{T}} A_R G(u) A_R^{\mathrm{T}} Q_C w_1 + Q_C^{\mathrm{T}} A_V w_3 + Q_C^{\mathrm{T}} A_S w_4 = 0.$$

Regarding the two relations for w_1 above and the fact that $G(u)$ is positive definite, we obtain

$$(A_C A_R A_V A_S)^{\mathrm{T}} Q_C w_1 = 0, \quad \text{i.e.,} \quad Q_C w_1 = 0, \tag{40}$$

since the network does not contain LI-cutsets. Furthermore, using the 2nd and the 5th–8th line of (38) as well as the fact that $L(u)$ is positive definite, we get

$$w_2 = 0, \quad w_8 = w_9 = 0, \quad w_6 = w_7 = 0.$$

The 1st line of (38) reads as

$$A_C C(u) A_C^{\mathrm{T}} w_1 + A_V w_3 + A_S w_4 = 0$$

now and implies

$$A_C^{\mathrm{T}} w_1 = 0, \quad w_3 = 0, \quad w_4 = 0,$$

since $C(u)$ is positive definite and the network does not contain CVS-loops with at least one voltage source or one semiconductor. Together with (40) and the 9th line of (38), this yields to

$$w_1 = 0, \quad \mathfrak{r}\,\mathrm{grad}\, w_5 = 0.$$

Regarding (39) and (40), we find

$$Q_{\mathfrak{r}\,\mathrm{grad}}\, w_5 = 0 \tag{41}$$

and, consequently, $w_5 = 0$. That means that $\mathcal{G}_1(u)$ is injective.

(iii) It remains to show that $\overline{\mathrm{im}\,\mathcal{G}_1(u)} = Y$ if the network contains neither LI-cutsets nor CVS-loops with at least one voltage source or one semiconductor. Since $C^\infty(\Omega)$ is dense in $L_2(\Omega)$, it is enough to show that

$$\mathbb{R}^{n-1} \times \mathbb{R}^{n_L} \times \mathbb{R}^{n_V} \times (C^\infty(\Omega))^5 \times \mathbb{R} \subseteq \mathrm{im}\,\mathcal{G}_1(u).$$

Let y belong to $\mathbb{R}^{n-1} \times \mathbb{R}^{n_L} \times \mathbb{R}^{n_V} \times (C^\infty(\Omega))^5 \times \mathbb{R}$. With similar arguments like above, one can show that the matrix

$$\begin{pmatrix} A_C C(u) A_C^{\mathrm{T}} + A_R G(u) A_R^{\mathrm{T}} Q_C & A_V & A_S \\ A_V^{\mathrm{T}} Q_C & 0 & 0 \\ A_S^{\mathrm{T}} Q_C & 0 & 0 \end{pmatrix}$$

is nonsingular because the network contains neither LI-cutsets nor CVS-loops with at least one voltage source or one semiconductor. Consequently, we find w_1, w_3 and w_4 such that

$$\begin{aligned} A_C C(u) A_C^{\mathrm{T}} w_1 + A_R G(u) A_R^{\mathrm{T}} Q_C w_1 & \\ + A_V w_3 + A_S w_4 &= y_1, \\ A_V^{\mathrm{T}} Q_C w_1 &= y_3, \\ A_S^{\mathrm{T}} Q_C w_1 &= -\tfrac{1}{\varepsilon} \int_0^l \int_0^\tau y_4(s)\, \mathrm{d}s\, \mathrm{d}\tau. \end{aligned} \tag{42}$$

Defining

$$\begin{aligned}
w_2 &= L^{-1}(u)y_2 + L^{-1}(u)A_L^{\mathrm{T}}Q_C w_1, \\
w_5(x) &= \tfrac{1}{\varepsilon}\int_0^x \int_0^\tau y_4(s)\,\mathrm{d}s\,\mathrm{d}\tau + (1-f(x))A_S^{\mathrm{T}}Q_C w_1 \\
&\qquad + \tfrac{x(l-x)}{\varepsilon\, l}(w_4 - y_9 - \mathfrak{r}\, y_7 - \mathfrak{r}\, y_8), \\
w_8 &= y_7 - \tfrac{q}{\varepsilon}\mu_n n \int_0^x y_4(s)\,\mathrm{d}s, \\
w_9 &= y_8 - \tfrac{q}{\varepsilon}\mu_p p \int_0^x y_4(s)\,\mathrm{d}s, \\
w_6 &= y_5 - \tilde{R}_V(u)\mathrm{grad}\, Q_{\mathfrak{r}\,\mathrm{grad}}\, w_5 - \tilde{R}_{J_n}(u)w_8 \\
&\quad + \tfrac{1}{q}\mathrm{div}\, w_8 - \tilde{R}_{J_p}(u)w_9, \\
w_7 &= y_6 - \tilde{R}_V(u)\mathrm{grad}\, Q_{\mathfrak{r}\,\mathrm{grad}}\, w_5 - \tilde{R}_{J_n}(u)w_8 \\
&\quad - \tilde{R}_{J_p}(u)w_9 - \tfrac{1}{q}\mathrm{div}\, w_9
\end{aligned}$$

successively, we obtain $w \in X$ and $\mathcal{G}_1(u)w = y$. Note that equation (42) ensures $w_5(l)$ to be equal zero in order to guarantee $w_5 \in H_0^1(\Omega)$. □

Conclusions

The further miniaturization demands a refined network analysis describing certain semiconductor elements by contributed models. Using the instationary drift-diffusion model, the device equations represent a system of elliptic and parabolic differential equations. The network is described by a differential-algebraic system. Both systems are mutually coupled via boundary conditions.

The coupled system can be analyzed as abstract differential algebraic system. For the one-dimensional case (with respect to position), network topological criteria are described that guarantee index-1 systems. It is still an open question for the higher-dimensional case. Furthermore, it will be of interest under which conditions the coupled system has index 2.

The presented index results form a basis for the discussion of suitable discretizations of the coupled system. Following the method of lines one should look for space discretizations that imply a DAE with the same index as the abstract DAE has.

Acknowledgment

The author wishes to thank Prof. R. März, Prof. G. Wachutka, Dr. U. Feldmann, Dr. M. Günther, Dr. R. Lamour and A. Bartel for very fruitful discussions.

References

[1] G. Albinus, H. Gajewski, and R. Hünlich. *Thermodynamic design of energy models of semiconductor devices.* Nonlinearity **15(2)** (2002), 367–383.

[2] G. Alì, A. Bartel, M. Günther, and C. Tischendorf. *Elliptic partial differential-algebraic multiphysics models in electrical network design.* Technical Report **02/05**, Institute of Scientific Computing and Mathematical Modeling, University of Karlsruhe, 2002. To appear in Math. Models Meth. Appl. Sci. 2003.

[3] M. Arnold and M. Günther. *Preconditioned dynamic iteration for coupled differential-algebraic systems.* BIT Numerical Mathematics **41** (2001), 1–25.

[4] S.L. Campbell and W. Marszalek. *The Index of an Infinite-Dimensional Implicit System.* Math. Comput. Model. Dyn. Syst. **5(1)** (1999), 18–42.

[5] W.L. Engl, R. Laur, and H.K. Dirks. *MEDUSA – A simulator for modular circuits.* IEEE Trans. CAD **1(2)** (1982), 85–93.

[6] D. Estévez Schwarz and C. Tischendorf. *Structural analysis of electric circuits and consequences for MNA.* Int. J. Circ. Theor. Appl. **28** (2000), 131–162.

[7] H. Gajewski. *Analysis und Numerik des Ladungstransports in Halbleitern.* GAMM Mitteilungen **16** (1993), 35–57.

[8] H. Gajewski and K. Gröger. *On the basic equations for carrier transport in semiconductors.* J. Math. Anal. Appl. **113** (1986), 12–35.

[9] H. Gajewski and K. Gröger. *Semiconductor equations for variable mobilities based on Boltzmann statistics or Fermi-Dirac statistics.* Math. Nachr. **140** (1989), 7–36.

[10] K. Gröger. *Initial boundary value problems from semiconductor device theory.* ZAMM **67(8)** (1987), 345–355.

[11] M. Günther. *Partielle differential-algebraische Systeme in der numerischen Zeitbereichsanalyse elektrischer Schaltungen.* Number 343 in Fortschritt-Berichte VDI, Reihe 20, Rechnerunterstützte Verfahren. VDI Verlag, Düsseldorf, 2001. Habilitation.

[12] M. Günther. *A PDAE model for interconnected linear RLC networks.* Mathematical and Computer Modelling of Dynamical Systems **7(2)** (2001), 189–203.

[13] A. Jüngel. *Quasi-hydrodynamic Semiconductor Equations.* Birkhäuser, Basel, 2001.

[14] R. Lamour, R. März, and C. Tischendorf. *PDAEs and further mixed systems as abstract differential algebraic systems.* Technical Report **01-11**, Institute of Mathematics, Humboldt-Univ. of Berlin, 2001.

[15] J. Litsios, B. Schmithüsen, U. Krumbein, A. Schenk, E. Lyumkis, B. Polsky, and W. Fichtner. *DESSIS 3.0 Manual.* ISE Integrated Systems Engineering, Zürich, 1996.

[16] W. Lucht, K. Strehmel, and C. Eichler-Liebenow. *Indexes and special discretization methods for linear partial differential algebraic equations.* BIT **39(3)** (1999), 484–512.

[17] P. A. Markowich, C. A. Ringhofer, and C. Schmeiser. Semiconductor equations. Springer Verlag Wien, 1990.

[18] F.M. Rotella. Mixed circuit and device simulation for analysis, design, and optimization of opto-electronic, radio frequency, and high speed semiconductor devices. PhD thesis, Stanford University, 2000.

[19] S. Scharfenberg. Mixed-Level Circuit-Device Simulation auf einem heterogenen Workstation-Cluster. Shaker Verlag, 1996. PhD thesis.

[20] D. Schroeder. Modelling of Interface Carrier Transport for Device Simulation. Computational Microelectronics. Springer-Verlag, Wien New York, 1990.

[21] S. Selberherr. Analysis and Simulation of Semiconductor Devices. Springer-Verlag, Wien New York, 1984.

[22] S. M. Sze. Physics of Semiconductor Devices. John Wiley & Sons, New York, 1981.

[23] C. Tischendorf. *Coupled systems of partial and differential algebraic equations in circuit and device simulation. Modeling and numerical analysis.* Submitted as habilitation thesis at Humboldt Univ. of Berlin, 2003.

[24] C. Tischendorf. *Topological index calculation of DAEs in circuit simulation.* Surv. Math. Ind. **8** (1999), 187–199.

[25] G. Wachutka. *Unified framework for thermal, electrical, magnetic, and optical semiconductor device modeling.* COMPEL **5** (1991), 311–321.

[26] G. Wachutka. *Consistent treatment of carrier emission and capture kinetics in electrothermal and energy transport models.* Microelectronics Journal **26** (1995), 307–315.

[27] C. Wasshuber, H. Kosina, and S. Selberherr. *SIMON – A simulator for single-electron tunnel devices and circuits.* IEEE CAD **16(9)** (1997), 937–944.

Caren Tischendorf
Humboldt-University of Berlin
Institute of Mathematics
10099 Berlin, Germany
email: `caren@mathematik.hu-berlin.de`

Circuits with Stochastic Sources

Modeling, Simulation and Optimization of Integrated Circuits
(Eds.) K. Antreich, R. Bulirsch, A. Gilg and P. Rentrop

International Series of Numerical Mathematics Vol. 146, 251–267

Transient Noise Simulation: Modeling and Simulation of $1/f$-Noise

Georg Denk, David Meintrup, and Stefan Schäffler

Abstract. In this paper we present a new approach for the transient noise simulation of electronic circuits with stochastic differential algebraic equations (SDAEs). The first part treats the modeling of noise in the time domain which is accomplished with generalized stochastic processes. This allows not only to model white noise like thermal and shot noise, but also $1/f$-noise or flicker noise. It is shown that fractional Brownian motion reflects the properties of $1/f$-noise, namely a spectrum proportional to $1/f$ with f denoting the frequency. Some consequences of this approach on the solvability of the circuit equations are presented. In the second part we give remarks on the implementation of numerical schemes for SDAEs. Besides the integration scheme itself the generation of appropriate random numbers is a major issue. Finally we present some numerical experiments.

1. Introduction

Integrated circuits can be found nearly everywhere in all-days life. The design of these chips is a difficult task which is facilitated by simulation. A crucial part of the design process is the verification of the physical behavior of the circuits at transistor level which is done with the help of analog simulation. One of the most important parts is the transient analysis, i.e., the simulation in the time domain. Normally this analysis does not take noise into account, but due to the decreasing device sizes and supply voltages the small signal-to-noise ratio demands the consideration of noise. Though noise models are well established in the frequency domain, the modeling in the time-domain requires some new approaches. This issue has been tackled for white noise sources (thermal noise, shot noise), see, e.g., [3], but not for the important $1/f$-noise, also known as flicker noise. In addition, the mathematical methods for solving the circuit equations differ significantly between noise-less and noisy simulations. One approach is the computation of expectation value and standard deviation as done by [3]. This method has two drawbacks: It requires

2000 *Mathematics Subject Classification.* Primary 65C30, 60H35; Secondary 60G18, 60G20, 65C20.
Key words and phrases. noise, circuit simulation, flicker noise, $1/f$-noise, modeling, fractional Brownian motion, stochastic differential-algebraic equations.

some linearizations which may not be acceptable in every situation and it does not consider $1/f$-noise directly. Therefore, we aim at the computation of paths of the solution which gives all information and enables the computation of expectation value and standard deviation as a postprocessing step.

In the next section we will state the circuit equations used for transient noise analysis. After giving an introduction to generalized stochastic processes in Section 3, we will present the modeling of thermal and shot noise resp. $1/f$-noise in Section 4 resp. Section 5. Some remarks on the numerical integration of the noisy circuit equations and numerical experiments conclude this paper.

2. Circuit simulation

In order to perform the (noiseless) transient simulation of an integrated circuit, the physical behavior of the circuit has to be transfered into mathematical equations. Most circuit simulators use the modified nodal analysis (MNA) as mathematical foundation which means that Kirchhoff's current law is applied to every node of the circuit. Hereby all branch currents are substituted by element relations in admittance form while branch voltages are expressed in terms of node voltages via Kirchhoff's voltage law. This set of equations is expanded by the element relations for voltage-defining elements. In order to get charge conservation, a charge-oriented formulation is used which leads to the following type of equation:

$$A \cdot \dot{q}(x) + f_{\text{stat}}(x) + s(t) = 0, \quad x(0) = x_0, \tag{2.1}$$

where $x \in \mathbb{R}^n$ is the vector of unknown nodal voltages and branch currents through voltage-defining elements, $t \in [0, t_{\text{end}}]$ is the time, the independent variable. $A \in \mathbb{R}^{n,m}$ is an incidence matrix describing the topology of the dynamic circuit parts, $q(x) : \mathbb{R}^n \to \mathbb{R}^m$ is the vector of terminal charges and branch fluxes. $f_{\text{stat}}(x) : \mathbb{R}^n \to \mathbb{R}^n$ describes the static elements, and $s(t) : \mathbb{R} \to \mathbb{R}^n$ is the vector of independent sources. x_0 are consistent initial values for Eq. (2.1). For details on the setup of equations see [6]. As the rank of matrix A is generally not full, Eq. (2.1) is a *differential-algebraic equation (DAE)*. In the following it will be assumed that the equations have an index of at most 1.

For a transient noise simulation all noise sources are modeled as appropriate current sources shunt in parallel to the ideal noiseless elements (for details see Section 4 and 5). This extends Eq. (2.1) to the *stochastic differential-algebraic equation (SDAE)*

$$A \cdot \dot{q}(x) + f_{\text{stat}}(x) + s(t) + g(t, x) \cdot \eta_t = 0, \tag{2.2}$$

with $g(t, x) : \mathbb{R} \times \mathbb{R}^n \to \mathbb{R}^{n,k}$ holding the noise densities, and $\eta_t \in \mathbb{R}^k$ is the vector of Gaussian white noise processes. Equation (2.2) is not a mathematically precise expression. The reason for this is that there is no proper stochastic process that has the properties of Gaussian white noise. Therefore, the mathematical interpretation of the differential equation (2.2) is the following integral equation:

$$A \cdot q(X_\tau)\Big|_0^t + \int_0^t f_{\text{stat}}(X_\tau)\,\mathrm{d}\tau + \int_0^t s(\tau)\,\mathrm{d}\tau + \int_0^t g(\tau, X_\tau)\,\mathrm{d}B_\tau = 0, \qquad (2.3)$$

with X_τ denoting the stochastic process describing the circuit behavior and B_τ is a k-dimensional Brownian motion process. The two tasks involved with Eq. (2.3), namely modeling the noise sources and building a numerical scheme for a solution, are investigated in the following.

3. General theory of stochastic processes

The integral equation (2.3) is mathematically well defined, since the integral with respect to Brownian motion $\int_0^t g(\tau, X_\tau)\,\mathrm{d}B_\tau$ has been given a precise mathematical meaning by the Japanese mathematician Itô and is called Itô-integral. Another question is, why does the integral equation (2.3) represent the stochastic differential equation (2.2). The reason for this is that formally differentiating Eq. (2.3) leads exactly to (2.2), provided that we know that

$$\dot{B} = \frac{\mathrm{d}B_t}{\mathrm{d}t} = \eta_t. \qquad (3.1)$$

We will explain in Section 4 that the relation (3.1) is a correct mathematical statement provided that one considers Brownian motion and white noise in a more general framework, as so-called generalized processes.

To sum up we can state the following: A stochastic differential equation (2.2) driven by white noise finds its mathematical interpretation in Eq. (2.3) because

- $\dot{B} = \dfrac{\mathrm{d}B_t}{\mathrm{d}t} = \eta_t$ as generalized Brownian resp. white noise processes,
- the integrals $\displaystyle\int_0^t g(\tau, X_\tau)\,\mathrm{d}B_\tau$ with respect to B_t are well defined.

Other components in electronic circuits generate noise that is not white. Typically it has a power spectrum $S(f)$ that is proportional to $(\frac{1}{|f|})^\beta$ with $\beta \approx 1$. We call it *1/f-noise* and denote the corresponding process by $\eta_{1/f}(t)$. Let us now consider a stochastic differential equation that is driven by 1/f-noise. In analogy to (2.2) we get

$$\frac{\mathrm{d}X_t}{\mathrm{d}t} = f_{\text{stat}}(t, X_t) + g(t, X_t) \cdot \eta_{1/f}(t), \quad X_0 = x_0. \qquad (3.2)$$

Again, in order to turn (3.2) into a mathematically well-defined equation, we would need a stochastic process, let us call it B_t^H, that fulfills

- $\dot{B}^H = \dfrac{\mathrm{d}B_t^H}{\mathrm{d}t} = \eta_{1/f}(t)$ as generalized processes,
- the integrals "$\displaystyle\int_0^t g(\tau, X_\tau)\,\mathrm{d}B_\tau^H$" with respect to B_t^H are well defined.

With these two properties one could turn the differential equation (3.2) into a stochastic integral equation

$$X_t = \int_0^t f_{\text{stat}}(\tau, X_\tau)\,\mathrm{d}\,\tau + \text{“}\int_0^t g(\tau, X_\tau)\,\mathrm{d}\,B_\tau^H\text{”}, \quad X_0 = x_0. \tag{3.3}$$

In the following we consider the problem posed by $\dot{B}^H = \eta_{1/f}(t)$. That is, we will introduce a stochastic process B_t^H, the so-called *fractional Brownian motion* whose derivative can be interpreted as $1/f$-noise process. To that end we have to introduce generalized stochastic processes and their spectra what will be done in this section. We will then demonstrate how this theory applies to the white noise case in Section 4. Fractional Brownian noise will play the role of Brownian motion for modeling $1/f$-noise, as carried out in Section 5.

Generalized Stochastic Processes: As mentioned above, we need a generalized concept to describe noise in a mathematically proper way. The main reference for this subject is [5]. We denote by K the space of “test functions”, such that a function $\chi : \mathbb{R} \to \mathbb{R}$ belongs to K if and only if χ is smooth and identically 0 outside of a compact interval.

Definition 3.1 (Generalized function)**.** A continuous linear functional

$$\Phi : K \to \mathbb{R}, \quad \chi \mapsto \Phi(\chi)$$

is called a *generalized function.*

Example 3.2. Every function $\psi \in K$ gives rise to a generalized function by setting

$$\Phi_\psi : K \to \mathbb{R}, \quad \chi \mapsto \Phi_\psi(\chi) := \int_{-\infty}^{\infty} \chi(t)\psi(t)\,\mathrm{d}\,t$$

Example 3.3. Let us consider the functional $\Phi : K \to \mathbb{R}$ defined by

$$\Phi(\chi) := \chi(t_0) \quad \text{for all } \chi \in K, t_0 \in \mathbb{R} \text{ fixed.}$$

This generalized function is called *Dirac Delta function* and usually denoted by $\delta(t - t_0)$.

Instead of assigning a real number to every test function we can assign a random variable what leads to the notion of generalized stochastic processes. Throughout this section we consider a fixed probability space $(\Omega, \mathcal{S}, P)$.

Definition 3.4 (Generalized stochastic process)**.** Let Φ be a function on K that assigns to every $\chi \in K$ a random variable,

$$\chi \mapsto \Phi(\chi) \text{ random variable on } (\Omega, \mathcal{S}, P).$$

Φ is called a *generalized stochastic process*, if it is linear and continuous.

Example 3.5. Every stochastic process X_t with continuous paths gives rise to a generalized stochastic process X defined by

$$X(\chi) := \int_{-\infty}^{\infty} \chi(t) X_t\,\mathrm{d}\,t, \quad \chi \in K.$$

As for stochastic processes, we can define expectation and covariance by taking them for each test function.

Definition 3.6 (Expectation and covariance functional). Let Φ be a generalized stochastic process. Then its *expectation functional* $m_\Phi : K \to \mathbb{R}$ is given by

$$m_\Phi(\chi) := E[\Phi(\chi)] = \int_\Omega \Phi(\chi) \, \mathrm{d} P,$$

and its *covariance functional* $C_\Phi : K \times K \to \mathbb{R}$ is given by

$$C_\Phi(\chi, \psi) := E([\Phi(\chi) - m_\Phi(\chi)][\Phi(\psi) - m_\Phi(\psi)]).$$

An important class of examples are Gaussian processes. A generalized stochastic process Φ is called *Gaussian*, if for any finite number of linearly independent test functions $\chi_1, \ldots, \chi_n \in K$ the process $(\Phi(\chi_1), \ldots, \Phi(\chi_n))$ has a multivariate Gaussian distribution. Every generalized Gaussian process is characterized by its expectation and covariance functional, more precisely we have:

Proposition 3.7. *Let $m : K \to \mathbb{R}$ be a continuous linear functional and $C : K \times K \to \mathbb{R}$ a continuous bilinear positive semi-definite functional. Then there is up to equivalence exactly one generalized Gaussian process Φ with*

$$m_\Phi = m \text{ and } C_\Phi = C.$$

We will use this proposition in Section 4 to construct generalized Gaussian processes by prescribing expectation and covariance.

It is well known that the canonical Brownian motion has almost sure (a. s.) continuous paths, but is a. s. nowhere differentiable. One of the main advantages of the generalized stochastic processes is that they are always differentiable and the derivative is again a generalized stochastic process.

Definition 3.8 (Derivative of a generalized process). Let Φ be a generalized stochastic process. Then its *derivative* $\dot{\Phi}$ is generalized stochastic process given by

$$\dot{\Phi}(\chi) := -\Phi(\dot{\chi}), \quad \chi \in K.$$

The derivative of a Gaussian process is again a Gaussian process, and we can even describe it in terms of the original process, as stated in the next proposition.

Proposition 3.9. *Let Φ be a generalized Gaussian process with expectation functional m_Φ and covariance functional C_Φ. Then its derivative $\dot{\Phi}$ is a generalized Gaussian process with expectation and covariance:*

$$m_{\dot{\Phi}}(\chi) := -m(\dot{\chi}) \quad \text{and} \quad C_{\dot{\Phi}}(\chi, \psi) = C(\dot{\chi}, \dot{\psi}).$$

Spectra: A key point in simulating noise is to establish a valid connection between the measurable data and the stochastic process that is supposed to model the noise. It is obvious that the better the stochastic process is chosen, the better the results of the simulation will be. The measurable data are usually expressed in form of the power spectrum $S(f)$. The mathematical relation between the noise and the stochastic process is given by the following definition.

Definition 3.10 (Spectra of stochastic processes)**.** Let X_t be a stationary stochastic process and denote by $R_X(h) := C(X_{t+h}, X_t)$ its auto-correlation function. Then the *power spectrum* or *spectral density* S_X of X_t is the Fourier transform of R_X, i.e.,

$$\begin{aligned} S_X : \mathbb{R} &\rightarrow \mathbb{C} && (3.4)\\ f &\mapsto S_X(f) := \mathrm{Fou}(R_X) := \int_{-\infty}^{\infty} R_X(h) \exp^{-ifh} \mathrm{d}\, h. && (3.5) \end{aligned}$$

If Φ is a stationary generalized stochastic process and its covariance functional can be written as

$$C_\Phi(\chi, \psi) = \int_{-\infty}^{\infty} \int_{-\infty}^{\infty} R_\Phi(t - \tau)\chi(t)\psi(\tau)\, \mathrm{d}\, t\, \mathrm{d}\, \tau,$$

then the *spectral density of* Φ is defined as

$$\begin{aligned} S_\Phi : \mathbb{R} &\rightarrow \mathbb{C}\\ f &\mapsto S_\Phi(f) := \mathrm{Fou}(R_\Phi) := \int_{-\infty}^{\infty} R_\Phi(h) \exp^{-ifh} \mathrm{d}\, h. \end{aligned}$$

Remark 3.11. Note that we defined spectra only for stationary processes, because they give rise to an auto-correlation function. There are ways of defining spectra for non-stationary processes which leads to time-dependent spectra (see [4]). Their physical and mathematical interpretation is less clear and subject of further investigation.

In Def. 3.10 we got the spectrum from the auto-correlation function of a given process. The situation in practice is the converse. A power spectrum is given by measurements, and the task is to find a stationary (generalized) stochastic process, whose auto-correlation function fulfills Eq. (3.5). In other words, the auto-correlation function of the process we are looking for is the Fourier inverse of the power spectrum, given by

$$R : \mathbb{R} \rightarrow \mathbb{C} \text{ with } t \mapsto R(t) = \frac{1}{2\pi} \int_{-\infty}^{+\infty} S(f) \exp^{ift} \mathrm{d}\, f. \qquad (3.6)$$

The shape of the power spectrum also determines the name of the noise that is observed. The most important examples are:

White noise: Let assume that $S(f) = c$ is constant for all $f \in \mathbb{R}$. Without loss of generality we can assume that $c = 1$. Hence every frequency f is equally represented in $S(f)$, and in analogy to white light this noise is also called white. We will study the corresponding process in the next section.

1/*f*-noise: Let assume that a power spectrum is given that is proportional to an inverse power of f, i.e., $S(f) \sim (\frac{1}{|f|})^\beta$, with $\beta > 0$. This noise is called $1/f$-noise. We will study the process needed to simulate this noise in Section 5.

1/*f*²-noise: This is a special case of $1/f$-noise with $\beta = 2$. It is sometimes called Brownian noise.

4. Modeling of thermal and shot noise

Thermal noise has its foundation in the thermal motion of electrons. It is most important within resistors. The noise density $g(t, X_t)$ is given by

$$g(t, X_t) \equiv g() = \sqrt{\frac{4\,k\,T}{R}},$$

with k denoting Boltzmann's constant, T is the temperature and R is the resistance (cf. Fig. 1). As the noise density does not depend on X_t, this type of noise is called *additive noise*.

Shot noise is present in semiconductors and is due to the discrete character of the elementary charge. The noise density depends on the current and is given by

$$g(t, X_t) \equiv g(X_t) = \sqrt{2\,q\,|I_{\mathrm{det}}(t)|},$$

with q is the elementary charge and $I_{\mathrm{det}}(t)$ is the deterministic current through the device (cf. Fig. 1). As g depends on the stochastic process X_t, the shot noise is called *multiplicative noise.*

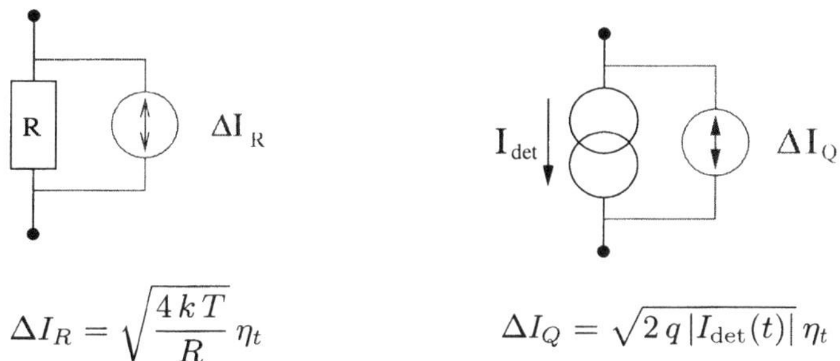

$$\Delta I_R = \sqrt{\frac{4\,k\,T}{R}}\,\eta_t \qquad \Delta I_Q = \sqrt{2\,q\,|I_{\mathrm{det}}(t)|}\,\eta_t$$

FIGURE 1. Modeling of thermal noise (left) and shot noise (right)

Both thermal and shot noise have a constant power spectrum $S(f) = 1$. Let us recall that our aim is to find a process whose Fourier transform of the auto-correlation function is equal to the power spectrum $S(f) = 1$. It is well known that the Fourier inverse of 1 is the Dirac Delta function $\delta(t)$. Therefore our task is to find a generalized process whose auto-correlation functional is equal to the Delta function. As we will show now this is the case for the derived process of the generalized Brownian motion.

A Gaussian process is up to equivalence uniquely determined by its expectation and its covariance function. For example, the standard Brownian motion $(B_t)_{t\geq 0}$ is given by the expectation

$$E(B_t) = 0, \quad t \geq 0, \tag{4.1}$$

and covariance

$$\mathrm{Cov}(\tau, t) = \min(\tau, t), \quad \tau, t \geq 0. \tag{4.2}$$

3.5, we can define the corresponding generalized Gaussian process.

Definition 4.1 (Generalized Brownian motion). The *generalized Brownian motion process* B is given by

$$B(\chi) := \int_{-\infty}^{\infty} \chi(t) B_t \,\mathrm{d}\,t.$$

It follows immediately from (4.1) and (4.2) that the expectation functional m_B and the covariance functional C_B are:

$$\begin{aligned} m_B(\chi) &= 0, \quad \chi \in K, \\ C_B(\chi, \psi) &= \int_0^{\infty}\int_0^{\infty} \min(\tau, t)\chi(\tau)\psi(t) \,\mathrm{d}\,t \,\mathrm{d}\,\tau, \quad \chi, \psi \in K. \end{aligned}$$

As pointed out in Section 3 the derivative of a generalized stochastic process always exists. In case of a generalized Gaussian process like B we know by Prop. 3.9 that the generalized process $W := \dot{B}$ is Gaussian and

$$\begin{aligned} m_W(\chi) &= 0, \quad \chi \in K, \\ C_W(\chi, \psi) &= C_B(\dot{\chi}, \dot{\psi}) = \int_0^{\infty}\int_0^{\infty} \min(\tau, t)\dot{\chi}(\tau)\dot{\psi}(t) \,\mathrm{d}\,t \,\mathrm{d}\,\tau, \quad \chi, \psi \in K. \end{aligned} \tag{4.3}$$

To detect the covariance kernel in Eq. (4.3) we apply partial integration which leads to

$$C_W(\chi, \psi) = \int_0^{\infty} \chi(t)\psi(t) \,\mathrm{d}\,t$$

what we can rewrite as

$$C_W(\chi, \psi) = \int_0^{\infty}\int_0^{\infty} \delta(t-\tau)\chi(t)\psi(\tau) \,\mathrm{d}\,t \,\mathrm{d}\,\tau. \tag{4.4}$$

We can conclude from Eq. (4.4) that the covariance kernel of W is $R_W(h) = \delta(h)$, hence the Fourier transform of the given spectrum is

$$1 = S(f) = \mathrm{Fou}(R_W) = \int_{-\infty}^{\infty} R_W(h) \exp^{-ihf} \,\mathrm{d}\,h.$$

Therefore W is called the white noise process. We sum up our results in the next proposition.

Proposition 4.2. *The white noise process W is a generalized Gaussian process with the following properties:*

$$\begin{aligned} m_W(\chi) &= 0, \quad \chi \in K, \\ C_W(\chi, \psi) &= \int_0^{\infty}\int_0^{\infty} \delta(t-\tau)\chi(\tau)\psi(t) \,\mathrm{d}\,t \,\mathrm{d}\,\tau, \quad \chi, \psi \in K, \\ \dot{B} &= W, \ \textit{where } B \textit{ is generalized Brownian motion}, \\ S_W(f) &= \mathrm{Fou}(\delta(h)) = 1. \end{aligned}$$

From a numerical point of view, the Gaussian white noise processes can be simulated with the help of N(0,1)-distributed pseudo-random numbers (algorithms for generating such pseudo-random numbers can be found in, e.g., [7]).

The algebraic part of Eq. (2.3) leads to a problem not present in the deterministic transient analysis, namely the so-called *direct noise*. Consider the circuit from [12] given in Fig. 2. The modified nodal analysis leads to

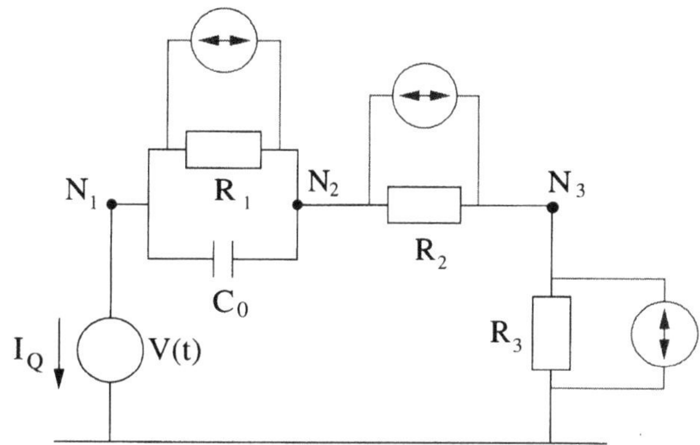

FIGURE 2. Schematic of an example circuit

$$\begin{pmatrix} C_0 & -C_0 & 0 & 0 \\ -C_0 & C_0 & 0 & 0 \\ 0 & 0 & 0 & 0 \\ 0 & 0 & 0 & 0 \end{pmatrix} \cdot \begin{pmatrix} \dot{u}_1 \\ \dot{u}_2 \\ \dot{u}_3 \\ \dot{I}_Q \end{pmatrix} + \begin{pmatrix} \frac{1}{R_1} & -\frac{1}{R_1} & 0 & 1 \\ -\frac{1}{R_1} & \frac{1}{R_1} + \frac{1}{R_2} & -\frac{1}{R_2} & 0 \\ 0 & -\frac{1}{R_2} & \frac{1}{R_2} + \frac{1}{R_3} & 0 \\ 1 & 0 & 0 & 0 \end{pmatrix} \cdot \begin{pmatrix} u_1 \\ u_2 \\ u_3 \\ I_Q \end{pmatrix}$$

$$+ \begin{pmatrix} \sqrt{\frac{4kT}{R_1}} & 0 & 0 \\ -\sqrt{\frac{4kT}{R_1}} & \sqrt{\frac{4kT}{R_2}} & 0 \\ 0 & -\sqrt{\frac{4kT}{R_2}} & \sqrt{\frac{4kT}{R_3}} \\ 0 & 0 & 0 \end{pmatrix} \cdot \begin{pmatrix} \nu_1 \\ \nu_2 \\ \nu_3 \end{pmatrix} = \begin{pmatrix} 0 \\ 0 \\ 0 \\ V(t) \end{pmatrix}$$

This test circuit has been simulated without considering the noise sources (deterministic transient analysis) and with the noise sources active (transient noise analysis). The difference between these simulations are given for the nodal voltages at nodes 2 and 3 in Fig. 3. While the result for node 2 is as expected, the difference for node 3 is much larger. Re-simulating the circuit with smaller step-sizes yields even larger differences.

As the equation for node 3 is purely algebraic, we do not have a smoothing integration for the stochastic process. Gaussian white noise processes have an unlimited variance which is mapped to the large differences for node 3. Only the step-sizes used for the simulation restrict the variance, if the step-size approaches 0, the waveform would blacken the plot. The equation for node 2 has some differential parts in it, so the stochastic process is integrated resulting in the continuous Brownian motion. Smaller step-sizes lead to a more notched waveform but not to larger differences between deterministic and stochastic simulation.

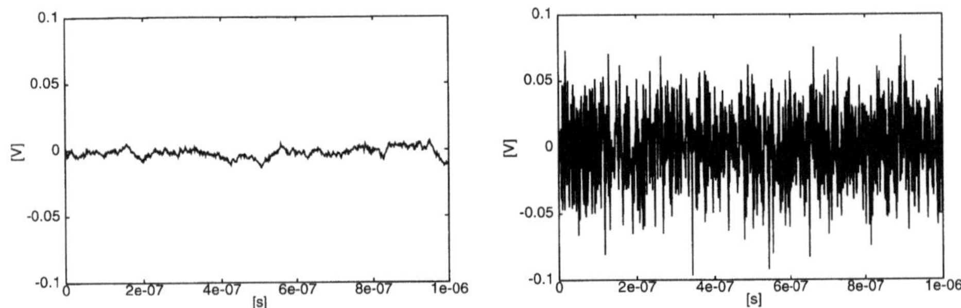

FIGURE 3. Difference between noiseless and noisy simulation of the test circuit: node 2 (left) and node 3 (right)

For a more mathematical interpretation of the situation we transform the SDAE into a stochastic differential equation (SDE) and an algebraic stochastic equation:

$$\begin{aligned} \dot{y}_1(t) + f_1(y_1, y_2) + s_1(t) + g_1(y_1, y_2, t) \cdot \eta_{t,1} &= 0, \\ f_2(y_1, y_2) + s_2(t) + g_2(y_1, y_2, t) \cdot \eta_{t,2} &= 0. \end{aligned}$$

This is always possible due to the index-1 assumption. If $g_2(y_1, y_2, t) \neq 0$, the SDAE is called *SDAE with direct noise* leading to the situation described above. If $g_2(y_1, y_2, t) = 0$, the SDAE is called *SDAE without direct noise.* It has been shown in [12] for linear SDAEs with index 1 that the solution process exists and is unique iff there is no direct noise.

There are two remedies for this situation: a restriction of the transient noise analysis to circuits without direct noise or a different modeling of the noise sources. Direct noise can be avoided, if there are always capacitances in parallel to noise sources. Due to parasitics present in real-world circuits this is true for many applications. It is also possible to relax the restriction when the direct noise only enters into algebraic components without interaction with dynamical ones, see [15] for details.

Though the modeling of thermal and shot noise with Gaussian white noise processes is well accepted, it is somehow physically incorrect, see, e.g., [14, 12]. Starting with a corpuscular model for thermal noise, we arrive at a stochastic process described by the Ornstein-Uhlenbeck (OU) equation

$$X_t \Big|_0^{t_{\text{end}}} = -\int_0^{t_{\text{end}}} a\, X_t\, dt + \int_0^{t_{\text{end}}} \sigma\, dB_t$$

with some physical parameters $a \in \mathbb{R}^+$, $\sigma \in \mathbb{R}$ depending on the device. For $\sigma/a = \text{const}$, $a \to \infty$ the OU-process converges to the Wiener process, so for large a this model is quite similar to the Gaussian white noise model used traditionally (for typical materials $a \approx 10^{13}$). The model according to the OU-process is physically more acceptable and it can be shown for linear SDAEs of index 1 with additive noise that there exists a unique Gaussian solution process for every consistent initial value for Eq. (2.3). For details see [12].

The issue of direct noise is of great theoretical interest and requires future investigations. From a numerical point of view, however, the computed stochastic process does not behave as a Gaussian white noise process but possesses a finite band-width due to the finite step-sizes. This facilitates the numerical treatment of white-noise models used in transient noise simulations.

5. Modeling of $1/f$-noise

Another important noise source is $1/f$-noise which is also called flicker noise. One of its foundations are crystal defects in the semiconductor. The noise density $g(t, X_t)$ is given by

$$g(t, X_t) \equiv g(X_t, f) = \sqrt{\frac{K|I(t)|^\alpha}{f^\beta}},$$

with K denoting a model parameter (typical range $10^{-25} \dots 10^{-19}$), α is a physical constant $(0.5 \dots 2.0)$, β is in most cases 1, $I(t)$ is the current through the device, and f is the frequency. The density depends on the frequency f, a quantity normally not present in transient analysis.

For a mathematical treatment of $1/f$-noise we introduce a new process, called *fractional Brownian motion*. It will play the role in simulating $1/f$-noise that Brownian motion had in simulating white noise.

Fractional Brownian Motion: Fractional Brownian motion $(B_t^H)_{t\geq 0}$ (fBm for short) is defined as a Gaussian process with expectation

$$E(B_t^H) = 0 \text{ for all } t \geq 0, \tag{5.1}$$

and covariance function

$$C(\tau, t) = \frac{1}{2}(\tau^{2H} + t^{2H} - |\tau - t|^{2H}), \quad \tau, t \geq 0. \tag{5.2}$$

We put $B_t^H = 0$ for $t < 0$. The parameter H is a fixed real number with $0 < H < 1$ and is called the *Hurst parameter*. Note that for $H = \frac{1}{2}$ the fractional Brownian motion is simply the standard Brownian motion B_t.

Fractional Brownian motion has been introduced for the first time by Mandelbrot and van Ness [8]. They were motivated by the self-similarity of fBm, i.e., we have for every $\alpha > 0$

$$B_{\alpha t}^H = \alpha^H B_t^H \text{ in distribution.}$$

The variance of a fBm is given by

$$E[(B_t^H)^2] = t^{2H}, \quad t \geq 0.$$

In analogy to Brownian motion the fBm is not a stationary process, but its increments $(B_{t+h}^H - B_t^H)$ are stationary. This is a first indication that, once introduced as generalized process, the derived process could be stationary and henceforth has a spectral density. We will make these considerations precise in the following.

Finally, we want to mention that one has to distinguish the three cases $H = 1/2$, $0 < H < 1/2$, and $1/2 < H < 1$. As mentioned above the first case reduces to standard Brownian motion. For $0 < H < 1/2$ the increments of fBm are negatively correlated, whereas for $1/2 < H < 1$ the increments are positively correlated. This changes dramatically the behavior of the process. In particular, fBm B_t^H with $0 < H < 1/2$ is used to model systems with strong intermittency, whilst fBm B_t^H with $1/2 < H < 1$ is used to model systems with strong aftereffect.

1/*f*-noise: Let us assume that a power spectrum is given that is proportional to an inverse power of f, i.e., $S(f) \sim (\frac{1}{|f|})^\beta$, with $\beta > 0$. Again, our task is to find a stochastic process with an auto-correlation function that is the Fourier inverse of $S(f)$. Now the Fourier inverse of $S(f)$ is proportional to $|t|^{\beta-1}$. As this is supposed to be an auto-correlation function of a stochastic process, $R(0)$ corresponds to the variance of the process. For $\beta > 1$ the variance would be equal to 0, which implies that the process is 0 almost sure, which is a contradiction. A similar argument holds for $\beta = 1$. Therefore we get:

Remark 5.1. There is no stochastic process that simulates $(\frac{1}{|f|})^\beta$-noise for $\beta \geq 1$. Therefore from now on we restrict β to $0 < \beta < 1$.

For $0 < \beta < 1$ we proceed in analogy to the white noise case replacing Brownian motion by fractional Brownian motion. Hence as a first step we have to introduce fBm as a generalized process.

Definition 5.2 (Fractional Brownian motion as generalized process). The *generalized fractional Brownian motion process* B^H is given by

$$B^H(\chi) := \int_{-\infty}^{\infty} \chi(t) B_t^H \, \mathrm{d}t, \quad \chi \in K.$$

It follows from (5.1) resp. (5.2) that the expectation functional resp. the covariance functional are

$$\begin{aligned} m_{B^H}(\chi) &= 0, \quad \chi \in K, \\ C_{B^H}(\chi, \psi) &= \int_{-\infty}^{\infty} \int_{-\infty}^{\infty} \frac{1}{2}(\tau^{2H} + t^{2H} - |\tau - t|^{2H}) \chi(\tau) \psi(t) \, \mathrm{d}t \, \mathrm{d}\tau, \ \chi, \psi \in K. \end{aligned}$$

The derived process $W^H := \dot{B}^H$ is by Proposition 3.9 known to be a generalized Gaussian process with expectation functional

$$m_{W^H}(\chi) = 0, \quad \chi \in K,$$

and covariance functional $(\chi, \psi \in K)$

$$C_W(\chi, \psi) = C_{B^H}(\dot{\chi}, \dot{\psi}) = \int_0^{\infty} \int_0^{\infty} \frac{1}{2}(\tau^{2H} + t^{2H} - |\tau - t|^{2H}) \dot{\chi}(\tau) \dot{\psi}(t) \, \mathrm{d}t \, \mathrm{d}\tau. \quad (5.3)$$

To detect the covariance kernel in (5.3) we have to apply partial integration twice which leads to the following expression for $1/2 < H < 1$:

$$C_{W^H}(\chi, \psi) = \int_0^\infty \int_0^\infty R_{W^H}(t-\tau)\chi(t)\psi(\tau)\,\mathrm{d}\,t\,\mathrm{d}\,\tau$$

where we abbreviated the covariance kernel of W^H as

$$R_{W^H}(h) := H(2H-1)|h|^{2H-2}.$$

Hence with $\beta = 2H - 1$ we have the following relation between W^H and $1/f$-noise:

$$S(f) = \left(\frac{1}{|f|}\right)^\beta \sim \mathrm{Fou}(R_{W^H}) = \int_{-\infty}^\infty R_{W^H}(h) \exp^{-ihf} \mathrm{d}\,h.$$

This relation justifies to call W^H the fractional noise process. We sum up our results in the next proposition.

Proposition 5.3. *The fractional noise process W^H, $1/2 < H < 1$, is a generalized Gaussian process with the following properties:*

$$\begin{aligned} m_{W^H}(\chi) &= 0, \quad \chi \in K, \\ C_{W^H}(\chi, \psi) &= \int_0^\infty \int_0^\infty H(2H-1)|t-\tau|^{2H-2}\chi(\tau)\psi(t)\,\mathrm{d}\,t\,\mathrm{d}\,\tau, \quad \chi, \psi \in K, \\ \dot{B}^H &= W, \text{ where } B^H \text{ is generalized fBm}, \\ S_{W^H(f)} &= \mathrm{Fou}(R(h)) \sim \left(\frac{1}{|f|}\right)^{2H-1}. \end{aligned}$$

For a numerical simulation of fBm we need appropriate random numbers. These can be generated starting with N(0,1)-distributed random numbers which are converted according to Eq. (5.2). The covariance matrix has the dimension of the already computed timesteps.

6. Numerical integration of SDAEs

For an efficient implementation of a transient noise analysis in a circuit simulator several issues have to be resolved:

- efficient computation of appropriate pseudo-random numbers
- development of numerical integration schemes for SDAEs
- development of a step-size control
- adaption of the nonlinear solver

While there are many efficient implementations available for the computation of N(0,1)-distributed pseudo-random numbers, the direct implementation of random numbers for the fBm leads to a very inefficient transient noise analysis. The reason is the dimension of the necessary covariance matrix, it increases with the number of computed time steps. Even with efficient Schur-complement techniques the computational effort increases quadratically with the number of time steps.

This means for simulations which require a huge number of integration steps that for most of the time the simulator is just computing random numbers instead of evaluating the circuit equations. As the influence of time steps decreases with the number of intermediate points, it seems reasonable to omit time steps which are more than a user defined number of time steps beyond the current timepoint. Thus a window is defined which – after the initialization process – is moved across the covariance matrix. With the Sherman-Morrison-Woodbury formula this approach helps to keep the computational effort, which is necessary for the computation of pseudo-random numbers for the fBm, below a certain level, especially it does no longer depend on the number of already computed timesteps. The error introduced by the windowing technique is user controllable, but for reasonable window sizes (say 100) it is reasonable small.

The development of an efficient integration scheme for SDAEs is quite difficult, as the formula has to cope with the charge-oriented formulation of Eq. (2.3), with the DAE index issue, and with high-dimensional equation systems (up to 100,000). Additionally the numerical scheme has to deal with different types of noise (multiplicative, non-commutative, ...). The first step for the numerical treatment of $1/f$-noise was to find the stochastic process B_t^H, the fractional Brownian motion. We proved that its derived process W^H has an auto-correlation function whose Fourier transform is the power spectrum of $1/f$-noise. Therefore, fBm is the right process to simulate $1/f$-noise. Now, as a second step one needs to define integrals with respect to fBm, i.e., expressions of the type

$$\text{“}\int_0^t g(\tau, X_\tau)\,\mathrm{d}\,B_\tau^H\text{”}. \tag{6.1}$$

The theory of stochastic integration is well established for Brownian motion, more generally for martingales or semi-martingales. Unfortunately fBm is not a semi-martingale. Hence a completely new approach is needed to give sense to the expression (6.1). There are recent publications with first attempts of defining integrals with respect to fBm (e.g., [1, 2]). However it is at this time unclear which approach is suited for SDEs resp. SDAEs. This is a major challenge for future work.

It is beyond the scope of this paper to present all the details about schemes for simulating white noise, so we restrict ourself to an overview of recent research. The development of such schemes for charge-oriented SDAEs started with Schein [12, 13] who investigated schemes for linear SDAEs with additive noise. A more general approach was done by Penski [10] which allowed the computation of multiplicative noise. This formula, however, requires a large number of function evaluation which restricts its usage to small-sized circuits. As the noise sources are relatively small compared to the deterministic signals, this can be exploited in the development of new schemes for SDEs [9]. Promising results which combine the exploitation of small noise with the application to charge-oriented SDAEs can be found in [15].

The adaptive step-size control is a very essential point in the efficient solution of DAEs. Though some interesting first results can be found in [11], there exists no robust and efficient step-size control for charge-oriented SDAEs yet. Therefore, we use the step-size control for DAEs in our test implementation with some minor modifications in the nonlinear solver regarding convergence control. As the noise sources are relatively small, this approach works quite well even if it is not theoretically founded.

7. Numerical experiments

For the transient noise simulation of real-world circuits, a modified version of the numerical scheme given in [10] was implemented in Infineon's in-house simulator TITAN. The transient noise analysis was performed for a cross-section of a 256 Mb DRAM chip which contains approximately 37,000 elements and 150,000 noise sources leading to a system with nearly 17,000 equations. The window size for the computation of flicker noise sources has been set to 100 timesteps. The overhead necessary for the noise analysis is about 30 % compared to the simulation without noise. In Fig. 4 the path of a nodal voltage is given. The simulation with noise sources gives a zig-zag line around the straight line resulting from the noiseless simulation. The peaks arising from some switching events within the circuits are quite similar between both simulations, the deviation from the deterministic waveform is small. For this digital circuit the transient noise simulation shows that no problems due to noise are to be expected.

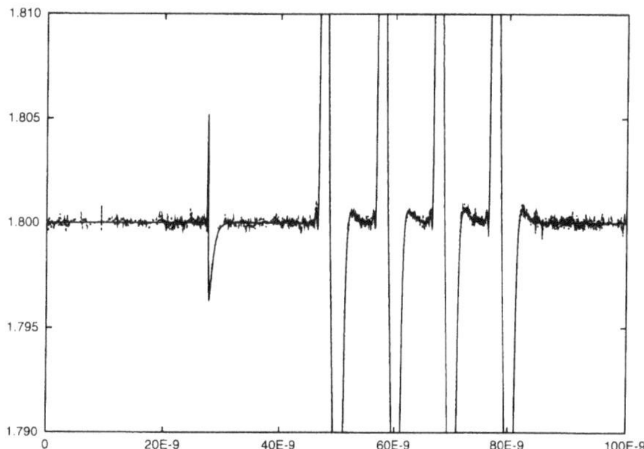

FIGURE 4. Simulation result for a cross section of a 256 Mb DRAM: without noise (*solid line*), with noise (*dotted line*)

References

[1] T.E. Duncan, Y.Z. Hu, B. Pasik-Duncan, *Stochastic calculus for fractional Brownian motion. I. Theory.* SIAM J. Control Optim. **38** (2000), 582–612.

[2] L. Decreusefond, A.S. Ustunel, *Stochastic analysis of the fractional Brownian motion,* Potential Anal. **10** (1999), 177–214.

[3] A. Demir, A. Sangiovanni-Vincentelli, *Analysis and Simulation of Noise in Nonlinear Electronic Circuits and Systems.* Kluwer Academic Publishers, 1998.

[4] P. Flandrin, *On the spectrum of fractional Brownian motions.* IEEE Trans. Inform. Theory **35** (1989), 197-199.

[5] I. M. Gelfand, J. Wilenkin, *Verallgemeinerte Funktionen.* VEB Deutscher Verlag der Wissenschaften, 1964.

[6] M. Günther, U. Feldmann, *CAD-based electric-circuit modeling in industry I: Mathematical structure and index of network equations.* Surv. Math. Ind. **8** (1999), 97–129.

[7] D.E. Knuth, *The Art of Computer Programming I.* Addison-Wesley (1973).

[8] B.B. Mandelbrot. J.W. van Ness, *Fractional Brownian motions, fractional noises and applications.* SIAM Rev. **10** (1968), 422–437.

[9] G.N. Milstein, M.V. Tretyakov, *Mean-square numerical methods for stochastic differential equations with small noise.* SIAM J. Sci. Comput. **18** (1997), 1067–1087.

[10] Chr. Penski, *Numerische Integration stochastischer differential-algebraischer Gleichungen in elektrischen Schaltungen.* Shaker Verlag (2002) (Ph. D. thesis, Technische Universität München).

[11] W. Römisch, R. Winkler, *Stepsize control for mean-square numerical methods for SDEs with small noise.* In preparation.

[12] O. Schein, *Stochastic differential algebraic equations in circuit simulation.* Ph. D. thesis, Technische Universität Darmstadt (1999).

[13] O. Schein, G. Denk, *Numerical solution of stochastic differential-algebraic equations with applications to transient noise simulation of microelectronic circuits.* J. Comput. Appl. Math. **100** (1998), 77–92.

[14] L. Weiß, W. Mathis, *A thermodynamical approach to noise in nonlinear networks.* Int. J. Circ. Theor. Appl. **26** (1998), 147–165.

[15] R. Winkler, *Stochastic DAEs in transient noise simulation.* To appear in the Proceedings of *Scientific Computing in Electrical Engineering*, June, 23rd–28th, 2002, Eindhoven, Springer Series *Mathematics in Industry.* Proceedings of the SCEE (2002).

Acknowledgment

The authors acknowledge the valuable discussions and fruitful contributions given by R. Winkler (HU Berlin) and C. Hillermeier (UniBw München).

Georg Denk
Infineon Technologies
Memory Products
Balanstr. 73
D-81541 München, Germany
email: georg.denk@infineon.com

David Meintrup
Universität der Bundeswehr
Lehrstuhl für Mathematik und Projektmanagement
Werner-Heisenberg-Weg 39
D-85577 Neubiberg, Germany
email: David.Meintrup@UniBw-Muenchen.de

Stefan Schäffler
Universität der Bundeswehr
Lehrstuhl für Mathematik und Projektmanagement
Werner-Heisenberg-Weg 39
D-85577 Neubiberg, Germany
email: stefan.schaeffler@UniBw-Muenchen.de

Modeling, Simulation and Optimization of Integrated Circuits
(Eds.) K. Antreich, R. Bulirsch, A. Gilg and P. Rentrop

International Series of Numerical Mathematics Vol. 146, 269–282

Noise Analysis of Nonlinear Electrical Circuits and Devices

Wolfgang Mathis and Laurens Weiss

Abstract. In this paper we discussed the mathematical modelling of thermal noise in linear and nonlinear electric networks and its physical interpretation. A new approach of thermal noise analysis in nonlinear networks with a well-defined physical interpretation is presented. It is illustrated by means of a few network examples.

1. Introduction

Since around 1913 first nonlinear electrical circuits containing electronic tubes were developed by Meissner, Armstrong and others (see Mathis [27] for further details) the problem of noise became a main subject of circuit designers. In 1918 Schottky [33] published a first fundamental paper where he investigated the physical aspects of thermal and shot noise. In 1928 thermal noise was studied in more detail for linear time-invariant circuits by Johnson [18] and Nyquist [30] where the famous Johnson-Nyquist formula was derived using equilibrium thermodynamical arguments. Basically all these considerations applied to the so-called Langevin approach that was developed by Langevin [25] in 1908 for linear physical systems. A mathematical reasoning of the concept of Langevin can be given if the mathematical concepts of stochastic processes and stochastic differential equations are used. Although this approach became very successful in the analysis of noisy linear electrical circuits and systems (see, e.g., Haus and Adler [17]) it was observed by McDonald and van Kampen around 1957 (for an overview see van Kampen [21]) that one runs into difficulties if one tries to formulate a noise description for certain nonlinear physical systems.

The main problem behind these difficulties was that it is very essential to distinguish between external and internal noise in physical systems. In the first case there are no problems because it deals with the nonlinear transformation of stochastic processes. In the latter case new physical aspects arise with respect to the interpretation of the results since in contrast to the first case the origin

---

Key words and phrases. Dynamical Systems, nonlinear, noise.
We thank Dipl.-Ing. B. Beute for programming of stochastic integration methods and numerous calculations.

for deterministic behavior and the stochastic disturbance is the same. If, e.g., an electrical current is considered we have a flow of quantized charges that are responsible for the deterministic as well as the stochastic behavior; the stochastic dynamics cannot be suppressed without a breakdown of the deterministic current. Fortunately internal noise can be described by external noise sources in the case of linear systems which is the reason why the Langevin approach is successful. In case of noisy nonlinear systems it was pointed out by van Kampen [22] that moments of different orders are coupled. As a consequence one cannot start with a deterministic description of a system and add the stochastic disturbance – just like in the linear case.

The paper is organized in the following manner: In Section 2 we discuss some general aspects of physical dynamical systems. In the following section a theory for a large class of noisy nonlinear electrical circuits is presented. This theory was developed by Weiss and Mathis (see, e.g., [35], [36], [34]) during the last few years based on Stratonovich's concept of nonlinear nonequilibrium statistical thermodynamics. Unfortunately the theory is restricted to so-called reciprocal circuits and therefore no amplifying circuit elements (transistors or tubes) can be included in a circuit under consideration. Especially noisy nonlinear oscillators cannot be analyzed from this point of view. Therefore in Section 2 some aspects of stochastic differential equations with bifurcations are discussed.

2. General stochastic dynamical systems

In the state space theory of deterministic dynamical systems the traditional approach emphasizes trajectories in the state space $\mathcal{S}$ where trajectories can be interpreted as solutions of differential equations with prescribed initial states. In theoretical mechanics the state space corresponds to the space of all positions and momentums; it possesses the structure of a cotangent bundle $\mathcal{T}^*\mathcal{S}$ (see, e.g., Abraham, Marsden [1]). In contrast to mechanics a state space of an electrical circuit possesses at most the structure of a differentiable manifold (see, e.g., Mathis [28]). In certain cases the trajectory approach of dynamical systems is not useful. Boltzmann observed that the dynamics of sets of initial states instead of a single initial state is suitable to calculate interesting physical quantities (see van Kampen [21], Chapter XIV). Today similar methods can be applied to study so-called chaotic systems. In the following we restrict ourselves to the case of differential equations on $\mathbb{R}^n$ (index 0 from a DAE point of view) and omit discussions about differential equations on more general sets (e.g., with the structure of a differentiable manifold). Therefore we only proceed with stochastic differential equations based on this class of deterministic differential equations. Note that stochastic differential equations on differentiable manifolds were studied by Elworthy [14], [15] and Kunita [24] as well as more recently by Arnold and his group at the University of Bremen [7]. Recent progress in stochastic bifurcations is discussed by Arnold [4]. Mathis and Prochaska discussed some aspects of stochastic bifurcations in [29]. We only want to emphasize that there are two concepts of bifurcation in stochastic systems (Arnold [4]): 1) P-bifurcation, where probability density functions change its form. 2) D-bifurcation, where Ljapunov exponents change its signs and as a conclusion invariant measures that replace equilibrium points in deterministic systems bifurcate. Mathis and Prochaska illustrate in [29] these aspects in the case of stochastic Andronov-Hopf bifurcation using results from Schenk-Hoppe and Arnold [9]. In deterministic analysis of electrical oscillators this bifurcation theorem was developed by Andronov et al. in 1935 (see, e.g., Andronov, Witt, Chaikin [3]; first published in 1937) based on results of Henri Poincaré.

For his statistical mechanics Boltzmann used (scalar) density functions of positions and momentums and derived corresponding dynamical equations. The deterministic behavior of a system is described by a set of differential equations on $\mathbb{R}^n$ (or state space equations)

$$\frac{dx}{dt} = F(x) \tag{1}$$

where $F : \mathbb{R}^n \to \mathbb{R}^n$. If we are interested in the dynamics of a suitable class of density functions $f : \mathbb{R} \to \mathbb{R}$ where an evolution equation with an associated Frobenius-Perron-Operator $\mathcal{P}^t$ can be formulated

$$f(x,t) = \mathcal{P}^t\left(f(x)\right); \tag{2}$$

this equation is closely related to the generalized Liouville equation

$$\frac{\partial f}{\partial t} = -\mathrm{div}\left\{f \cdot F\right\} = -\sum_{i=1}^{n} \frac{\partial (f \cdot F_i)}{\partial x_i}. \tag{3}$$

In the Langevin approach of stochastic systems we start with such a deterministic description and add a stochastic process ξ

$$\frac{dx}{dt} = F(x) + \sigma(x)\,\xi, \tag{4}$$

where the coefficient $\sigma(x)$ characterizes the coupling of the noise sources. The first term on the right should be interpreted as dissipation whereas the second term corresponds to fluctuations.

Using the concept of stochastic differential equations ξ has to be interpreted as a generalized white noise process but in order to solve these equations a more generalized concept of integration (e.g., Kunita [24]) is needed. Essentially there are two concepts of stochastic integration which are due to Ito and Stratonovich, respectively, and associated types of stochastic differential equations

$$dx = F(x)dt + \sigma(x)\,dw \tag{5}$$

where w is the so-called Wiener process. Both concepts are mathematically equivalent to a partial differential equation for the probability density P of Fokker-Planck type which generalizes in some sense the concept of Liouville equations (see Arnold

[5], Section 4.2)

$$\frac{\partial P}{\partial t} = -\sum_{i=1}^{n} \frac{\partial (P \cdot F_i)}{\partial x_i} + \sum_{i,j=1}^{n} \frac{\partial^2 (\sigma^2 P)}{\partial x_i \partial x_j}. \tag{6}$$

In the case of linear stochastic differential equations – the original subject of Langevin – there is no difference between Ito's and Stratonovich's type. Unfortunately stochastic differential equations (of Ito or Stratonovich type) are sound concepts only from a mathematical point of view if we consider nonlinear Langevin equations. The reason is that each type corresponds to a certain interpretation rule; otherwise its meaning is not well defined. It is interesting to see that for nonlinear Langevin equations in contrast to linear ones the deterministic equation (without noise) does not correspond to the averaged equation (see van Kampen's paper for further details [22])

$$\left\langle \frac{dx}{dt} \right\rangle = \frac{d \langle x \rangle}{dt} = \langle F(x) \rangle + \langle \sigma(x)\, \xi \rangle \,. \tag{7}$$

Even if σ is constant we note that the function F and the statistical average operator $\langle \cdot \rangle$ do not commute in every case. Only if $\langle F(x) \rangle = F(\langle x \rangle)$ is valid the averaged equation for the first moment of the stochastic process x fulfills the deterministic equation $\dot{x} = F(x)$. Therefore like van Kampen [19] [20] we come to the conclusion that there is no good reason why the dissipation term should be identical to the vector field of the deterministic equation. In order to obtain a sound description of physical systems additional considerations are needed. In the next section we will discuss some ideas in this direction.

3. A physical theory of noisy nonlinear electrical networks

It was discussed in the previous section that in the nonlinear Langevin approach deterministic differential equations are extended by a stochastic part with an additive white noise process. Although the stochastic part can be cancelled if the coupling coefficient $\sigma(x)$ between the deterministic system and the white noise process will be set to zero the equivalence of the deterministic equation and the differential equation for the first moment of the describing quantity get lost. The reason is simply that in nonlinear stochastic systems a coupling of moments occurs that is similar to the coupling of frequencies in nonlinear deterministic systems (e.g., distortion in almost nonlinear amplifiers or nonlinear oscillators). Therefore the mathematical procedure of the Langevin approach for nonlinear deterministic systems can be justified but there are problems with its physical interpretations. Following van Kampen [20] and Stratonovich [32] a physical reasonable approach for a large class of nonlinear noisy electrical circuits was presented by Weiss and Mathis [35], [36], [34]. In the following we will give a short survey of the main ideas of this approach.

Since in the nonlinear extension of the Langevin approach serious problems with its physical interpretation occur one must describe the evolution of the entire system as a stochastic process (see van Kampen [20], p. 184). If it is reasonable to associate the Markov property to the describing variable $x(t)$ then the evolution equation is a so-called master equation having the general form

$$\frac{\partial P(x,t)}{\partial t} = \int \{W(x|x')P(x',t) - W(x'|x)P(x,t)\}\, dx' \tag{8}$$

where $P(x,t) := P(x,t|x_0,t_0)$ is the transition probability between t_0 and t, while $W(x|x')\triangle t$ is so small that P does not vary much, but large enough for the Markov property (see van Kampen [21], p. 184). For solving the integro-differential equation – equivalent with the so-called Chapman-Kolmogorov equation – two approaches are available which result in approximative solutions near the thermodynamic equilibrium. Following van Kampen a suitable system parameter Ω which measures the size of the system under consideration is chosen and a so-called Ω-expansion can be derived. This approach which is explained in every detail in [21] can be helpful in simple physical systems. For more complex systems it is more suitable to use Stratonovich's approach who solved the master equation with the so-called Kramers-Moyal expansion and derived the following equation

$$\frac{\partial P(x,t)}{\partial t} = \sum_{m=1}^{\infty} \frac{(-1)^m}{m!} \sum_{\alpha_1 \cdots \alpha_m = 1}^{r} \frac{\partial^m}{\partial x_{\alpha_1} \cdots x_{\alpha_m}} \left[K_{\alpha_1 \cdots \alpha_m}(x) P(x,t)\right] . \tag{9}$$

In order to apply this equation to a noisy physical system the coefficients $K_{\alpha_1 \cdots \alpha_m}(x)$ have to be determined as functions of the deterministic system parameters. Using ideas of Stratonovich a corresponding procedure for nonlinear noisy electrical networks is presented by Weiss and Mathis [34], [35], [36]. In the following the main steps are described.

It is known from equilibrium thermodynamics that the m-fold moments are proportional to kT where k is the Boltzmann constant and T the absolute temperature. Stratonovich assumed that the same must hold for (conditional) nonequilibrium moments (in a "certain neighborhood" of equilibrium) and therefore for the unknown coefficients $K_{\alpha_1 \cdots \alpha_m} \sim (kT)^{m-1}$. Hence Stratonovich used an expansion in the small parameter kT instead of Ω like in van Kampen's approach.

Since the equilibrium case must be enclosed in the non-equilibrium theory, the equilibrium distribution (Gibbs distribution)

$$P_{eq}(x) = C \cdot e^{-\frac{\Psi(x)}{kT}} \tag{10}$$

must be a stationary solution of the master equation (and its Kramers-Moyal expansion) where $\Psi(x)$ denotes Helmholtz free energy. Near the equilibrium a similar relation between $P(x)$ and $\Psi(x)$ should be valid. Following Stratonovich we choose

$$P(x) = C \cdot e^{-\frac{\Psi(x) - y \cdot x}{kT}} \tag{11}$$

where $y := \partial\Psi(x)/\partial x$ denotes the so-called conjugate thermodynamic forces.

The deterministic equations impose additional constraints on the coefficients $K_{\alpha_1 \cdots \alpha_m}$. Finally the Onsager-Casimir reciprocity relations of linear nonequilibrium thermodynamics (see Casimir [13] and Anderson [2]), which are a consequence of time reversal symmetry, are used to reduce the number of unknown coefficients.

In a series of papers Weiss and Mathis (see, e.g., also the dissertation of Weiss [34]) adapted Stratonovich's approach to nonlinear noisy electrical networks. They showed that the so-called topologically complete electrical networks represents the most comprehensive class of networks that can be described by Stratonovich's approach. This class of networks was characterized mathematically for the first time by Brayton and Moser [11] (see Fig. 1)

$$L_\rho(i_\rho)\frac{di_\rho}{dt} = \frac{\partial B(i,u)}{\partial i_\rho} \quad (\rho = 1,\dots,r), \tag{12}$$

$$C_\sigma(u_\sigma)\frac{du_\sigma}{dt} = \frac{\partial B(i,u)}{\partial u_\sigma} \quad (\sigma = 1,\dots,s) \tag{13}$$

where the so-called (Brayton-Moser) mixed potential $B(i,u) := \Sigma(i) - \Pi(u) + (i, \Gamma u)$ is related to the dissipative parts Σ (current potential) and Π (voltage potential) as well as Γ (interconnection) of a network.

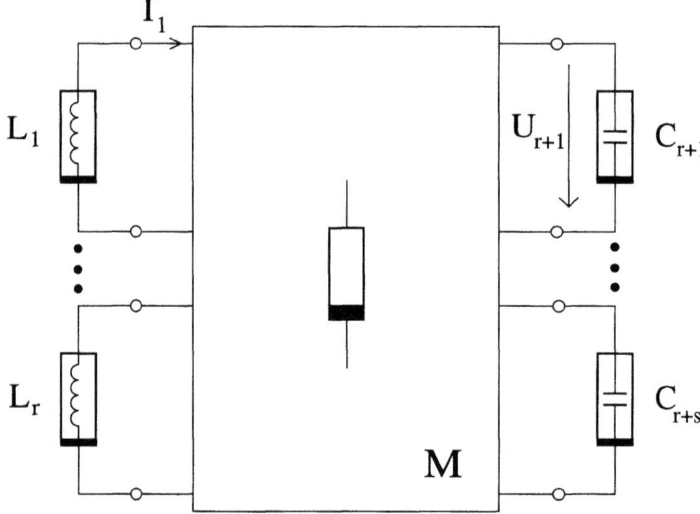

FIGURE 1. Block Diagram of Network represented by Brayton-Moser Equations

If currents i_ρ through inductors and voltages u_σ across the capacitors of topologically complete electrical networks are used as state space variables x_α of the Stratonovich approach different approximations of the master equation with Kramers-Moyal expansion can be derived for networks of this class. As free energy $\Psi(x)$ the energy relation for all r inductors and s capacitors

$$\Psi(i,u) := \frac{1}{2}\sum_{\rho=1}^{r} L_\rho(i_\rho)i_\rho^2 + \frac{1}{2}\sum_{\sigma=r+1}^{r+s} C_\sigma(u_\sigma)u_\sigma^2 \tag{14}$$

is used. In the case of linear noisy electrical networks a Fokker-Planck type equation for the probability density $P(i, u)$ has been derived by Weiss and Mathis

$$\frac{\partial P(i,u)}{\partial t} = \mathcal{L}_1 P(i,u) \tag{15}$$

where

$$\mathcal{L}_1 := - \sum_{\rho_1,\rho_2} \frac{\Sigma_{,\rho_1,\rho_2}(0)}{L_{\rho_1}} i_{\rho_2} \frac{\partial}{\partial i_{\rho_1}} - \sum_{\rho_1,\sigma_2} \frac{\gamma_{\rho_1,\sigma_2-r}}{L_{\rho_1}} u_{\sigma_2} \frac{\partial}{\partial i_{\rho_1}} \tag{16}$$

$$+ \sum_{\sigma_1,\rho_2} \frac{\gamma_{\rho_2,\sigma_1-r}}{C_{\sigma_1}} i_{\rho_2} \frac{\partial}{\partial u_{\sigma_1}} - \sum_{\sigma_1,\sigma_2} \frac{\Pi_{,\sigma_1,\sigma_2}(0)}{C_{\sigma_1}} u_{\sigma_2} \frac{\partial}{\partial u_{\sigma_1}} \tag{17}$$

$$- kT \sum_{\rho_1,\rho_2} \frac{\Sigma_{,\rho_1,\rho_2}(0)}{L_{\rho_1} L_{\rho_2}} \frac{\partial^2}{\partial i_{\rho_1} i_{\rho_2}} - kT \sum_{\sigma_1,\sigma_2} \frac{\Pi_{,\sigma_1,\sigma_2}(0)}{C_{\sigma_1} C_{\sigma_2}} \frac{\partial^2}{\partial u_{\sigma_1} u_{\sigma_2}} \tag{18}$$

and $\Sigma_{,\rho_1,\rho_2}$ means $\partial^2 \Sigma / \partial i_{\rho_1} \partial i_{\rho_2}$, etc.

It can be shown by inspection that

$$P_{eq}(i,u) = C \exp\left(-\frac{\Psi(i,u)}{kT}\right) \tag{19}$$

is a stationary solution of Eq. (15), where $\Psi(i, u)$ is the free energy of an electrical network. Furthermore this is a partial differential equation for the probability density $P(i, u)$ of a Fokker-Planck type and therefore a set of mathematical equivalent stochastic differential equations (with Stratonovich's interpretation rule) exists ($\rho = 1, \ldots, r, \sigma = 1, \ldots, s$)

$$di_\rho = \left(\sum_{\rho_2} \frac{\Sigma_{,\rho,\rho_2}(0)}{L_\rho} i_{\rho_2} + \sum_{\sigma} \frac{\gamma_{\rho,\sigma-r}}{L_\rho} u_\sigma \right) dt \tag{20}$$

$$+ \sum_{k=r+s+1}^{r+s+t} \pm \sqrt{\frac{2kTR_k}{L_\rho^2}} dw_k \tag{21}$$

$$du_\sigma = \left(\sum_{\sigma_2} \frac{\Pi_{,\sigma,\sigma_2}(0)}{C_\sigma} u_{\sigma_2} - \sum_{\rho} \frac{\gamma_{\rho,\sigma-r}}{C_\sigma} i_\rho \right) dt \tag{22}$$

$$+ \sum_{l=r+s+t+1}^{r+s+t+u} \pm \sqrt{\frac{2kTG_l}{C_\sigma^2}} dw_l. \tag{23}$$

These relationships correspond to multi-dimensional Nyquist formulas for linear noisy electrical networks.

If we consider nonlinear noisy networks in the linear-quadratic series approximation of their nonlinearities, compared to equation (15) an additional term occurs in the evolution equation of the probability density

$$\frac{\partial P(i,u)}{\partial t} = \mathcal{L}_1 P(i,u) + \mathcal{L}_2 P(i,u), \tag{24}$$

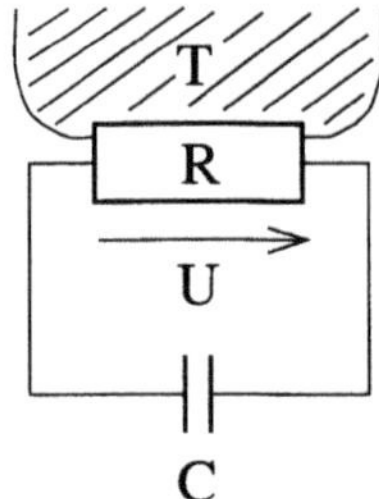

FIGURE 2. Linear RC network with thermal bath

where $\mathcal{L}_2$ is a third-order differential operator. Due to the third-order derivatives (24) is more general than a Fokker-Planck equation, hence there is no corresponding stochastic differential equation for the state space variables i and u and the Gaussian white noise source model approach fails. But even if no third-order derivatives occur just like in networks without inductors additional deterministic (drift) terms (dt-part of stochastic differential equations) arise. If these terms are omitted we get classical nonlinear Fokker-Planck equations but then the Brillouin paradoxon [12] can be constructed. This means that the description does not satisfy the second law of thermodynamics which is acceptable at most as approximation under certain conditions (see also Wyatt and Coram [38]). It was emphasized by Weiss and Mathis that their noise theory for nonlinear electrical networks can be used to explain and to quantify the restrictions of the classical noise source picture.

4. Application to simple noisy electrical networks

In the last section a framework for the analysis of noisy electrical networks was presented if networks possess the property of reciprocity and can be described by Brayton-Moser equations [11]. For illustrating this framework we consider some simple examples.

As a first example the simple linear noisy RC network in Fig. 2 is discussed that is coupled to a thermal bath and therefore noise occur. In contrast to the description in conventional textbooks a Brayton-Moser formulation is chosen in order to apply our framework. If u is the voltage across the capacitor C the corresponding mixed potential function can be formulated as ($G := 1/R$)

$$B(u) = \int Gu\,du = G\,u^2 = -\Pi(u) \tag{25}$$

since the current term $\Sigma(i)$ and the crossing term $(i, \Gamma u)$ are missing. For calculating the linear approximation, that is in cases of linear noisy networks already the complete noise description, the second derivative of $-\Pi(u)$ at $u = 0$ is needed. As a result we get

$$\frac{\partial P(u)}{\partial t} = \mathcal{L}_1 P(u) = \frac{\partial}{\partial u}\left(\frac{G}{C} uP(u)\right) + \frac{\partial^2}{\partial u^2}\left(\frac{GkT}{C^2} P(u)\right). \tag{26}$$

The partial differential equation for the probability density $P(u)$ is linear but with a non-constant drift coefficient, that is a Fokker-Planck equation. In this simple example the Fokker-Planck equation can be solved by numerical methods in an acceptable time. For more complex networks the solution process for the corresponding multi-dimensional Fokker-Planck equation is very time-consuming.

The corresponding stochastic differential equation – in Stratonovich interpretations – is

$$du = -\frac{G}{C} u \, dt \pm \sqrt{\frac{2kTG}{C}} \, dw \quad u(t_0) = u_0 \tag{27}$$

where the RC network with noise source and the corresponding noise current spectrum $S_I := 2kTG = 2kT/R$ is shown in Fig. 3.

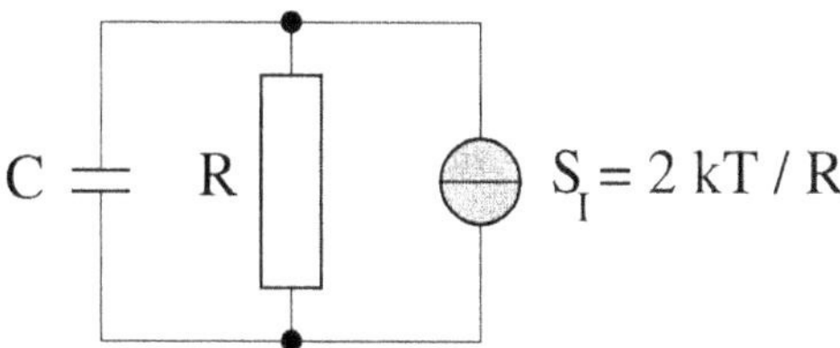

FIGURE 3. Lineare RC network with Noise Source

In correspondence to the non-stochastic case numerical integration methods for stochastic differential equations are available (see, e.g., Kloeden, Platen, Schurz [23]). Using, e.g., stochastic trapezoidal methods a set of stochastic trajectories can be calculated. After a suitable averaging process over the set of trajectories and the extraction of moments for each time the dynamics of the associated probability density function of this simple network results; in Fig. 4 the probability density function is shown.

Finally we remark that the mixed potential can be used in order to derive the usual (state) differential equation for the network

$$\frac{du}{dt} = -\frac{G}{C} u. \tag{28}$$

Now we consider an RC network with a nonlinearity in cubic approximation coupled to a thermal bath with the following mixed potential function

$$B(u) = \int \left(Gu + \frac{1}{2}\beta u^2 + \frac{1}{2}\gamma u^3 \right) du = -\Pi(u). \tag{29}$$

Using our framework for the linear-quadratic approximation (or $\gamma = 0$) a Fokker-Planck type equation for the density function $P(u)$ can be derived

$$\frac{\partial P(u)}{\partial t} = \frac{\partial}{\partial u} \left(\frac{G}{C} u P(u) \right) + \frac{\partial^2}{\partial u^2} \left(\frac{GkT}{C^2} P(u) \right) + \mathcal{L}_2 P(u), \tag{30}$$

where

$$\mathcal{L}_2 = \frac{\partial}{\partial u} \left(\frac{\beta}{2C} u^2 - \frac{\beta kT}{2C^2} \right) + \frac{\partial^2}{\partial u^2} \left(\frac{\beta kT}{2C^2} u \right). \tag{31}$$

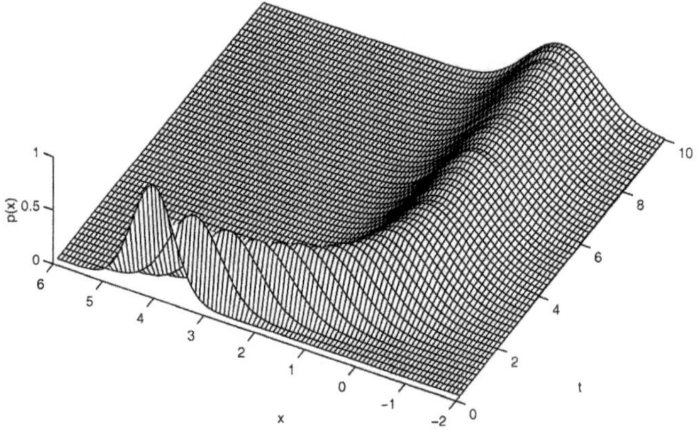

FIGURE 4. Solution of the Fokker-Planck Equation

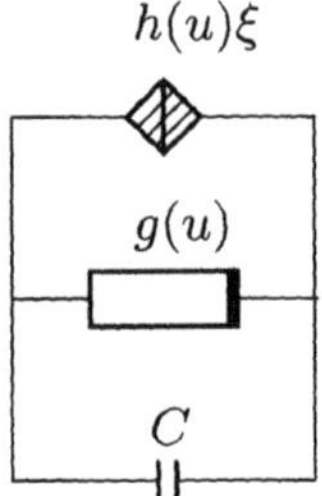

FIGURE 5. Nonlinear RC network with thermal bath

Also in this case the density equation is of the Fokker-Planck type from a mathematical point of view but in contrast to the classical case a correction term of the order 1 in kT occurs in the drift term. Therefore a physical interpretation in the usual sense is not available. On the other hand the correction term prevents that the Brillouin paradoxon [12] arises. The Fokker-Planck equation (29) is equivalent to the following stochastic differential – or Langevin – equation

$$\frac{du}{dt} = -\frac{1}{C}\left(Gu + \frac{1}{2}\beta u^2 - \frac{\beta kT}{4C}\right) + \sqrt{1 + \frac{\beta u}{2G}}\,\xi(t) \quad u(t_0) = u_0 \tag{32}$$

where $\langle \xi(t)\xi(t')\rangle = (2GkT/C^2)\delta(t-t')$. The Langevin equation corresponds to a network model with noise sources shown in Fig. 5 where the following aspects have to be included: 1) $g(u) = Gu + 1/2\beta u^2$ is the nonlinear resistor. 2) $\beta kT/(4C)$ corresponds to an independent current source. 3) $h(u) := \sqrt{1 + \beta u/(2G)}$ corresponds to a controlled current source. Since the average $\langle u\rangle$ is of the order 0 in kT it is not influenced by the correction term (order 1 in kT). Therefore the correction term is not included in the deterministic network model.

If the mixed potential function $B(u)$ is extended by a term with $\gamma \neq 0$ then a now differential operator $\mathcal{L}_3$ arises in the density equation

$$\frac{\partial P(u)}{\partial t} = (\mathcal{L}_1 + \mathcal{L}_2 + \mathcal{L}_3)\, P(u). \tag{33}$$

It was shown by Weiss [34] that derivatives higher than two and an additional parameter ϵ occur, and that this parameter cannot be determined as a function of the deterministic network parameters from general physical considerations. Higher derivatives lead to the conclusion that an exact correspondence to a stochastic differential equation does not exist but only an approximative one. Further discussion about fundamental aspects of this new approach can be found in several papers of Weiss and Mathis [35], [36] as well as Weiss [34].

Some more interesting results in noise modelling that can be used for the analysis of semiconductor devices were presented in Weiss and Mathis [37]. Since with the exception of diodes and tunnel diodes the other device models examined in [37] make use of controlled sources the latter elements are non-reciprocal and cannot be fully analyzed within the framework presented here. If, on the other hand, only the channel resistors of FETs and diodes are analyzed our framework is applicable. As our approach works only with thermodynamical quantities certain assumptions have to be fulfilled that the results are valid. The limitations of our approach are discussed in details by Weiss [34]. If the assumptions are fulfilled it is possible to calculate the noise properties of a channel resistor by means of resistor characteristics in a certain order of approximation.

As an example we discuss the thermal noise of J-FETs – a rather complicated case. The drain current I_D of an n-channel J-FET can be described as a function of the Drain-Source voltage U_{DS} in second-order approximation as

$$I_D \approx g_0 \left(\left(1 - \sqrt{\frac{U_{diff} - U_{GS}}{U_0}} \right) U_{DS} - \frac{1}{4} \frac{U_{DS}^2}{\sqrt{(U_{diff} - UGS)U_0}} \right) + O(U_{DS}^3), \tag{34}$$

where the physical quantities $U_0, U_{diff}, U_{GS}, g_0$ can assumed to be known and to be independent of U_{DS}. The current density spectrum S_I that is calculated with our concept in the second-order is precisely the same as the one determined by van der Ziel's S_I from microscopic considerations [39] (in the same order of approximation)

$$S_I(U_{DS}) = 2kTg_0 \left(1 - \sqrt{\frac{U_{diff} - U_{GS}}{U_0}} - \frac{U_{DS}}{4\sqrt{(U_{diff} - U_{GS})U_0}} \right) + O(U_{DS}^2). \tag{35}$$

This equality of results cannot be coincidence. In the cases of pn-diodes, tunnel diodes and MOSFETs we derived the same agreement in the corresponding order of approximation.

5. Conclusions

In this paper we discussed the mathematical modelling of thermal noise (including shot noise of diodes) in linear and nonlinear electric networks and its physical interpretation. It was reasoned by means of arguments from van Kampen, who was one of the most eminent researchers in stochastic processes in physics, that the famous Ito-Stratonovich dilemma, which became an essential subject in the mathematical literature, is no problem in the case of physical systems. In contrast to this the substantial difference between external and internal noise in physical systems is really a problem. Whereas an external (input) noise source can be treated with the theory of (nonlinear) transformation of stochastic processes internal noise is much more difficult. In this paper a new approach of the authors for noise analysis in nonlinear networks with a well-defined physical interpretation is presented. It is illustrated by means of a few network examples. This new approach is based on ideas of Stratonovich that have been published in the field of nonlinear nonequilibrium statistical thermodynamics. Although some interesting results in the area of semiconductor noise models were presented by means of this new concept, non-reciprocal networks – also in the linear case – cannot be treated. Since this is necessary for analyzing active networks (e.g., amplifiers and oscillators with transistors) our approach has to be extended.

However, it should be emphasized that it is a substantial result of our studies that noise analysis of nonlinear network models cannot be exact if noisy devices are described by noiseless device models and (Gaussian white) noise sources, or, in other words, using the Langevin approach. Furthermore within our framework one can understand to which extend approximations are used if the conventional white noise source concept is applied.

References

[1] Abraham, R.; J.E. Marsden, Foundations of Mechanics, Second edition, The Benjamin/Cummings Publishing Company, 1978

[2] Anderson, B.D.O., Nonlinear networks and Onsager-Casimir reversibility, IEEE Trans. Circuits and Systems 27, 1980, 1051–1058

[3] Andronov, A.A., A.A. Witt, S.E. Khaikin, Theory of Oscillators. Pergamon, Oxford, 1966

[4] Arnold, L., Random dynamical systems. Springer-Verlag, Berlin-Heidelberg-New York 1998

[5] Arnold, L., The unfoldings of dynamics in stochastic analysis. Comput. Appl. Math. 16, 1997, 3–25

[6] Arnold, L.; P. Imkeller, Normal forms for stochastic differential equations. Probab. Theory Relat. Fields 110, 1998, 559–588

[7] Arnold, L., Trends and open problems in the theory of random dynamical systems. in: Probability Towards 2000, L. Accardi, C.C. Heyde, eds., Springer Lecture Notes in Statistics Vol. 128, 1998, 34–46.

[8] Arnold, L.; Xu Kedai, Normal forms for random diffeomorphismen. J. Dynamics and Differential Equations 4, 1992, 445–483

[9] Arnold, L.; N. Sri Namachchivaya, K.R. Schenk-Hoppe, Toward an understanding of stochastic Hopf Bifurcation: a case study. Intern. Journ. Bifurcation and Chaos, 6, 1996, 1947–1975

[10] Boxler, P., A stochastic version of center manifold theory. Probab. Th. Rel. Fields 83, 1989, 509–545

[11] Brayton, R.K.; J.K. Moser, A theory of nonlinear networks – I and II. Quart. Appl. Math. 22, 1964, 1–33, 81–104

[12] Brillouin, L., Can the rectifier be a thermodynamical demon. Phys. Rev. 78, 1950, 627–628

[13] Casimir, H.B.G., On Onsager's principle of microscopic reversibility. Rev. Mod. Phys. 83, 1951, 34–40

[14] Elworthy, K.D., Geometric aspects of diffusions on manifolds. Ecole d'Eté de Prob. de Saint-flour, pp. 277–425, 1988. Springer Lecture Notes in Mathematics, Vol. 1362

[15] Elworthy, K.D., Stochastic differential equations on manifolds. Cambridge University Press, Cambridge 1982

[16] Guckenheimer, J.; P. Holmes, Nonlinear oscillations, dynamical systems, and bifurcation of vector fields. Springer-Verlag, Berlin-Heidelberg 1983

[17] Haus, H.A.; R.B. Adler, Circuit Theory of Linear Noisy Networks. John Wiley & Sons, New York 1959

[18] Johnson, J.B., Thermal agitation of electric charge in conductors. Phys. Rev. 32, 1928, 97–109

[19] van Kampen, N.G., The Validity of Nonlinear Langevin Equations. Journ. Statist. Physics 25, 1981, 431–442.

[20] van Kampen, N.G., Ito versus Stratonovich. Journ. Statist. Physics 24, 1981, 175–187.

[21] van Kampen, N.G., Stochastic Processes in Physics and Chemistry. North Holland, Amsterdam 1992

[22] van Kampen, N.G., Thermal fluctuations in nonlinear systems. Journ. Math. Phys. 4, 1963, 190–194

[23] Kloeden, P.E., E. Platen, H. Schurz: Numerical Solution of SDE Through Computer Experiments. Springer-Verlag, Berlin Heidelberg 1994

[24] Kunita, H., Stochastic flows and stochastic differential equations. Cambridge 1990 University Press, Cambridge

[25] Langevin, P., Sur la théorie du mouvement brownien, Comptes Rendus Acad. Sci. (Paris) 146, 1908, 530–533

[26] Mathis, W., Theory of Nonlinear Circuits, German language. Springer-Verlag, Berlin-Heidelberg, 1987

[27] Mathis, W., Historical remarks to the history of electrical oscillators (invited). In: Proc. MTNS-98 Symposium, July 1998, IL POLIGRAFO, Padova 1998, 309–312.

[28] Mathis, W., Nonlinear Networks as dynamical systems on differentiable manifolds. Proc. 3rd MATHMOD, Wien, 2–4 Februar 2000, Vol. 1, S. 209–212.

[29] Mathis, W.; M. Prochaska, Deterministic and stochastic Andronov-Hopf bifurcation in nonlinear electronic oscillators. Proc. 11th NDES 2003, Scuol, Switzerland, May

[30] Nyquist, H., Thermal agitation of electric charge in conductors. Phys. Rev. 32, 1928, 110–113

[31] Pecseli, H.L., Fluctuations in Physical Systems. Cambridge University Press, Cambridge 2000

[32] Stratonovich, R.L., Nonlinear Nonequilibrium Thermodynamics I + II. Springer-Verlag, Berlin-Heidelberg, 1992 + 1994

[33] Schottky, W., Über spontane Stromschwankungen in verschiedenen Elektrizitätsleitern, Ann. d. Phys. 57, 1918, 541–567

[34] Weiss, L., Rauschen in nichtlinearen elektronischen Schaltungen und Bauelementen – ein thermodynamischer Zugang. Dissertation, University of Magdeburg, Germany 1999

[35] Weiss, L.; W. Mathis, Irreversible Thermodynamics and thermal noise of nonlinear networks. COMPEL 17, 1998, 635–648

[36] Weiss, L.; W. Mathis, A Thermodynamical Approach to Noise in Nonlinear Networks. Int. Journ. Circ. Theor. Appl. 26, 1998, 147–165

[37] Weiss, L.; W. Mathis, A thermodynamic noise model for nonlinear resistors. IEEE Trans. Electr. Dev. Lett. 20, 1999, 402–404

[38] Wyatt, J.L.; G.J. Coram, Nonlinear device noise models: Satisfying the thermodynamic requirements. IEEE Trans. Electr. Dev. 46, 1999, 184–193

[39] van der Ziel, A.: Thermal noise in field-effect transistors. Proc. IRE 50, 1962, 1808–1812

Wolfgang Mathis
Institute of Theoretical Electrical Engineering (TET)
Department of Electrical Engineering
University of Hannover
Appelstr. 9A
D-30167 Hannover, Germany
email: `mathis@tet.uni-hannover.de`

Laurens Weiss
Infineon Technologies AG
Balanstr. 73
D-81541 München, Germany
email: `laurens.weiss@infineon.com`

Modeling, Simulation and Optimization of Integrated Circuits
(Eds.) K. Antreich, R. Bulirsch, A. Gilg and P. Rentrop

International Series of Numerical Mathematics Vol. 146, 283–301

On Parametric Test Design for Analog Integrated Circuits considering Error in Measurement and Stimulus

Michael Pronath, Helmut Graeb, and Kurt Antreich

Abstract. An overview of the problem of go/no-go testing for parametric faults in analog integrated circuits is given. In a linear model that includes both test stimulus and measurement errors, the influence of such errors on the test decision is analyzed in order to calculate trade-off curves for the probabilities of misclassification. This is achieved by an optimization of the correlation between test and performance.

1. Problem formulation

1.1. Specification

The specification of analog circuits is usually described by the type of circuit (e.g., operating amplifier, data converter, oscillator,...) and by a type-specific vector $\mathbf{f} = (f_1, \ldots, f_{n_f})$ of scalar performance variables like phase margin, delay, slew rate, power consumption, DNL, etc.

The values of these performances determine if a circuit will operate correctly when used in a certain application or not. For example, a low-power filter application may require that the power consumption of the used operational amplifiers be less than 1 mW and the transit frequency be larger than 1 MHz. Such requirements are collected in an electrical specification

$$\underset{1 \le i \le n_f}{\forall} f_i \ge b_i \tag{1}$$

or in vector notation

$$\mathbf{f} \ge \mathbf{b} \,. \tag{2}$$

From the tester's point of view, the performance specification bounds $\mathbf{b} = (b_1, \ldots, b_{n_f})$ are fixed limits that should be guaranteed to hold for every sold circuit.

An upper bound on a performance f_i can be easily included by defining a new performance $f_j := -f_i$ and setting a lower bound on f_j. Therefore it is sufficient to consider exclusively lower bounds, which will simplify the notation of our concepts.

1.2. Parametric faults

The production of integrated circuits involves many complex physical and chemical processing steps. Inevitable fluctuations in the manufacturing process make the quality of the produced devices vary, i.e., each specified performance f_i varies a little from chip to chip. Manufacturing is therefore a stochastic process. Knowledge of this process helps to design better (=more robust) products as well as better (=more sensitive) tests. A simulation model of the performance variation induced by process fluctuation is therefore needed for design as well as for test design.

It turned out, that variation of device model parameters $\mathbf{s}$, like oxide thickness or threshold voltage of transistors, yields a suitable and efficient formulation, because it is process-specific and independent from the circuit. This means that the same model can be applied without adaption to all different kinds of circuits that are manufactured in the same process. Random variation of these parameters $\mathbf{s}$ leads to a random variation of the circuit performances $\mathbf{f}$, since $\mathbf{s}$ determines the electrical behavior of the individual devices which constitute the circuit. For each value of $\mathbf{s}$, circuit simulation yields the corresponding circuit performances $\mathbf{f}$. Since circuit simulation is deterministic, it is a function that maps $\mathbf{s}$ to $\mathbf{f}$. To keep the notation simple, we will write $\mathbf{f}(\mathbf{s})$ to denote this function in the following, without making a notational difference between the performance vector $\mathbf{f}$ and the performance function $\mathbf{f}(\mathbf{s})$.

Therefore, the simulation model of parametric variation is determined by two factors: The performance function $\mathbf{f}(\mathbf{s})$, and the probability distribution of $\mathbf{s}$. The performance function is usually not available in an explicit form, but rather as a set of implicit nonlinear algebro-differential equations that describe the electrical behavior of the circuit. The evaluation of $\mathbf{f}(\mathbf{s})$ for a given value of $\mathbf{s}$ is done by solving this set of equations numerically, which is called "circuit simulation". Circuit simulation is done by specialized software and is beyond the scope of this text, see [9] or [5] for details. Evaluation of $\mathbf{f}(\mathbf{s})$ by circuit simulation is an extremely time-consuming task, which dominates the computational costs of design and test. Therefore, efficient algorithms for design and test have to aim at evaluating $\mathbf{f}(\mathbf{s})$ as few times as possible. In the following, we assume that $\mathbf{f}(\mathbf{s})$ is not explicitly known and has to be evaluated by circuit simulation, so that the number of evaluations should be kept small.

Process characterization estimates parameters of the joint probability function of device model parameters from wafer measurement data [8]. In practice, only the multinormal Gaussian distribution and the log-normal distribution are used, and the latter is easily transformed into a Gaussian distribution by taking the logarithm of the random variable.[1] Therefore, only the normal distribution needs to be considered for the individual device parameters.

In this text, the vector of the random device parameters is denoted as $\mathbf{s}$. To simplify the notation, we do not distinguish between a random variable and its realizations. The probability density function of a p-dimensional multinormal

[1] $X \sim \exp(N(\mu, \sigma^2)) \Leftrightarrow \ln X \sim N(\mu, \sigma^2)$.

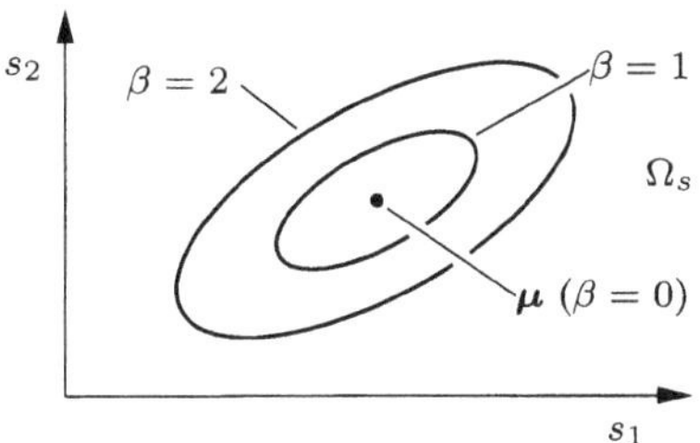

FIGURE 1. Contour lines of the probability density function of a multinormal distribution.

distribution with parameters $\boldsymbol{\mu}$ and $\boldsymbol{\Sigma}_s$ is

$$pdf(\mathbf{s}) = (2\pi)^{-\frac{p}{2}} (\det \boldsymbol{\Sigma}_s)^{-\frac{1}{2}} \exp\left(-\tfrac{1}{2}(\mathbf{s}-\boldsymbol{\mu})^{\mathrm{T}} \boldsymbol{\Sigma}_s^{-1} (\mathbf{s}-\boldsymbol{\mu})\right) . \tag{3}$$

This distribution is completely characterized by its mean vector $\boldsymbol{\mu}$ and its symmetric, positive-definite covariance matrix $\boldsymbol{\Sigma}_s$. In analogy to the notation $N(\mu, \sigma^2)$ for a normally distributed random variable, the symbol $N_p(\boldsymbol{\mu}, \boldsymbol{\Sigma}_s)$ denotes a p-dimensional random variable vector that conforms to a multinormal distribution with parameters $\boldsymbol{\mu}$ and $\boldsymbol{\Sigma}_s$. Therefore, using the symbol "$\sim$" ("is distributed like"), we can write

$$\mathbf{s} \sim N_p(\boldsymbol{\mu}, \boldsymbol{\Sigma}_s) . \tag{4}$$

As can be seen from (3), the probability density for a point $\mathbf{s}$ depends on its distance from the mean $\boldsymbol{\mu}$ in a norm defined by the covariance matrix $\boldsymbol{\Sigma}_s$. We will use this distance more often in the following, using the letter β:

$$pdf(\mathbf{s}) = C \cdot \exp\left(-\tfrac{1}{2}\beta^2(\mathbf{s})\right) \tag{5}$$

with

$$\beta(\mathbf{s}) = \sqrt{(\mathbf{s}-\boldsymbol{\mu})^{\mathrm{T}} \boldsymbol{\Sigma}_s^{-1} (\mathbf{s}-\boldsymbol{\mu})} \text{ , and}$$
$$C = (2\pi)^{-\frac{p}{2}} (\det \boldsymbol{\Sigma}_s)^{-\frac{1}{2}} . \tag{6}$$

Therefore, the contour lines $\{\mathbf{s} \in \Omega_s \mid pdf(\mathbf{s}) = const.\}$ of the probability density function can be expressed in terms of β: $\{\mathbf{s} \in \Omega_s \mid \beta(\mathbf{s}) = const.\}$. Figure 1 shows an example of the contour lines of the probability density function of a multinormal distribution with some correlation between s_1 and s_2. They are ellipses that are centered at the mean value.

Figure 2 illustrates, how the performance function $\mathbf{f}(\mathbf{s})$ connects the parameter space Ω_s and the performance space Ω_f in two important aspects. At first, the distribution of the performance values of the produced devices results from mapping the parameter distribution via the performance function to the performance space. As said before, the performance function is not explicitly known, and therefore neither is the distribution of $\mathbf{f}$. Secondly, the performance bounds in the performance space define, which devices satisfy the specification and which

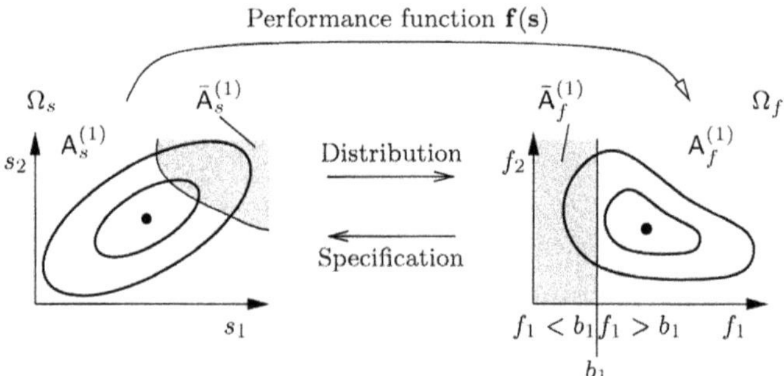

FIGURE 2. The performance function maps the device model parameter variation to the performance space, which defines the distribution of the performance values $\mathbf{f}$. Specifications in the performance space define the acceptance region in the parameter space.

do not. In the following, we will use the performance bounds $\mathbf{f} \geq \mathbf{b}$ in the form of a set they define

$$\mathsf{A}_f = \{\mathbf{f} \in \Omega_f \mid \mathbf{f} \geq \mathbf{b}\} , \tag{7}$$

which will be called the *acceptance region* in the performance space. Since each performance f_i has its own limit b_i, and every limit must be kept, the whole acceptance region is the intersection of the individual acceptance regions

$$\mathsf{A}_f^{(i)} = \{\mathbf{f} \in \Omega_f \mid f_i \geq b_i\} , \tag{8}$$

$$\mathsf{A}_f = \bigcap_i \mathsf{A}_f^{(i)} . \tag{9}$$

The remaining performance space is then characterized by

$$\bar{\mathsf{A}}_f^{(i)} = \Omega_f \setminus \mathsf{A}_f^{(i)} = \{\mathbf{f} \in \Omega_f \mid f_i < b_i\} \tag{10}$$

$$\bar{\mathsf{A}}_f = \Omega_f \setminus \mathsf{A}_f = \bigcup_i \bar{\mathsf{A}}_f^{(i)} . \tag{11}$$

Figure 2 shows the acceptance region in performance space for performance f_1 (the second performance has no bound in this example).

Every vector $\mathbf{s} \in \Omega_s$ corresponds to only one performance vector $\mathbf{f}$, which either fulfills the specification or not. This allows to clearly separate good from faulty devices in the parameter space:

$$\mathsf{A}_s = \{\mathbf{s} \in \Omega_s \mid \mathbf{f}(\mathbf{s}) \in \mathsf{A}_f\} , \tag{12}$$

with the corresponding individual spaces $\mathsf{A}_s^{(i)}$ and the remaining space $\bar{\mathsf{A}}_s$ defined accordingly. We may take for granted that $\boldsymbol{\mu}$ is inside the acceptance region, because else the nominal design violated the specification which is a design error.

There usually is no explicit form for the acceptance region A_s or its boundary $\partial\mathsf{A}_s$. A good linear approximation to the individual acceptance regions $\mathsf{A}_s^{(i)}$ is however at hand: Due to the probability distribution of $\mathbf{s}$, the production samples will be concentrated in a small area around the mean $\boldsymbol{\mu}$, whereas the largest part of the parameter space is empty. For many applications, it is reasonable to use a good (in a probabilistic sense) linear approximation of $\partial\mathsf{A}_s^{(i)}$, which is accurate at the *worst-case point* $\mathbf{s}_{\mathrm{wc}}^{(i)}$. The worst-case point of a specification is the parameter set that has the highest probability density among all vectors on the boundary of the acceptance region, i.e.,

$$\mathbf{s}_{\mathrm{wc}}^{(i)} = \operatorname*{argmax}_{\mathbf{s}\in\partial\mathsf{A}_s^{(i)}} pdf(\mathbf{s}) \,. \tag{13}$$

Focusing on the worst-case point is reasonable, because most of the faulty elements lie close to it in Ω_s. The worst-case point shows the place in Ω_s, where most parametric faults will occur.

As seen in (5), the probability density depends on the distance measure β. Therefore, an equivalent definition is

$$\mathbf{s}_{\mathrm{wc}}^{(i)} = \operatorname*{argmin}_{\mathbf{s}\in\partial\mathsf{A}_s^{(i)}} \beta^2(\mathbf{s}) \,. \tag{14}$$

Figure 3 illustrates this definition. The smallest distance β in (14) corresponds to the smallest contour ellipsis. Therefore, the worst-case point lies on the smallest contour ellipsis, which just touches $\bar{\mathsf{A}}_s^{(i)}$ in one point, that is the worst-case point. Finding the worst-case point is a nonlinear optimization problem and lies beyond the scope of this text; algorithms that perform this task are discussed for instance in [2] and [1].

The distance between the worst-case point and the mean is called *worst-case distance*

$$\beta_{\mathrm{wc}}^{(i)} = \beta(\mathbf{s}_{\mathrm{wc}}^{(i)}) \,. \tag{15}$$

In the example of Fig. 3, $\beta_{\mathrm{wc}}^{(i)} = 1$.

Every specification has its own worst-case point and distance, so that the symbols $\mathbf{s}_{\mathrm{wc}}$ and β_{wc} always have the specification index i. To simplify the notation, the specification index will be omitted in the following whenever there is only one specification being considered. Then, $\mathsf{A}_s^{(i)}$ and $\mathsf{A}_f^{(i)}$ will be written without the index, too.

The worst-case distance can be written explicitly for a specified performance f that depends linearly on $\mathbf{s}$:

$$\beta_{\mathrm{wc}} = \frac{f_0 - b}{\sigma_f} \,. \tag{16}$$

For linear performances, the worst-case distance is the distance between the specification bound b and the nominal value $f_0 = f(\boldsymbol{\mu})$, in multiples of the standard deviation σ_f of the performance.

The partial yield Y of a specification is the percentage of circuits which fulfill the specification. The yield is therefore equal to the probability, that a produced

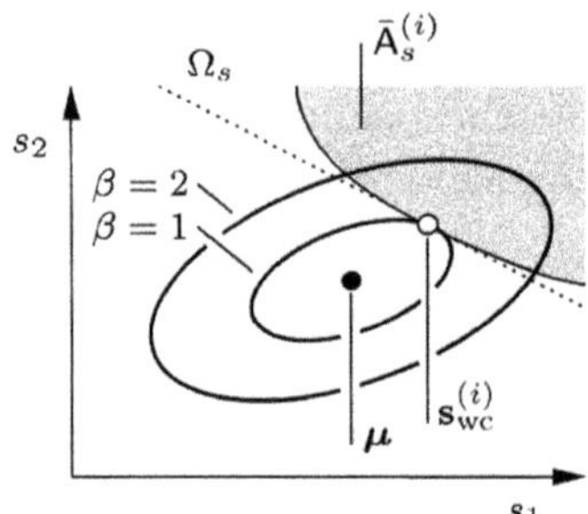

FIGURE 3. The worst-case point $\mathbf{s}_{\text{wc}}$ is part of the boundary of $\mathsf{A}_s^{(i)}$ and has the highest probability density among all boundary points. The dotted line in this figure is a linear approximation to $\partial\mathsf{A}_s^{(i)}$ through the worst-case point.

circuit will satisfy the specification:

$$Y = \mathcal{P}\{f \geq b\} \ . \tag{17}$$

The linear approximation to A_s shown in Figure 3 can be used to estimate Y:

$$Y \approx \Phi_1(\beta_{\text{wc}}) = \frac{1}{\sqrt{2\pi}} \int_{-\infty}^{\beta_{\text{wc}}} e^{-\xi^2/2} d\xi \ . \tag{18}$$

This approximation is exact for linear performance functions $f(\mathbf{s})$ and has been shown to be accurate enough for many circuit examples [2].

1.3. Test

1.3.1. TEST PROCEDURE In the context of electrical testing of integrated circuits, the term "test" refers to a plan $(\mathbf{c}, M, d)$ of instructions that is carried out to obtain a go/no-go decision for each device under test (DUT):

1. **Operating conditions c** lay down the environment that the DUT is exposed to during the test. Operating conditions include for instance the temperature of the DUT during test, or resistive and capacitive loads that must be attached to the output nodes of the DUT during test. They also comprise voltage level and driver strength of the power supply as well as other constant or time-varying input voltages to the DUT, which are usually called **test stimuli**.
2. **Measurement instruction** M. After the environment of the DUT is set up, electrical measurements are taken at the circuit as prescribed in a measurement plan M. The results of these measurements are for instance time-series of voltages and currents, pole-zero plots, power spectra or histograms. To be able to process these unstructured data in a common, systematic way, a set of scalar **observations** $\mathbf{o} = (o_1, \ldots, o_{n_o})$ are extracted from them. Observations are for example the real part of a certain pole, the 3rd coefficient of the Fourier series of the output voltage, the input-output delay, the output

current at a certain moment in time etc. The observation vector $\mathbf{o}$ usually includes (but does not consist of exclusively) circuit performances that are part of the circuit specification.

3. **Decision rule** d. Based on these observations, the go/no-go decision is made by evaluation of a pre-defined decision rule d for the observations made:

$$d : \Omega_o \rightarrow \{0, 1\} \tag{19}$$

If $d(\mathbf{o}) = 1$, then the DUT is accepted and passes the test; if $d(\mathbf{o}) = 0$, then the DUT is rejected.

The decision rule separates the space of observations Ω_o into the sets of measurement results that pass the test

$$\mathsf{P} = \{\mathbf{o} \in \Omega_o \mid d(\mathbf{o}) = 1\} \tag{20}$$

and those that do not and are rejected

$$\bar{\mathsf{P}} = \{\mathbf{o} \in \Omega_o \mid d(\mathbf{o}) = 0\} = \Omega_o \setminus \mathsf{P} \,. \tag{21}$$

For testing for parametric faults it is usual, that specifications are tested separately, and the DUT is accepted only if it passes every single test. This does not mean that the specified performances are directly measured, but that for every specified performance there is a test decision function $d^{(i)}$ that is constructed in order to indicate as good as possible if the respective specification is fulfilled or not. Then,

$$\mathsf{P} = \bigcap_i \mathsf{P}^{(i)} \qquad \bar{\mathsf{P}} = \bigcup_i \bar{\mathsf{P}}^{(i)} \,, \tag{22}$$

since a DUT is accepted if it passes every test, and rejected, if it fails any test. A set of devices coming from production undergoes the test procedure and each device is either accepted and sent to the customer, or is rejected. The test should accept good devices and reject bad devices, but practical tests are unable to perfectly separate good from faulty devices due to systematic error and measurement error.

It is the task of test design to find decision functions $d^{(i)}$ that distinguish good from faulty devices as accurately as possible, i.e., that minimize the probabilities of misclassification. Misclassification of faulty devices leads to test escapees that lower the average product quality perceived by the customer. Misclassification of good devices lowers the overall yield of the process and increases production cost per device.

1.3.2. Testing for parametric faults For testing parametric faults, the relations between the vectors of statistical parameters $\mathbf{s}$, performances $\mathbf{f}$, and test observations $\mathbf{o}$ need to be examined. As shown in Section 1.2, the specifications $\mathbf{f} \geq \mathbf{b}$ in performance space define an acceptance region A_s in the space of statistical parameters.[2] Under the assumption that there is a bijective observation

[2]The term "acceptance region" was coined by research on design and is *not* the set of devices that are accepted by the test. A_s is rather the set of parameter vectors that make *acceptable*, i.e., good, devices.

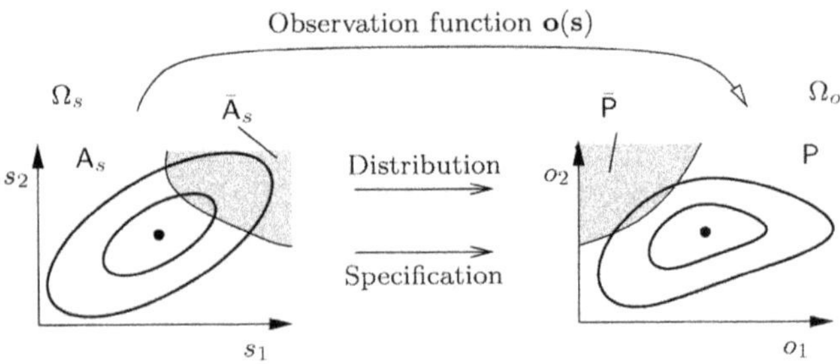

FIGURE 4. Ideal mapping from $\mathbf{s}$ to $\mathbf{o}$.

function $\mathbf{o}(\mathbf{s})$ and its inverse $\mathbf{s}(\mathbf{o})$, a perfect test is given by

$$\mathsf{P} = \{\mathbf{o} \in \Omega_o \mid \mathbf{s}(\mathbf{o}) \in \mathsf{A}_s\} = \{\mathbf{o} \in \Omega_o \mid \mathbf{f}(\mathbf{s}(\mathbf{o})) \geq \mathbf{b}\}\,, \tag{23}$$

because all good circuits pass the test and all faulty circuits are rejected. Figure 4 illustrates this concept.

Unfortunately, this ideal test does not exist in practice for a couple of reasons. As noted before, A_s is not explicitly given, which impedes the evaluation of (23), and every approximation introduces an error that may lead to misclassification. To achieve a bijective observation function $\mathbf{o}(\mathbf{s})$ with an existing inverse $\mathbf{s}(\mathbf{o})$, the number of observations must be equal to or exceed the number of statistical parameters. Requiring bijectivity therefore impedes testing or leads to large test sets even for specifications that could be tested easily by a small number of observations.

Furthermore, the variation of the statistical parameters is not the only factor that influences the test observations. Measurement error and variation in the test environment have a significant impact, too. Considering these factors, there is no observation function $\mathbf{o}(\mathbf{s})$ at all, because the same DUT (=the same parameter vector $\mathbf{s}$) may show slightly different test results $\mathbf{o}$ when a test is repeated. Quantization of measurement data in the test equipment is a special case of measurement error that inhibits an inversion of $\mathbf{o}(\mathbf{s})$, since different parameter values $\mathbf{s}$ lead to the same quantized observation vector $\mathbf{o}$.

Especially the last argument shows, that test design for parametric faults must consider the statistical properties of device parameters, performances, observations and measurement error. In contrast, equation (23) is a non-statistical, purely geometrical approach that wants to map A_s from the parameter space to P in the observation space.

Mapping the observation vectors $\mathbf{o}$ directly onto the performance vectors $\mathbf{f}$ by looking for a function $\mathbf{f}(\mathbf{o})$ required that

$$\mathbf{o}(\mathbf{s}_1) = \mathbf{o}(\mathbf{s}_2) \Rightarrow \mathbf{f}(\mathbf{s}_1) = \mathbf{f}(\mathbf{s}_2) \text{ for all } \mathbf{s}_1, \mathbf{s}_2 \in \Omega_s, \tag{24}$$

an assumption that is rarely justifiable and usually fails in presence of measurement error.

In the following, finding a test criterion d and its corresponding set P is always guided by the principle that optimality is defined by misclassification errors, which are to be optimized under consideration of systematic error and measurement error.

2. Sources of error in testing for parametric faults

For manual as well as for automatic test generation, it is important to know how accurately a given test strategy will classify good and faulty circuits. Measurement error plays an important role here and must be taken into account to assess the precision of a test strategy. A second, often underestimated cause of error is the limited degree of accuracy of analog test stimulus generation.

2.1. Test model

2.1.1. SOURCES OF VARIATION The observations $\mathbf{o}$ and the performance f vary between circuits due to statistical fluctuations of device and process parameters like t_{ox} or V_{th} variation. These variations are modeled by a statistical distribution of device and process parameters $\mathbf{s} \sim N(\mathbf{0}, \mathbf{\Sigma}_s)$.

There are random variations in the test equipment that cause the test environment to vary between measurements at different DUTs. For instance, the rise time or the frequency of a test input signal are subject to small statistical variations that are determined by the accuracy of the input signal generation. The same is valid for temperature, supply voltage, and so on. These variations are modeled by an additional random vector of test environment parameters $\varepsilon_c \sim N(\mathbf{0}, \mathbf{\Sigma}_c)$.

Another cause of variation of $\mathbf{o}$ is the finite precision of the measuring process which is determined by limitations in speed or resolution of the applied measurement equipment. This variation is modeled by a random vector $\varepsilon_m \sim N(\mathbf{0}, \mathbf{\Sigma}_m)$. In most cases, the measurement errors of the observations are independent, and therefore $\mathbf{\Sigma}_m$ is a diagonal matrix. We may assume, that every measurement is affected by measurement error and that in practice there is nothing like a perfect measurement with infinite accuracy. Therefore $\mathbf{\Sigma}_m$ is positive definite.

2.1.2. A LINEAR MODEL OF TEST A first order approximation of the observations $\mathbf{o}$ and one specified performance f is given by

$$\mathbf{o} = \mathbf{o}_0 + \mathbf{G} \cdot \mathbf{s} + \mathbf{H} \cdot \varepsilon_c + \varepsilon_m \tag{25}$$

$$f = f_0 + \mathbf{k}^{\mathrm{T}} \mathbf{s} \, . \tag{26}$$

This linearization is calculated at the worst-case point, because the boundary of the acceptance region is of interest here, not the nominal behavior.

Due to the nonlinearity of the real performance and observations with respect to the statistical parameters, this linear approximation introduces a linearization error that is neglected in the following. Although such a linear model is not always sufficient for design of an optimum test, the results show that in most cases it is a good enough approximation for analysis of the effects of changing error variances on the test decisions.

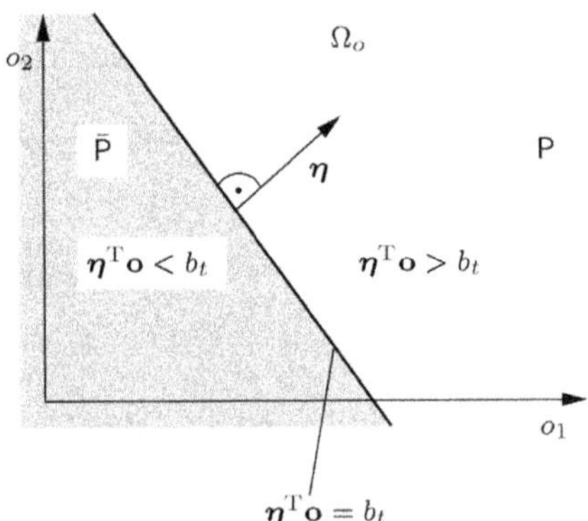

FIGURE 5. The set P of accepted devices for the linear model.

To define a test criterion, the test variable t is calculated from $\mathbf{o}$ via a linear function

$$t(\mathbf{o}) = \boldsymbol{\eta}^{\mathrm{T}}\mathbf{o}\,. \tag{27}$$

The test criterion is then defined by a test specification bound b_t on t:

$$d(\mathbf{o}) = \begin{cases} 1 & \text{if } t(\mathbf{o}) \geq b_t \\ 0 & \text{else} \end{cases} \tag{28}$$

in accordance with Eq. (19). The test decision rule is therefore fully specified by two parameters: The coefficient vector $\boldsymbol{\eta}$ and the test specification bound b_t. Since t is a linear function of $\mathbf{o}$, the set of accepted devices P is a linearly bounded half-space (see Fig. 5).

The performance f, the observations $\mathbf{o}$ and the test variable t are all normal Gaussian distributed with $f \sim N(f_0, \sigma_f^2)$, $\mathbf{o} \sim N(\mathbf{o}_0, \boldsymbol{\Sigma}_o)$ and $t \sim N(t_0, \sigma_t^2)$. The parameters of these distributions are

$$\sigma_f^2 = \mathbf{k}^{\mathrm{T}}\boldsymbol{\Sigma}_s\mathbf{k} \tag{29}$$

$$\boldsymbol{\Sigma}_o = \mathbf{G}\boldsymbol{\Sigma}_s\mathbf{G}^{\mathrm{T}} + \mathbf{H}\boldsymbol{\Sigma}_c\mathbf{H}^{\mathrm{T}} + \boldsymbol{\Sigma}_m\,, \tag{30}$$

and

$$t_0 = \boldsymbol{\eta}^{\mathrm{T}}\mathbf{o}_0 \qquad \sigma_t^2 = \boldsymbol{\eta}^{\mathrm{T}}\boldsymbol{\Sigma}_o\boldsymbol{\eta}\,. \tag{31}$$

The joint probability distribution of f and t is a multinormal distribution.

The correlation factor ρ between the performance f and the test variable t is

$$\rho = \frac{\mathbf{k}^{\mathrm{T}}\boldsymbol{\Sigma}_s\mathbf{G}^{\mathrm{T}}\boldsymbol{\eta}}{\sqrt{\boldsymbol{\eta}^{\mathrm{T}}\boldsymbol{\Sigma}_o\boldsymbol{\eta}\mathbf{k}^{\mathrm{T}}\boldsymbol{\Sigma}_s\mathbf{k}}}\,. \tag{32}$$

There is a performance specification (see Eq. (1))

$$f \overset{!}{\geq} b \tag{33}$$

with a worst-case distance $\beta_{\rm wc} = (f_0 - b)/\sigma_f$ (see Section 1.2). In analogy to $\beta_{\rm wc}$, we define the test distance

$$\beta_{\rm t} = \frac{t_0 - b_t}{\sigma_t} . \tag{34}$$

These two numbers $\beta_{\rm wc}$ and $\beta_{\rm t}$ are the distances between the performance specification bound and the nominal value of the performance (resp. the test specification bound and the nominal value of the test variable), expressed in multiples of the standard deviation.

2.1.3. MISCLASSIFICATION IN THE LINEAR MODEL Misclassification is caused by the three sources of variation:

1. Measurement error ε_m and
2. stimulus error ε_c cause a variation in the observations $\mathbf{o}$ (and consequently in the test variable t), that misleadingly indicates a change in the specified performance although there is none.
3. Random parameter fluctuations $\mathbf{s}$ influence both, test variable t and performance f. However, if the gradients $\nabla_{\mathbf{s}} t = \mathbf{G}^{\rm T}\boldsymbol{\eta}$ and $\nabla_{\mathbf{s}} f = \mathbf{k}$ differ, then there is always some kind of variation $\delta\mathbf{s}$ possible that will influence f ($\nabla_{\mathbf{s}}^{\rm T} f \cdot \delta\mathbf{s} \neq 0$), but not t ($\nabla_{\mathbf{s}}^{\rm T} t \cdot \delta\mathbf{s} = 0$). These undetectable variations increase the misclassification probability, too.

Fig. 6 shows one contour ellipsis of the joint probability density function of f and t, and the mean (f_0, t_0). There are four regions of interest with regard to classification:

1. In region I are all circuits where $f \geq b$, i.e., that fulfill the specification, and $t \geq b_t$, i.e., the test will accept them.
2. In region II, there are all circuits that violate their specification and are correctly rejected by the test.
3. Region III consists of all circuits that fulfill the specification but are wrongly rejected by the test.
4. In region IV are all circuits that violate the specification and yet pass the test.

Whereas the circuits in regions I and II are correctly classified, circuits in regions III and IV are either good but rejected or faulty but accepted. If we denote the probability volumes of these regions $\mathcal{P}_{\rm I}$, $\mathcal{P}_{\rm II}$, $\mathcal{P}_{\rm III}$, and $\mathcal{P}_{\rm IV}$, then these probability volumes can be calculated from

$$\begin{aligned} \mathcal{P}_{\rm I} + \mathcal{P}_{\rm III} &= \Phi_1(\beta_{\rm wc}) = Y & \mathcal{P}_{\rm II} + \mathcal{P}_{\rm IV} &= 1 - Y \\ \mathcal{P}_{\rm I} &= \Phi_2(\beta_{\rm wc}, \beta_{\rm t}, \rho) & \mathcal{P}_{\rm II} &= \Phi_2(-\beta_{\rm wc}, -\beta_{\rm t}, \rho) \end{aligned} \tag{35}$$

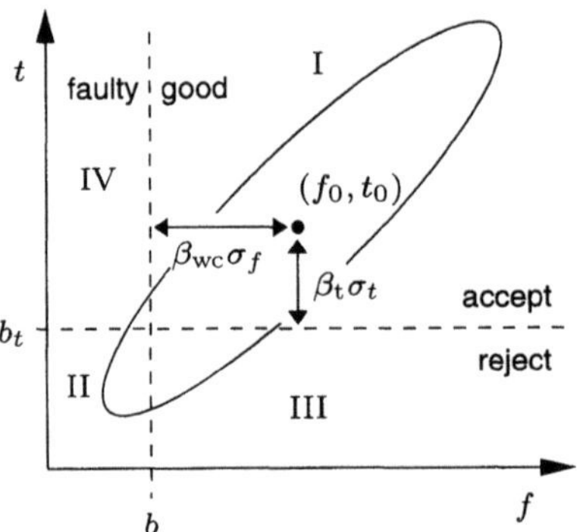

FIGURE 6. The probability densitiy function of the test variable t and the specified performance f.

with Φ_n being the n-dimensional Gaussian normal distribution functions

$$\Phi_1(x) = \frac{1}{\sqrt{2\pi}} \int_{-\infty}^{x} e^{-\xi^2/2} d\xi \tag{36}$$

$$\Phi_2(x, y, \rho) = \frac{1}{2\pi\sqrt{1-\rho^2}} \int_{-\infty}^{x} \int_{-\infty}^{y} e^{-(\xi^2 - 2\xi\zeta\rho + \zeta^2)/2(1-\rho^2)} \, d\zeta \, d\xi \tag{37}$$

and ρ being the correlation factor between f and t (Eq. (32)). There are fast and precise algorithms for a numerical evaluation of Φ_1 and Φ_2, see [3, 4].

We define *yield loss* by

$$y_l = \mathcal{P}\{\text{rejected}|\text{good}\} = \frac{\mathcal{P}\{\text{rejected} \wedge \text{good}\}}{\mathcal{P}\{\text{good}\}} = \frac{\mathcal{P}_{\text{III}}}{\mathcal{P}_{\text{I}} + \mathcal{P}_{\text{III}}} \tag{38}$$

as the portion of mistakenly rejected circuits in all good circuits, and *fault loss* by

$$f_l = \mathcal{P}\{\text{accepted}|\text{faulty}\} = \frac{\mathcal{P}\{\text{accepted} \wedge \text{faulty}\}}{\mathcal{P}\{\text{faulty}\}} = \frac{\mathcal{P}_{\text{IV}}}{\mathcal{P}_{\text{II}} + \mathcal{P}_{\text{IV}}} \tag{39}$$

as the portion of mistakenly accepted circuits in all bad circuits. Fault loss is nearly proportional to the *defect level*

$$d_l = \mathcal{P}\{\text{faulty}|\text{accepted}\} = \frac{\mathcal{P}\{\text{accepted} \wedge \text{faulty}\}}{\mathcal{P}\{\text{accepted}\}} = \frac{\mathcal{P}_{\text{IV}}}{\mathcal{P}_{\text{I}} + \mathcal{P}_{\text{IV}}}, \tag{40}$$

which is the portion of faulty devices among all bought devices.

2.2. Choice of test specification bound

Yield loss and fault loss should be as small as possible, but cannot be minimized independent from each other. Instead, there is a trade-off between both probabilities of misclassification, and any point on the trade-off curve may be preferrable.

There are four rather obvious, reasonable goals for selecting a test specification bound β_t, that are

1. minimize the sum of yield loss and fault loss:
$$\beta_t^* = \underset{\beta_t}{\operatorname{argmin}}(y_l + f_l) = \frac{1-\sqrt{1-\rho^2}}{\rho} \cdot \beta_{wc} . \tag{41}$$
2. minimize the larger one of both:
$$\min_{\beta_t} \max(y_l, f_l) , \tag{42}$$
Optimization of this min max-goal is equivalent to solving $y_l(\beta_t) = f_l(\beta_t)$ for β_t, because $y_l(\beta_t)$ is monotonic decreasing over β_t and $f_l(\beta_t)$ is monotonic increasing, both are continuous and bounded between 0 and 1. From a first order approximation of y_l and f_l at β_t^*, we get
$$\begin{aligned}\beta_t^{**} &= \underset{\beta_t}{\operatorname{argmin}} \max(y_l, f_l) \\ &\approx \beta_t^* + \sqrt{\frac{\pi}{2}} e^{\beta_t^{*2}/2}(y_l(\beta_t^*) - f_l(\beta_t^*)) .\end{aligned} \tag{43}$$
3. minimize the total probability of misclassification
$$\beta_t^g = \underset{\beta_t}{\operatorname{argmin}} Y \cdot y_l + (1-Y) \cdot f_l = \frac{1}{\rho} \cdot \beta_{wc} . \tag{44}$$
The total probability of misclassification is heavily biased towards yield coverage, because the parametric yield Y is maximized during the design of the circuit, and therefore $Y \gg 1 - Y$.
4. minimize yield loss for a given upper bound on defect level, for instance
$$\beta_t^d = \underset{\beta_t}{\operatorname{argmin}}\, y_l \quad \text{s.t. } d_l \leq 750\text{ppm} . \tag{45}$$

As we can see from Figure 6, a change in b_t will always decrease one probability of misclassification (P_{III} or P_{IV}) while increasing the other one. Hence, there is a trade-off between y_l and f_l over b_t (resp. over the normalized value β_t, which is used in the following as the trade-off parameter). As we can see from Eqs. (38), (39) and (35), the shape and position of a $(y_l(\beta_t), f_l(\beta_t))$ trade-off curve is fully specified by only two parameters: β_{wc} and ρ. As the former is constant for a given specification, ρ remains the only position parameter of these trade-off curves.

Fig. 7 shows some trade-off curves for $\beta_{wc} = 2.5$ and different values of ρ. It is a property of this linear model, that the trade-off curves over β_t intersect only at $(0,1)$ and $(1,0)$, and that a curve with a larger value of ρ always allows a more preferrable choice of misclassification errors than a curve with a smaller value of ρ. In Fig. 6, a larger value of ρ means a taller, thinner ellipsis, which reduces the volumes of both areas III and IV.

This trade-off is invariant to scale and position of t, i.e., a linearly transformed test variable $t' = c_0 + c_1 \cdot t$ has exactly the same y_l-f_l trade-off as t.

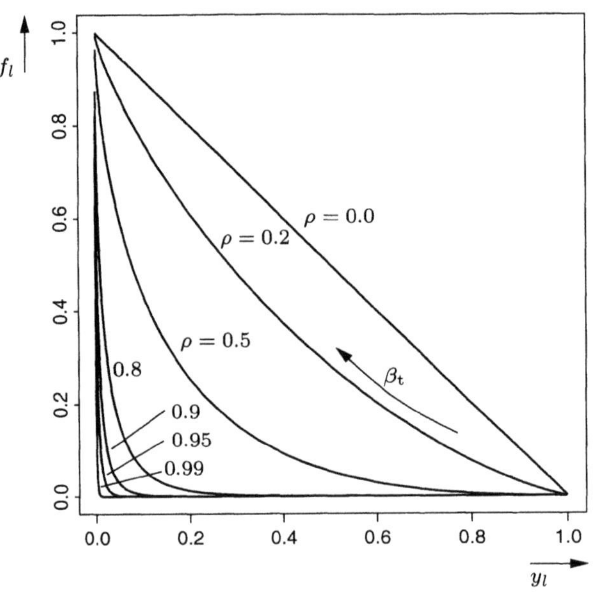

FIGURE 7. y_l-f_l trade-off for different values of ρ and $\beta_{\mathrm{wc}} = 2.5$.

Example: Consider the following model:

$$f = \mathbf{k}^{\mathrm{T}} \cdot \mathbf{s}$$
$$\mathbf{o} = \mathbf{G} \cdot \mathbf{s} + \boldsymbol{\varepsilon}$$
$$f \geq b = -4.0 \Rightarrow \text{fault-free}$$

with three statistical parameters s_1, s_2 and s_3, four measurements $m_1, \ldots, m_4$ and one performance f. The coefficients of this linear model are

$$\mathbf{k} = (-0.80, 6.00, 0.05)^{\mathrm{T}}$$

$$\text{and } \mathbf{G} = \begin{pmatrix} -10.65 & 10.65 & 0.00 \\ 6.00 & 13.50 & 1.50 \\ 12.00 & -9.00 & 0.00 \\ 0.00 & 14.85 & 1.50 \end{pmatrix} .$$

The model is already normalized with respect to the covariance matrices, that is $\mathbf{\Sigma}_s = \mathbf{I}_3$ and $\mathbf{\Sigma}_m = \mathbf{I}_4$. The condition number of $\mathbf{G}$ is 88 here. Since $\beta_{\mathrm{wc}} = 0.66$, the yield is approximately 75%.

A geometrical mapping of A_s to P in the sense of Eq. (23) is given by

$$t = \boldsymbol{\eta}^{\mathrm{T}} \mathbf{o} \geq b_t \Rightarrow \text{accept}$$
$$\text{with } \boldsymbol{\eta} = \mathbf{G}^{\mathrm{T}+} \mathbf{k} \,.$$

for the linear model of Eqs. (25) and (26).

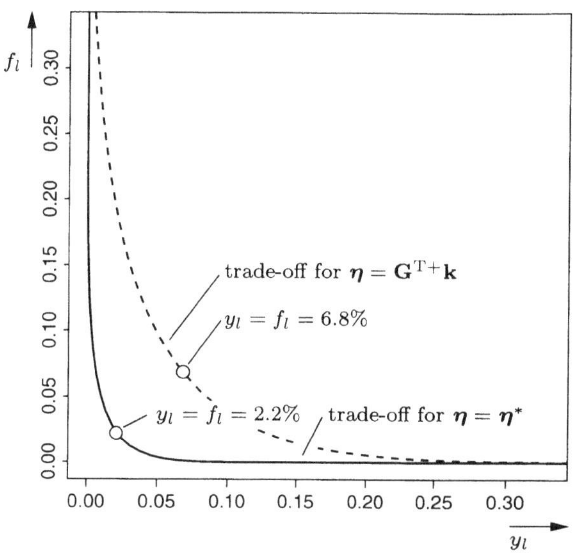

FIGURE 8. y_l-f_l trade-off for Example 1.

In this example, $\mathbf{G}^{\mathrm{T}+}\mathbf{k} = (0.8954, 0.8004, 0.3278, -0.7670)^{\mathrm{T}}$. Since there are more measurements than statistical parameters ($n_o > n_s$), $\mathbf{G}^{\mathrm{T}+}\mathbf{k}$ is the minimum norm solution to $\mathbf{G}^{\mathrm{T}}\boldsymbol{\eta} = \mathbf{k}$. For this value of $\boldsymbol{\eta}$, $\rho = 0.972$ from Eq. (32) and $\beta_{\mathrm{t}}^* = 0.520$ from Eq. (41). For $\beta_{\mathrm{t}} = \beta_{\mathrm{t}}^*$, $y_l = 8.13\%$ and $f_l = 5.31\%$ from Eqs. (38) and (39). From Eq. (43), $\beta_{\mathrm{t}}^{**} = 0.561$ and for $\beta_{\mathrm{t}} = \beta_{\mathrm{t}}^{**}$, $y_l = f_l = 6.8\%$. The dashed line in Figure 8 shows the y_l-f_l trade-off curve of this vector $\boldsymbol{\eta}$.

2.3. Pareto-optimal tests

The term "Pareto-optimality" of a configuration means that for both criteria that should be minimized simultaneously (y_l and f_l here), an improvement of one goal can only be achieved by increasing the other one. The trade-off curves in Fig. 7 contain all Pareto-optimal points of the optimization "minimize y_l and f_l for a given $\boldsymbol{\eta}$".

The choice of $\boldsymbol{\eta}$ determines the correlation factor ρ between test variable and specified performance. A high correlation yields a test with a better y_l-f_l trade-off than a low correlation. There is at least one value $\boldsymbol{\eta}^*$ that maximizes ρ and therefore has a trade-off curve that lies beneath all others. Hence, all tests with $\boldsymbol{\eta} = \boldsymbol{\eta}^*$ are Pareto-optimal with respect to the optimization "minimize y_l and f_l".

There are no restrictions to the choice of $\boldsymbol{\eta}$ and it is therefore reasonable to assume that the optimal value of $\boldsymbol{\eta}$ will always be preferred to any other choice. For further analysis, we will at first determine this optimum by maximizing $\rho(\boldsymbol{\eta})$.

Maximizing ρ is equivalent to maximizing ρ^2 under the constraint $\rho \geq 0$, which gives the problem

$$\max_{\boldsymbol{\eta}} \frac{(\mathbf{k}^{\mathrm{T}}\boldsymbol{\Sigma}_s\mathbf{G}^{\mathrm{T}}\boldsymbol{\eta})^2}{\boldsymbol{\eta}^{\mathrm{T}}\boldsymbol{\Sigma}_o\boldsymbol{\eta}} , \tag{46}$$

omitting the constant positive term $\mathbf{k}^{\mathrm{T}}\boldsymbol{\Sigma}_s\mathbf{k}$.

With the abbreviation $\mathbf{a} := \mathbf{G}\boldsymbol{\Sigma}_s\mathbf{k}$, the optimality condition is

$$\nabla_{\boldsymbol{\eta}} \frac{(\mathbf{a}^{\mathrm{T}}\boldsymbol{\eta})^2}{\boldsymbol{\eta}^{\mathrm{T}}\boldsymbol{\Sigma}_o\boldsymbol{\eta}} = \frac{\boldsymbol{\eta}^{\mathrm{T}}\boldsymbol{\Sigma}_o\boldsymbol{\eta}\cdot 2(\mathbf{a}^{\mathrm{T}}\boldsymbol{\eta})\mathbf{a} - (\mathbf{a}^{\mathrm{T}}\boldsymbol{\eta})^2\cdot 2\boldsymbol{\Sigma}_o\boldsymbol{\eta}}{(\boldsymbol{\eta}^{\mathrm{T}}\boldsymbol{\Sigma}_o\boldsymbol{\eta})^2} \stackrel{!}{=} 0 , \tag{47}$$

which is solved by either $\mathbf{a}^{\mathrm{T}}\boldsymbol{\eta} = 0$ or by

$$\boldsymbol{\eta}^{\mathrm{T}}\boldsymbol{\Sigma}_o\boldsymbol{\eta}\cdot\mathbf{a} = \mathbf{a}^{\mathrm{T}}\boldsymbol{\eta}\cdot\boldsymbol{\Sigma}_o\boldsymbol{\eta} . \tag{48}$$

Eq. (32) shows, that $\mathbf{a}^{\mathrm{T}}\boldsymbol{\eta} = 0$ implies $\rho = 0$ and is therefore a minimum.

From Eq. (48) we see that $\mathbf{a}$ and $\boldsymbol{\Sigma}_o\boldsymbol{\eta}$ must be collinear, because $\boldsymbol{\eta}^{\mathrm{T}}\boldsymbol{\Sigma}_o\boldsymbol{\eta}$ and $\mathbf{a}^{\mathrm{T}}\boldsymbol{\eta}$ are scalars. Applying the collinearity $\boldsymbol{\Sigma}_o\boldsymbol{\eta} = \alpha\cdot\mathbf{a}$ to (48) gives

$$\alpha\cdot\boldsymbol{\eta}^{\mathrm{T}}\mathbf{a}\cdot\mathbf{a} = \mathbf{a}^{\mathrm{T}}\boldsymbol{\eta}\cdot\alpha\cdot\mathbf{a} , \tag{49}$$

which is true for every α. Therefore any $\boldsymbol{\eta}$ that solves $\alpha\cdot\boldsymbol{\Sigma}_o\boldsymbol{\eta} = \mathbf{a}$ for any value of $\alpha \neq 0$ will satisfy (47). To satisfy the constraint $\rho \geq 0$, we see from (32) that α must be positive. Since the actual choice of α is equivalent to scaling t, which doesn't have any influence on the y_l-f_l trade-off, we may arbitrarily choose $\alpha = 1$ and therefore solve

$$\boldsymbol{\Sigma}_o\boldsymbol{\eta}^* = \mathbf{G}\boldsymbol{\Sigma}_s\mathbf{k} . \tag{50}$$

There is always a well-defined unique solution to Eq. (50), because $\boldsymbol{\Sigma}_o$ is square and positive definite. Since $\mathbf{a}^{\mathrm{T}}\boldsymbol{\eta}^* = \boldsymbol{\eta}^{*\mathrm{T}}\boldsymbol{\Sigma}_o\boldsymbol{\eta}^* > 0$, the correlation $\rho(\boldsymbol{\eta}^*)$ is positive. Therefore, $\boldsymbol{\eta}^*$ must be a maximum, because ρ^2 is bounded between 0 and 1, and there are no other extrema of ρ^2 but multiples of $\boldsymbol{\eta}^*$ and the minima that solve $\mathbf{a}^{\mathrm{T}}\boldsymbol{\eta} = 0$.

The right-hand side of (50), $\mathbf{G}\boldsymbol{\Sigma}_s\mathbf{k}$, is the covariance between f and $\mathbf{o}$. Equation (50) can be interpreted as a linear least squares regression of f over $\mathbf{o}$ if the sample size approaches infinity.

Example (continued): Consider $\boldsymbol{\eta}^*$ for the test decision, "$t = \boldsymbol{\eta}^{\mathrm{T}}\mathbf{o} \geq b_t \Rightarrow$ accept" with $\boldsymbol{\eta} = \boldsymbol{\eta}^* = (\mathbf{G}\mathbf{G}^{\mathrm{T}} + \mathbf{I})^{-1}\mathbf{G}\mathbf{k} = (0.1512, 0.2048, -0.0374, 0.0838)^{\mathrm{T}}$.

For this value of $\boldsymbol{\eta}$, $\rho = 0.997$ from Eq. (32) and $\beta_{\mathrm{t}}^* = 0.611$ from Eq. (41). For $\beta_{\mathrm{t}} = \beta_{\mathrm{t}}^*$, $y_l = 2.71\%$ and $f_l = 1.59\%$ from Eqs. (38) and (39). From Eq. (43), $\beta_{\mathrm{t}}^{**} = 0.628$ and for $\beta_{\mathrm{t}} = \beta_{\mathrm{t}}^{**}$, $y_l = f_l = 2.2\%$. The solid line in Figure 8 shows the y_l-f_l trade-off curve of this vector $\boldsymbol{\eta}^*$. Any multiple $c\cdot\boldsymbol{\eta}^*$ of $\boldsymbol{\eta}^*$ will lead to the same trade-off curve.

Figure 8 shows, that the geometrical mapping of A_s into the observation space leads to a significantly sub-optimal test. Norming the observations to make $\boldsymbol{\Sigma}_m = \mathbf{I}_4$ is necessary: a heteroscedastic $\boldsymbol{\Sigma}_m$ would make it even worse [7].

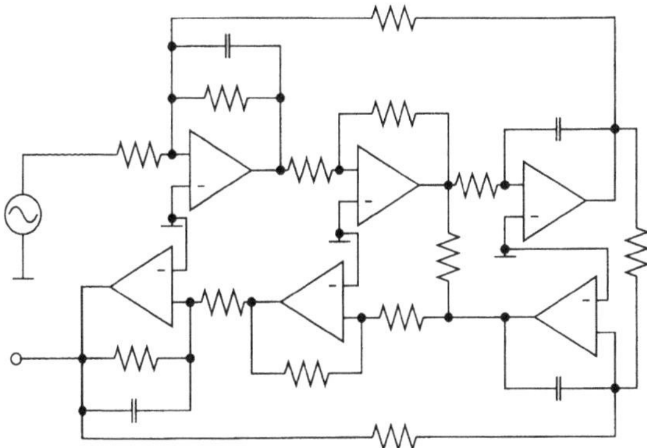

FIGURE 9. Mixed-signal test benchmark circuit from [6].

2.4. Simulation results

The leapfrog filter circuit shown in Figure 9 together with a simple test setup is used to demonstrate our approach. This circuit was proposed as a benchmark circuit for mixed-signal test in [6]. The specified performance of this filter is the 3dB cut-off frequency $f_{3\mathrm{dB}}$ with a specification $f_{3\mathrm{dB}} \geq 1300\mathrm{Hz}$ and a mean value of 1350Hz.

The test input signal is a square pulse. The parametric faults considered in this example are variations in the resistors and capacitances. The 10kΩ resistors R1–R13 vary independently with $3\sigma_R = 5\% \cdot R_{\mathrm{nom}} = 0.5\mathrm{k}\Omega$, and the four capacitors C1–C4 vary independently with $3\sigma_C = 5\% \cdot C_{\mathrm{nom}}$.

The test stimulus generation is subject to three different kinds of variations that are illustrated in Figure 10:

- $\varepsilon_{t,1}$ is a delay (jitter) of the signal, $\sigma(\varepsilon_{t,1}) = 5\mu\mathrm{s}$.
- $\varepsilon_{t,2}$ is a rise time variation, $\sigma(\varepsilon_{t,2}) = 3\mu\mathrm{s}$.
- $\varepsilon_{t,3}$ is a variation in the maximum input voltage, $\sigma(\varepsilon_{t,3}) = 10\mathrm{mV}$.

To take the precision of the output voltage measurement into account, an uncertainty of $\sigma_\varepsilon = 2\mathrm{mV}$ was introduced. The output signal was used to derive 20 different test measurement values $\mathbf{m}$: At first, the signal was sampled at 10 equidistant points to get 10 voltage values $v_1, \ldots, v_{10}$. Then, a Fourier transform was applied to the signal, and the magnitude of the first 10 coefficients $(c_1, \ldots, c_{10})$ were used for measurement.

Figure 11 compares the conventional approach and the optimal test combination $\boldsymbol{\eta}^*$ in how the yield loss depends on the variance of measurement error. Two cases are examined:

a) The stimulus error is not present, i.e., an absolutely perfect generation of test stimulus is assumed to be possible: $\mathbf{C}_c = \mathbf{0}$.
b) Stimulus error is present and considered during test design and validation.

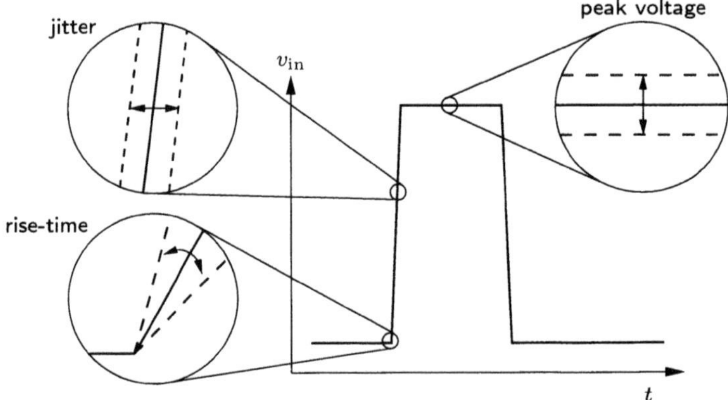

FIGURE 10. Variations of the input signal

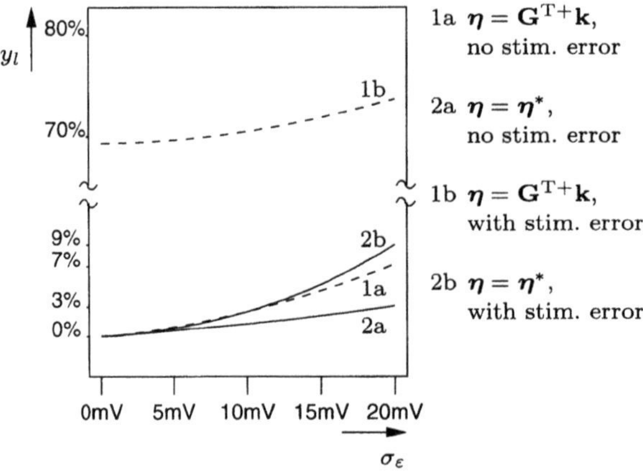

FIGURE 11. Yield loss y_l for a target defect level of 750ppm increases with measurement error.

The curves labeled "1a" and "2a" show the dependency of the yield loss y_l on the standard deviation of measurement error for a given maximum defect level $d_l \leq 750$ppm when there is no stimulus error present. The new approach (2a) decreases the demand of measurement precision, because the same yield loss is achieved at about twice the measurement error level than before (1a).

The curves labeled "1b" and "2b" show the dependency of the yield loss y_l on the standard deviation of measurement error for a given maximum defect level $d_l \leq 750$ppm when stimulus error is present. Since the conventional parameter estimation does not consider the moderate stimulus variation, the yield loss grows dramatically (1b).

The simulation results in Fig. 7 show that implicit test design is achievable at reasonable yield loss for a given defect level, if test environment error and measurement error are carefully considered.

3. Conclusion

We presented the problem of finding an optimal test decision vector for parametric faults in analog integrated circuits. In a linear model that includes both test stimulus and measurement errors, the influence of such errors on the test decision was analyzed in order to calculate trade-off curves for the probabilities of misclassification. Further possible applications are test selection and fault diagnosis.

References

[1] K. Antreich, J. Eckmueller, H. Graeb, M. Pronath, F. Schenkel, R. Schwencker, and S. Zizala. WiCkeD: Analog circuit synthesis incorporating mismatch. In *IEEE Custom Integrated Circuits Conference (CICC)*, pages 511–514, May 2000.

[2] K. Antreich, H. Graeb, and C. Wieser. Circuit analysis and optimization driven by worst-case distances. *IEEE Transactions on Computer-Aided Design of Circuits and Systems*, 13(1):57–71, January 1994.

[3] R.J. Bagby. Calculating normal probabilities. *Amer. Math. Monthly*, 102:46–49, 1995.

[4] Z. Drezner and G.O. Wesolowsky. On the computation of the bivariate normal integral. *Journal of Statistical Computation and Simulation*, 35:101–107, 1990.

[5] U. Feldmann, U. Wever, Q. Zheng, R. Schultz, and H. Wriedt. Algorithms for modern circuit simulation. *Archiv für Elektronik und Übertragungstechnik (AEÜ)*, 46:274–285, 1992.

[6] B. Kaminska et al. Analog and mixed-signal benchmark circuits – first release. In *IEEE International Test Conference (ITC)*, 1997.

[7] Walter M. Lindermeir, Helmut E. Gräb, and Kurt J. Antreich. Analog testing by characteristic observation inference. *IEEE Transactions on Computer-Aided Design of Circuits and Systems*, 18(9):1353–1368, September 1999.

[8] G. Müller-Liebler. PASTA – The characterization of the inherent fluctuations in the fabrication process for circuit simulation. *International Journal of Circuit Theory and Applications*, 23:413–432, 1995.

[9] Lawrence T. Pillage, Ronald A. Rohrer, and Chandramouli Visweswariah. *Electronic Circuit and System Simulation Methods*. McGraw-Hill, Inc., 1995.

Michael Pronath, Helmut Graeb, Kurt Antreich:
Technical University of Munich
Arcisstrasse 21
80333 Munich, Germany
email: {michael.pronath,helmut.graeb,kurt.antreich}@eda.ei.tum.de

Modeling, Simulation and Optimization of Integrated Circuits
(Eds.) K. Antreich, R. Bulirsch, A. Gilg and P. Rentrop

International Series of Numerical Mathematics Vol. 146, 303–318

Stochastic DAEs in Circuit Simulation

Werner Römisch and Renate Winkler

Abstract. Stochastic differential-algebraic equations (SDAEs) arise as a mathematical model for electrical network equations that are influenced by additional sources of Gaussian white noise. We sketch the underlying analytical theory for the existence and uniqueness of strong solutions, provided that the systems have noise-free constraints and are uniformly of DAE-index 1. In the main part we analyze discretization methods. Due to the differential-algebraic structure, implicit methods will be necessary. We start with a general p-th mean stability result for drift-implicit one-step methods applied to stochastic differential equations (SDEs). We discuss its application to drift-implicit Euler, trapezoidal and Milstein schemes and show how drift-implicit schemes for SDEs can be adapted to become directly applicable to stochastic DAEs. Test results of a drift-implicit Euler scheme with a mean-square step size control are presented for an oscillator circuit.

1. Introduction

Electrical noise limits the performance of electronic circuits and, hence, requires the analysis or simulation of its effects. Due to decreasing signal to noise ratios in special applications linear noise analysis around the deterministic solution is no longer satisfactory. The noise influences such systems in an essentially nonlinear way. We deal with two sources of electrical noise, namely, thermal noise of resistors and shot noise of pn-junctions. They are modelled as external Gaussian white noise sources in parallel to the original element (see Figures 1 and 2). Nyquist's theorem (see, e.g., [2, 4, 22]) states that the current through an arbitrary linear resistor having a resistance R, maintained in thermal equilibrium at a temperature T, can be described as the sum of the noiseless, deterministic current and a current due to a Gaussian white noise process with spectral density $S_{th} := \frac{2kT}{R}$, where k is Boltzmann's constant. Hence, the additional current is modelled as

$$I_{th} = \sigma_{th} \cdot \xi(t) = \sqrt{\tfrac{2kT}{R}} \cdot \xi(t),$$

where $\xi(t)$ is a standard Gaussian white noise process. In [21, 22] a thermodynamical foundation to apply this model to mildly nonlinear resistors and reciprocal networks is given.

This research has been supported by the BMBF-projekt 03-ROM3B3.

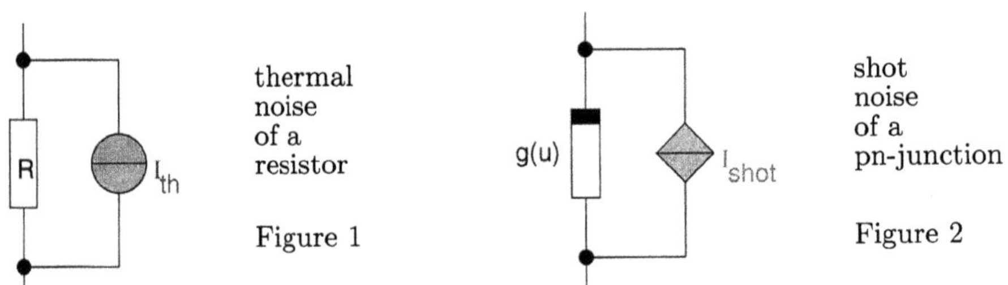

Figure 1

Figure 2

Shot noise of pn-junctions, caused by the discrete nature of current due to the elementary charge, is also modelled by a Gaussian white noise process, where the spectral density is proportional to the current I through the *pn*-junction: $S_{\text{shot}} := q|I|$, where q is the elementary charge. If the current through the *pn*-junction is described by a characteristic $I = g(u)$, where u is some voltage, the additional current is modelled by

$$I_{\text{shot}} = \sigma_{\text{shot}}(u) \cdot \xi(t) = \sqrt{q|g(u)|} \cdot \xi(t),$$

where $\xi(t)$ is a standard Gaussian white noise process. For a discussion of the model assumptions we refer to [2, 4, 21, 22].

The charge-oriented Modified Nodal Analysis (MNA) represents a standard tool in circuit simulation. The equations are generated automatically by combining the network topology, Kirchhoff's Current Law, and the characteristic equations describing the physical behaviour of the network elements. This results in large systems of DAEs, whose special structure was analyzed in a number of papers, e.g., [6, 8, 20]. We represent the topology of a network by means of the incidence matrix $(A_C, A_R, A_L, A_V, A_I, A_N)$, with indices referring to branches of capacitances, resistances, inductances, possibly controlled voltage and current sources, and n_N additional noise sources, respectively. Then the charge-oriented MNA system has the following structure (see [6, 8] for the deterministic case):

$$A_C q' + f_1(e, j_L, j_V, t) + A_N \text{diag}\,(\sigma(A_N^T e, t))\xi(t) = 0 \tag{1.1}$$

$$\phi' - A_L^T e = 0 \tag{1.2}$$

$$A_V^T e - v_s(e, j_L, t) = 0 \tag{1.3}$$

$$q - q_C(A_C^T e, t) = 0 \tag{1.4}$$

$$\phi - \phi_L(j_L, t) = 0, \tag{1.5}$$

where $f_1(e, j_L, j_V, t) := A_R g(A_R^T e, t) + A_L j_L + A_V j_V + A_I i_s(e, j_L, j_I, t)$, and q_C, g, ϕ_L, v_s, i_s, σ are given, noiseless functions. The vector of unknowns describing the system behaviour consists of all node potentials e, the branch currents of current-controlled elements (inductances and voltage sources) j_L, j_V, and the charges q of capacitances, and the fluxes ϕ of inductances. ξ denotes an n_N-dimensional vector of independent standard Gaussian white noise processes. In industry-relevant applications one has to deal with a large number of unknowns and noise sources.

The first block of equations (1.1) means a stochastic integral equation :

$$A_C q(s)\Big|_{t_0}^{t} + \int_{t_0}^{t} f_1(x(s),s)ds + \int_{t_0}^{t} A_N \text{diag}\,(\sigma)(x(s),s)dw(s) = 0,$$

where the second integral is an Itô integral, and w denotes an n_N-dimensional Wiener process (or Brownian motion) given on the probability space $(\Omega, \mathcal{F}, P)$ with a filtration $(\mathcal{F}_t)_{t\geq t_0}$ (see, e.g., [3, 9] for the stochastic background). A solution $x = x(t,\omega)$ is a stochastic process depending on the time t and the chance element $\omega \in \Omega$. The parameter ω is omitted in the notations above. The solution $x(t) = x(t,\cdot)$ for fixed time t is a vector-valued random variable in $L^p(\Omega)$, $p \geq 1$, a realization $x(\cdot,\omega)$ is called a path.

The equations (1.1)–(1.5) form a specially structured *Stochastic Differential Algebraic Equation* (SDAE) of the type

$$Ax(s)\Big|_{t_0}^{t} + \int_{t_0}^{t} f(x(s),s)ds + \int_{t_0}^{t} G(x(s),s)dw(s) = 0, \tag{1.6}$$

where A is a constant singular matrix, t varies over a compact interval $\mathcal{J}$. The short-hand notation

$$Ax'(t) + f(x(t),t) + G(x(t),t)\xi(t) = 0 \tag{1.7}$$

emphasizes the relations of (1.6) to its deterministic counterpart but may be misleading for readers who are less familiar with the stochastic background. Though the notation $x'(t)$ is used in (1.7), a typical realization $x(\cdot,\omega)$ of the solution is nowhere differentiable. A process $x(\cdot) = (x(t))_{t\in\mathcal{J}}$ is called a *strong solution* of (1.6) if it is adapted to the filtration (i.e., it does not depend on future information), and if, with probability 1, its sample paths are continuous, the integrals in (1.6) exist and (1.6) is satisfied.

In Section 2 we discuss some basics of an existence and uniqueness theory of strong solutions for SDAEs where we restrict to DAE-systems that have uniformly index 1 and noise-free constraints. In particular, we introduce the notion of an inherent regular SDE. The latter motivates to study discretization schemes first for SDEs. Hence, we provide in Section 3 a short introduction to p-th mean stability and convergence of general drift-implicit schemes. For the convenience of the reader the proof of the main stability result is shifted to the Appendix. In Section 4 we discuss several variants of drift-implicit schemes for SDAEs, namely, the drift-implicit Euler, trapezoidal and Milstein schemes. Special attention is paid to their convergence properties and to implementation issues. Finally, we report in Section 5 on numerical experience with the drift-implicit Euler scheme applied to the transient noise simulation in a ring-oscillator model.

2. Index 1 SDAEs

Due to the singularity of the matrix A the deterministic part of (1.6)

$$Ax'(t) + f(x(t),t) = 0, \tag{2.1}$$

where the solution x is now a deterministic function of t, forms a DAE. Solutions have to fulfil the constraints of the equation. The solution components belonging to $\ker A$ (we call them the algebraic components) do not occur under the differential operator d/dt, and the inherent dynamics live only in a lower-dimensional subspace. The DAE (2.1) is characterized as an index 1 DAE iff the constraints are locally solvable for the algebraic components. Solving an index 1 DAE involves a coupling of an integration task and a nonlinear equation solving task. If a DAE is of higher index, the constraints are not locally solvable for the algebraic components, and there exist solution components that are determined only by a hidden differentiation step, which may cause serious difficulties in the numerical solution of such problems (see, e.g., [1, 11]).

We assume here that the deterministic part (2.1) is globally an index 1 DAE in the sense that the constraints are regularly and globally uniquely solvable for the algebraic variables. The globally unique solvability is stronger than the deterministic index 1 condition, which requires only the non-singularity of the corresponding Jacobian and guarantees only local solvability of the constraints for the algebraic variables. The globally unique solvability holds for the MNA-system (1.1)–(1.5) if (see[23]) there are no loops of capacitances and voltage sources and no cut-sets of inductances or current sources, if the capacity, conductance, and inductance matrices are symmetric and uniformly positive definite, and if the controlled sources satisfy certain conditions described in [6] (see [6, 23]).

In [17, 18] it is shown that special conditions are needed to ensure solution processes that are not directly affected by white noise. Then the SDAEs are called SDAEs without direct noise, otherwise with direct noise. To avoid direct noise we have to assume that the noise sources do not appear in the constraints. This means that

$$\operatorname{im} G(x,t) \subseteq \operatorname{im} A \quad \forall (x,t) \in I\!R^n \times \mathcal{J}.$$

This is true for (1.7) if and only if there are always capacitances in parallel to a noise source. This is quite restrictive in the actual noise modelling (see also the example in Section 4). Nevertheless, one can also handle many situations where this condition is violated. Often noisy constraints are only needed for the determination of algebraic solution components that do not interact with the dynamical ones. Future work should be directed to a classification of such situations.
Under these conditions the constraints of the SDAE can be described by the deterministic equation

$$Rf(x(t),t) = 0,$$

where R is a projector along $\operatorname{im} A$, i.e., $R^2 = R$, $\ker R = \operatorname{im} A$. Solving the constraints for the algebraic components

$$Rf(u+v,t) = 0,\ Av = 0 \Longleftrightarrow v = \hat{v}(u,t),$$

inserting the result into the differential equations, and scaling the system by a pseudo-inverse A^- (with $AA^- = I - R, A^-A$ a projector along $\ker A$) leads to a so-called *inherent* regular SDE in the differential components u:

$$u' + A^- f(u + \hat{v}(u,t), t) + A^- G(u + v(u,t), t)\xi(t) = 0 . \tag{2.2}$$

It can be shown that (2.2), together with $x(t) = u(t) + \hat{v}(u(t), t)$, is equivalent to (1.6). Based on this, the following theorem on the existence and uniqueness of strong solutions of (1.6) is proved in [23]:

Theorem 2.1. *Let the above conditions be satisfied for* (1.6), *and assume that f and G are globally Lipschitz continuous with respect to x, continuous with respect to t, and that Ax^0 is $\mathcal{F}_{t_0}$-measurable, independent of the Wiener process w, and has finite p-th mean for some $p \geq 1$.*
Then there exists a strong solution $x(\cdot)$ of the initial value problem

$$Ax(t) - Ax^0 + \int_{t_0}^{t} f(x(s), s)ds + \int_{t_0}^{t} G(x(s), s)dw(s) = 0, \tag{2.3}$$

which is pathwise unique. Moreover, the solution $x(\cdot)$ has finite p-th mean.

Similarly, convergence properties of suitable drift-implicit discretization schemes for SDEs carry over to SDAEs. In the next section we therefore give some basic results for the discretization of SDEs.

3. Time discretization of stochastic differential equations

We consider the initial value problem for the SDE

$$x(s)\Big|_{t_0}^{t} + \int_{t_0}^{t} f(x(s), s)ds + \int_{t_0}^{t} G(x(s), s)dw(s) = 0, \; t \in \mathcal{J}, \; x(t_0) = x_0, \tag{3.1}$$

where $\mathcal{J} = [t_0, T]$, $f : I\!R^n \times \mathcal{J} \to I\!R^n$, $G : I\!R^n \times \mathcal{J} \to I\!R^{n\times m}$, w is an m-dimensional Wiener process on a given probability space $(\Omega, \mathcal{F}, P)$ with a filtration $(\mathcal{F}_t)_{t\in\mathcal{J}}$, and x_0 is a given $\mathcal{F}_{t_0}$-measurable initial value, which is independent of the Wiener process w. We assume that there exists a pathwise unique strong solution $x(\cdot)$.
Let us consider a generally drift-implicit discretization scheme of the form

$$x_\ell = x_{\ell-1} + \varphi(x_{\ell-1}, x_\ell; t_{\ell-1}, h_\ell) + \psi(x_{\ell-1}; t_{\ell-1}, h_\ell, I_{t_{\ell-1},h_\ell}), \; \ell = 1, \ldots, N, \tag{3.2}$$

on the deterministic grid $t_0 < t_1 < \ldots < t_N = T$ with stepsizes $h_\ell := t_\ell - t_{\ell-1}$, $\ell = 1, \ldots, N$. Here, φ and ψ are functions defined on $I\!R^n \times I\!R^n \times \mathcal{T}$ and $I\!R^n \times \mathcal{T} \times I\!R^M$ with $\mathcal{T} := \{(t,h) : t, t+h \in \mathcal{J}, h \in I\!R_+\}$, respectively, and mapping to $I\!R^n$. By $I_{t,h}$ we denote a vector of M multiple stochastic integrals having the form

$$I_{i_1,\ldots,i_k;t,h} = \int_t^{t+h} \int_t^{s_1} \cdots \int_t^{s_{k-1}} dw_{i_1}(s_1) dw_{i_2}(s_2) \cdots dw_{i_k}(s_k)$$

where the indices $i_1, \ldots, i_k$ are in $\{0, 1, \ldots, m\}$, k is bounded by certain finite order $k_{\max}$ and $dw_0(s)$ corresponds to ds.

For example, for the family of drift-implicit Euler schemes

$$x_\ell := x_{\ell-1} + h_\ell(\alpha f(x_\ell, t_\ell) + (1-\alpha) f(x_{\ell-1}, t_{\ell-1})) + G(x_{\ell-1}, t_{\ell-1})\Delta w_\ell, \; \ell = 1, \ldots, N,$$

where $\alpha \in [0,1]$, and $\Delta w_\ell := (w(t_\ell) - w(t_{\ell-1})) = (I_{i;t_{\ell-1},h_\ell})_{i=1}^m$, one has $k_{\max} = 1$, $M = m$, and

$$\begin{aligned} \varphi(z,x;t,h) &:= h(\alpha f(x,t+h) + (1-\alpha) f(z,t)), \\ \psi(z;t,h,I_{t,h}) &:= G(z,t)(w(t+h) - w(t)) = \sum_{i=1}^m g_i(z,t) \int_t^{t+h} dw_i(s), \end{aligned}$$

where $g_i(z,t)$, $i = 1,\dots,m$, are the columns of the matrix $G(z,t)$.
The family of drift-implicit Milstein schemes differs from the Euler schemes by an additional correction term for the stochastic part. The Milstein schemes are described by the same function φ, and $k_{\max} = 2$, $M = m + m^2$, and

$$\psi(z;t,h,I_{t,h}) := G(z,t)\Delta w_{t,h} + \sum_{j=1}^m (g_{j\,x} G)(z,t) I_{(j);t,h},$$

where $\Delta w_{t,h} := w(t+h) - w(t) = (I_{i;t,h})_{i=1}^m$, and $I_{(j);t,h} := (I_{j,i;t,h})_{i=1}^m$.

In [23], a result on numerical stability of drift-implicit schemes (3.2) in the mean-square sense has been derived which allows to study the behaviour of (3.2) under perturbations. Next we present a variant of such a stability result which supplements and extends Theorem 5 in [23].

Theorem 3.1. *Let $p \ge 1$ and x_0 have finite p-th mean. Assume that the scheme (3.2) satisfies the following properties:*

- *for all $z, \tilde z, x, \tilde x \in \mathbb{R}^n$, $(t,h) \in \mathcal{T}$, $h \le h^1$ we have*
  (A1) $|\varphi(z,x;t,h) - \varphi(\tilde z,\tilde x;t,h)| \le h(L_1|z-\tilde z| + L_2|x - \tilde x|)$
  for some positive constants h^1, L_1, L_2.
- *for all $(t,h) \in \mathcal{T}$, $h \le h^1$, and $\mathcal{F}_t$-measurable random vectors y, $\tilde y$ we have*
  (A2) $\mathbb{E}(\psi(y;t,h,I_{t,h}) - \psi(\tilde y;t,h,I_{t,h})|\mathcal{F}_t) = 0,$
  (A3) $\mathbb{E}(|\psi(y;t,h,I_{t,h}) - \psi(\tilde y;t,h,I_{t,h})|^p|\mathcal{F}_t) \le h^{\frac{p}{2}} L_3^p |y - \tilde y|^p,$
  (A4) $\mathbb{E}|\psi(0;t,h,I_{t,h})|^p < \infty,$
  for some constant $L_3 > 0$.

Then there exists constants $a \ge 1$, $h^0 > 0$ and a stability constant $S > 0$ such that the following holds true for each grid $\{t_0, t_1, \dots, t_N\}$ having the property $h := \max_{\ell=1,\dots,N} h_\ell \le h^0$ and $h \cdot N \le a \cdot (T - t_0)$:
For all $\mathcal{F}_{t_0}$-measurable random vectors x_0^*, $\tilde x_0$ having finite p-th mean, for all $\ell \in \{1,\dots,N\}$ and $\mathcal{F}_{t_\ell}$-measurable perturbations d_ℓ^*, $\tilde d_\ell$ having finite p-th mean the perturbed discrete system

$$\tilde x_\ell = \tilde x_{\ell-1} + \varphi(\tilde x_{\ell-1}, \tilde x_\ell; t_{\ell-1}, h_\ell) + \psi(\tilde x_{\ell-1}; t_{\ell-1}, h_\ell, I_{t_{\ell-1},h_\ell}) + \tilde d_\ell, \tag{3.3}$$

$\ell = 1,\dots,N$, has a unique solution $\{\tilde x_\ell\}_{\ell=0}^N$, and the following estimates are valid for any two solutions $\{x_\ell^*\}_{\ell=1}^N$ and $\{\tilde x_\ell\}_{\ell=0}^N$ of the perturbed discrete systems with

perturbations $\{d_\ell^*\}_{\ell=1}^N$ *and* $\{\tilde{d}_\ell\}_{\ell=1}^N$*:*

$$\mathbb{E}\max_{\ell=1,\dots,N}|x_\ell^*-\tilde{x}_\ell|^p \le S^p\Big(\mathbb{E}|x_0^*-\tilde{x}_0|^p+\frac{\max_{\ell=1,\dots,N}\mathbb{E}|s_\ell|^p}{h^{\frac{p}{2}}}+\frac{\mathbb{E}\max_{\ell=1,\dots,N}|r_\ell|^p}{h^p}\Big), \quad (3.4)$$

$$\max_{\ell=1,\dots,N}\mathbb{E}|x_\ell^*-\tilde{x}_\ell|^p \le S^p\Big(\mathbb{E}|x_0^*-\tilde{x}_0|^p+\frac{\max_{\ell=1,\dots,N}\mathbb{E}|s_\ell|^p}{h^{\frac{p}{2}}}+\frac{\max_{\ell=1,\dots,N}\mathbb{E}|r_\ell|^p}{h^p}\Big), \quad (3.5)$$

where $d_\ell := d_\ell^* - \tilde{d}_\ell$ *is splitted such that* $d_\ell = r_\ell + s_\ell$ *with* $\mathbb{E}(s_\ell|\mathcal{F}_{t_{\ell-1}}) = 0$.

The proof of Theorem 3.1 is given in the appendix.

Theorem 3.1 applies immediately to well-known schemes for SDEs. Here, we check the assumptions of Theorem 3.1 for the families of drift-implicit Euler and Milstein schemes. Condition (A1) follows from the Lipschitz continuity of the drift coefficient f, (A2) holds due to the explicit, non-anticipative discretization of the diffusion term, and the technical condition (A4) is satisfied since the function $G(0,\cdot)$ (and the functions $g_{j\,x}G(0,\cdot)$) are bounded on the compact interval $\mathcal{J}$. Condition (A3) is a consequence of standard properties of moments of stochastic integrals and the Lipschitz continuity of the diffusion coefficient G (and in case of the Milstein scheme of the functions $g_{j\,x}G$). For example, for the drift-implicit Euler scheme we obtain for any pair $(t,h)\in\mathcal{T}$ and any $\mathcal{F}_t$-measurable y, $\tilde{y}$

$$\begin{aligned}\mathbb{E}(|\psi(y;t,h,I_{t,h})-\psi(\tilde{y};t,h,I_{t,h})|^p|\mathcal{F}_t) &= |G(y,t)-G(\tilde{y},t)|^p\mathbb{E}(|\Delta w_{t,h}|^p|\mathcal{F}_t)\\ &\le L_G^p|y-\tilde{y}|^pC_ph^{\frac{p}{2}}\end{aligned}$$

where L_G is a Lipschitz constant of G and C_p a universal constant.

In the special case $x_\ell^* = x(t_\ell)$, the perturbations d_ℓ^* form the local discretization errors. We split them into

$$d_\ell^* = (d_\ell^* - \bar{d}_\ell^*) + \bar{d}_\ell^*, \quad \text{where } \bar{d}_\ell^* := \mathbb{E}(d_\ell^*|\mathcal{F}_{t_{\ell-1}}),$$

and obtain, in comparison with the exact solution of the numerical scheme x_ℓ,

$$\begin{aligned}\|\max_{\ell=1,\dots,N}|x(t_\ell)-x_\ell|\,\|_{L_p} &\le S_p\Big(\max_{\ell=1,\dots,N}\|d_\ell^*-\bar{d}_\ell^*\|_{L_p}/h^{\frac{1}{2}}+\|\max_{\ell=1,\dots,N}|\bar{d}_\ell^*|\,\|_{L_p}/h\Big),\\ \max_{\ell=1,\dots,N}\|x(t_\ell)-x_\ell\|_{L_p} &\le S_p\Big(\max_{\ell=1,\dots,N}\|d_\ell^*-\bar{d}_\ell^*\|_{L_p}/h^{\frac{1}{2}}+\max_{\ell=1,\dots,N}\|\bar{d}_\ell^*\|_{L_p}/h\Big),\end{aligned}$$

where $\|x\|_{L_p} := (\mathbb{E}|x|^p)^{1/p}$. If, by consistency arguments, the local error terms on the right-hand side are of order $\mathcal{O}(h^\gamma)$, we have global convergence of order γ.

4. Discretization schemes for index 1 SDAEs

Nowadays, a wide spectrum of discretization schemes for SDEs is available (cf. [3, 9, 12, 15]). However, SDAEs require special schemes. First decoupling the SDAE numerically and then applying a scheme to the resulting inherent SDE would be an inefficient procedure in general. We aim at numerical methods for SDAEs that work directly on the given implicit structure, as in the case of deterministic DAEs. Only little previous work has been done in this direction. In [17, 18] linear SDAEs are analyzed and the convergence of the drift-implicit Euler scheme is proved. In [14] a scheme with strong order 1 is developed for the specially structured SDAEs that arise in transient noise simulation for electronic circuits. Later we will point out its relation to the drift-implicit Milstein scheme.

Our approach also applies to nonlinear SDAEs. We present adaptations of known schemes for SDEs that are implicit in the deterministic and explicit in the stochastic part to the SDAE (1.6). Designing the methods such that the iterates x_ℓ have to satisfy the constraints of the SDAE at the current time-point t_ℓ

$$Rf(x_\ell, t_\ell) = 0,$$

is the key idea to adapt known methods for SDEs to (1.6).

The noise densities given in Section 1 contain small parameters. To exploit this in the analysis of the discretization errors we express the diffusion coefficient in the form

$$G(x,t) := \epsilon \tilde{G}(x,t), \quad \epsilon \ll 1. \tag{4.1}$$

4.1. Drift-implicit Euler scheme

On the deterministic grid $0 = t_0 < t_1 < \cdots < t_N = T$ the drift-implicit Euler scheme for (1.6) is given by

$$A\frac{x_\ell - x_{\ell-1}}{h_\ell} + f(x_\ell, t_\ell) + G(x_{\ell-1}, t_{\ell-1})\frac{1}{h_\ell}\Delta w_\ell = 0, \tag{4.2}$$

where $h_\ell = t_\ell - t_{\ell-1}$, $\Delta w_\ell = w(t_\ell) - w(t_{\ell-1})$. Realizations of Δw_ℓ can be simulated as $N(0, h_\ell I)$-distributed random variables. The Jacobian of 4.2 is the same as in the deterministic setting.

The scheme (4.2) for the SDAE (1.6) possesses the same convergence properties as the drift-implicit Euler scheme for SDEs. In general, its order of strong convergence is 1/2, i.e.,

$$\|x(t_\ell) - x_\ell\|_{L_p} = (E|x(t_\ell) - x_\ell|^p)^{1/p} \le c \cdot h^{1/2}, \quad h := \max_{\ell=1,\ldots,N} h_\ell,$$

holds for the p-th mean norm of the global errors for $p \ge 1$. For additive noise, i.e., $G(x,t) = G(t)$, the order of strong convergence is 1. For small noise, i.e., $G(x,t) = \epsilon\tilde{G}(x,t)$, the error is bounded by $O(h + \epsilon^2 h^{1/2})$ (see [16], or [13] for related results).

The smallness of the noise also allows special estimates of local error terms, which can be used to control the stepsize. The local error for the Euler scheme applied to SDEs with small noise is analyzed in [16]. As long as stepsizes with

$$h_\ell \gg \epsilon^2$$

are used, the dominating local error term of (4.2) is

$$\begin{aligned} \|\bar{d}_\ell^*\|_{L_p}/h_\ell = \mathcal{O}(h_\ell) &= \frac{1}{2}\|A^-(f(x_\ell, t_\ell) - f(x_{\ell-1}, t_{\ell-1}))\|_{L_p} + \mathcal{O}(\epsilon h_\ell^{1/2}) \\ &=: \eta_\ell + \mathcal{O}(\epsilon h_\ell^{1/2}), \end{aligned}$$

where A^- denotes a suitable pseudo-inverse of A. For $\epsilon \to 0$ it approaches the known error estimate in the deterministic setting. If an ensemble of solution paths is computed simultaneously, the estimate η_ℓ can be computed approximately and may be used to control the local error corresponding to a given tolerance. This results in an adaptive stepsize sequence that is uniform for all solution paths.

4.2. Drift-implicit Milstein scheme

We intend to design this method in such a way that it realizes the drift-implicit Milstein scheme for the inherent SDE $u' + \hat{f}(u,t) + \hat{G}(u,t)\xi(t) = 0$, i.e.,

$$\frac{u_\ell - u_{\ell-1}}{h_\ell} + \hat{f}(u_\ell, t_\ell) + \hat{G}(u_{\ell-1}, t_{\ell-1})\frac{1}{h_\ell}\Delta w_\ell - \sum_{j=1}^{m}(\hat{g}_{j\,u}\hat{G})(u_{\ell-1}, t_{\ell-1})\frac{1}{h_\ell}I_j^\ell = 0,$$

where

$$I_j^\ell = (I_{j,i}^\ell)_{i=1}^m, \; I_{j,i}^\ell = \int_{t_{\ell-1}}^{t_\ell}\int_{t_\ell}^{s} dw_i(\tau)dw_j(s), \; \hat{G} = (\hat{g}_1, \ldots, \hat{g}_m)$$

and

$$\hat{f}(u,t) := A^- f(u + \hat{v}(u,t), t), \; \hat{G}(u,t) := A^- G(u + \hat{v}(u,t), t).$$

The Milstein scheme is strongly convergent of order $\gamma = 1$. It differs from the Euler scheme by an additional correction term for the stochastic part, which includes double stochastic integrals. For additive noise the additional term vanishes and both schemes coincide.

The Milstein scheme for the inherent SDE is realized by

$$A\frac{x_\ell - x_{\ell-1}}{h_\ell} + f(x_\ell, t_\ell) + G(x_{\ell-1}, t_{\ell-1})\frac{\Delta w_\ell}{h_\ell} - \sum_{j=1}^{m}(g_{j\,x} x_u A^- G)(x_{\ell-1}, t_{\ell-1})\frac{I_j^\ell}{h_\ell} = 0,$$

where $G = (g_1, \ldots, g_m)$, which we call the drift-implicit Milstein scheme for (1.6).

We point out the explicit use of the inner derivative $x_u = I + \hat{v}_u$ and the scaling A^- in the last term. The inner derivative can be expressed as

$$x_u = I + \hat{v}_u = I - (A + \lambda R f_x)^{-1}\lambda R f_x = I - I + (A + \lambda R f_x)^{-1}A = (A + \lambda R f_x)^{-1}A$$

with a free parameter $\lambda \neq 0$. Choosing $\lambda = h$, it may be approximated via

$$\begin{aligned} x_u &= (A + hRf_x)^{-1}A &&= (A + hf_x)^{-1}A \quad + \mathcal{O}(h) \\ x_u A^- &= (A + hRf_x)^{-1}(I - R) &&= (A + hf_x)^{-1}(I - R) + \mathcal{O}(h) \end{aligned}$$

by means of the Jacobian of Newton's method. Hence, the term $x_u A^-$ can be substituted by $(A + hRf_x)^{-1}(I - R)$ without changing the order of the scheme. Penski's approach [14] results in a similar approximation to the Milstein scheme in a more specialized setting.

The higher order 1 of strong convergence of these schemes has to be paid for with the use of a large number of double stochastic integrals and the use of the derivatives of the diffusion coefficients. In an application with a large number of small noise sources one has to pay much for a mostly theoretical gain in accuracy.

4.3. Trapezoidal rule

The trapezoidal rule is widely used to integrate oscillatory solutions of ODEs. It is A-stable and convergent of order 2. It is also applied to index 1 DAEs of the form

$$Ax'(t) + f(x(t), t) = 0 \tag{4.3}$$

via the scheme

$$y_\ell := -y_{\ell-1} + 2\frac{x_\ell - x_{\ell-1}}{h}, \quad A(-y_{\ell-1} + 2\frac{x_\ell - x_{\ell-1}}{h}) + f(x_\ell, t_\ell) = 0 \tag{4.4}$$

that implicitly realizes the trapezoidal rule for the inherent regular ODE. This becomes clear by the following considerations: On the one hand (4.4) implies $Rf(x_\ell, t_\ell) = 0$, such that the iterates are forced to satisfy the constraints. On the other hand (4.4) implies $Ay_\ell + f(x_\ell, t_\ell) = 0$, such that $A(-y_{\ell-1}) = f(x_{\ell-1}, t_{\ell-1})$. Implementing the scheme (4.4) requires only residuals.

A stochastic counterpart of the trapezoidal rule for the integration of SDEs (3.1) is given by

$$\frac{x_\ell - x_{\ell-1}}{h_\ell} = \frac{1}{2}\{f(x_\ell, t_\ell) + f(x_{\ell-1}, t_{\ell-1})\} + G(x_{\ell-1}, t_{\ell-1})\frac{1}{h_\ell}\Delta w_\ell. \tag{4.5}$$

It is strongly convergent of order $\gamma = 1/2$ like the other Euler schemes. For small noise the error is bounded by $O(h^2 + \epsilon h + \epsilon^2 h^{1/2})$ (see [16], or [13] for related results).

An adaptation of this scheme to SDAEs, analogously to (4.4) in the deterministic case, would lead to an implicit discretization of the diffusion term. An appropriate implicit realization of (4.5) for the inherent SDE requires some more available information concerning the structure of the SDAE. Given the projector R and a separate evaluation of the drift term, one can use the scheme

$$A\frac{x_\ell - x_{\ell-1}}{h_\ell} + \frac{1}{2}\big(f(x_\ell, t_\ell) + (I-R)f(x_{\ell-1}, t_{\ell-1})\big) + G(x_{\ell-1}, t_{\ell-1})\frac{1}{h_\ell}\Delta w_\ell = 0, \tag{4.6}$$

which implicitly realizes (4.5) for the inherent SDE (cf. [10] for the deterministic case). Since the differential equations and the constraints are now treated differently, it is possible to use a different scaling for both parts, which leads to a better conditioned system:

$$\begin{aligned} A(x_\ell - x_{\ell-1}) + \frac{h_\ell}{2}(I-R)\big(f(x_\ell, t_\ell) + f(x_{\ell-1}, t_{\ell-1})\big) + G(x_{\ell-1}, t_{\ell-1})\Delta w_\ell& \\ +Rf(x_\ell, t_\ell) &= 0 \end{aligned} \tag{4.7}$$

After creating explicit constraints via scaling by a suitable non-singular matrix $\tilde{D}$ with

$$\tilde{D}A = \begin{pmatrix} \tilde{D}_1 A \\ \tilde{D}_2 A \end{pmatrix} = \begin{pmatrix} \tilde{A}_1 \\ 0 \end{pmatrix}, \ \operatorname{rank} \tilde{A}_1 = \operatorname{rank} A, \quad \tilde{D}f =: \begin{pmatrix} \tilde{f}_1 \\ \tilde{f}_2 \end{pmatrix}, \ \tilde{D}G =: \begin{pmatrix} \tilde{G}_1 \\ \tilde{G}_2 \end{pmatrix},$$

the suggested scheme (4.7) corresponds to

$$\begin{aligned}\tilde{A}_1 x_\ell - x_{\ell-1} + \frac{h_\ell}{2}\{\tilde{f}_1(x_\ell,t_\ell)+\tilde{f}_1(x_{\ell-1},t_{\ell-1})\} + \tilde{G}_1(x_{\ell-1},t_{\ell-1})\Delta w_\ell &= 0\\ \tilde{f}_2(x_\ell,t_\ell) &= 0.\end{aligned}$$

The iterates satisfy the constraints at the current time-point, and the trapezoidal rule for the inherent SDE is realized. The Jacobian with respect to the new iterate is

$$\begin{pmatrix}\tilde{A}_1 + h_\ell \tilde{f}'_{1x}(x,t)/2\\ \tilde{f}'_{2x}(x,t)\end{pmatrix}.$$

It is non-singular for sufficiently small stepsizes and its condition number is bounded independently of the stepsizes.

5. Numerical results

The drift-implicit Euler scheme has been used to simulate a ring-oscillator built of three coupled inverter steps with simple mosfet-models. Such an oscillator was also used for test runs in [14]. Thermal noise in the mosfets and in the resistors are modelled by multiplicative and additive white noise sources. The circuit diagram is given in Fig. 3. The corresponding noise-free circuit is a free running oscillator.

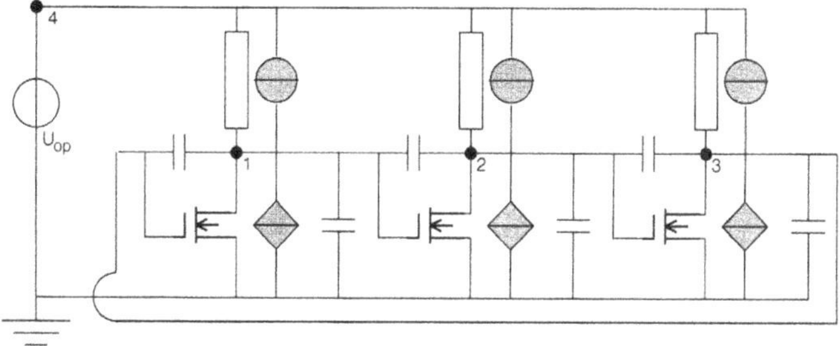

Figure 3: Thermal noise sources in a mosfet ring-oscillator model

The unknowns in the MNA system are the charges for the six capacities, the four nodal potentials and the current through the voltage source. The system is of index 1, but, formally, has direct noise. The three thermal resistance noise sources directly affect the current through the voltage source. However, the direct noise occurring in this current does not influence other variables. Omitting the corresponding variable together with the nodal equation for node 4 would lead to a system without direct noise. The diffusion coefficients have been scaled (by a factor 10^3) to make the noise effects more visible.

In Fig. 4 we present numerical results obtained with the drift-implicit Euler scheme. A mean-square estimate of the dominating local error term was used to control the stepsize according to the relative tolerance 10^{-4}. Realizations of the Wiener increments Δw_ℓ were simulated by a normal random number generator of the RANLIB library (of Fortran routines for random number generation). Fig. 4

shows the nodal potential at node 1: the dark solid lines correspond to two different paths of the stochastic potential and the dashed line to the noise-free potential. The solid grey lines give the mean function μ of 100 sample paths and the boundaries of the interval $[\mu - 3\sigma, \mu + 3\sigma]$, where σ denotes the standard deviation. The paths exhibit a highly visible phase noise and, hence, can hardly be considered as small perturbations of the deterministic potential. The mean function appears damped and differs considerably from the noise-free potential.

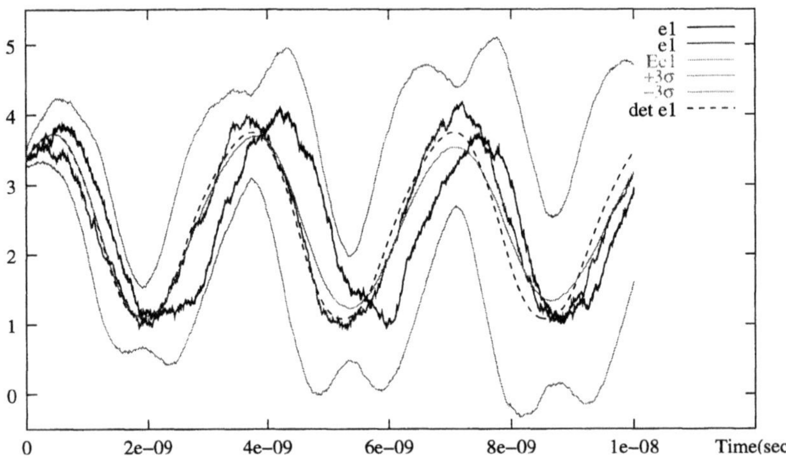

Figure 4: 2 sample paths of the voltage in node 1 (e1), the mean over 100 sample paths (E e1), the 3σ range ($\pm 3\sigma$), and the noiseless voltage (det e1)

Appendix

For the proof of Theorem 3.1 we need a discrete analogue of Gronwall's inequality.
Lemma: Let a_ℓ, $\ell = 1, \ldots, N$, and C_1, C_2 be nonnegative real numbers and assume that the inequalities

$$a_\ell \leq C_1 + C_2 \frac{1}{N} \sum_{i=1}^{\ell-1} a_i, \quad \ell = 1, \ldots, N,$$

are valid. Then we have $\max\limits_{\ell=1,\ldots,N} a_\ell \leq C_1 \exp(C_2)$.

Proof: (of Theorem 3.1)
Let $\tilde{d}_\ell$ be $\mathcal{F}_{t_\ell}$-measurable having a p-th order moment for each $\ell = 1, \ldots, N$. If the function φ does not depend on the variable x, the discretization scheme is explicit and the new iterate $\tilde{x}_\ell$ is given by

$$\tilde{x}_\ell = \tilde{x}_{\ell-1} + \varphi(\tilde{x}_{\ell-1}; t_{\ell-1}, h_\ell) + \psi(\tilde{x}_{\ell-1}; t_{\ell-1}, h_\ell, I_{t_{\ell-1}, h_\ell}) + \tilde{d}_\ell$$

for $\ell = 1, \ldots, N$. Otherwise, the scheme is implicit and the new iterate $\tilde{x}_\ell$ is given by the implicit equation (3.3). We assume that $h^0 > 0$ is chosen such that

$h^0 L_2 < 1$. Due to the global Lipschitz condition (A1), the equation

$$x = \tilde{x}_{\ell-1} + \varphi(\tilde{x}_{\ell-1}, x; t_{\ell-1}, h_\ell) + b_\ell$$

is uniquely solvable by the contraction principle since $h_\ell L_2 \leq h^0 L_2 < 1$. Moreover, the solution $\tilde{x}_\ell$ depends on $\tilde{x}_{\ell-1}$ and on b_ℓ in a Lipschitz continuous way (with a constant $L_4 > 0$). Since $b_\ell := \psi(\tilde{x}_{\ell-1}; t_{\ell-1}, h_\ell, I_{t_{\ell-1},h_\ell}) + \tilde{d}_\ell$ is a $\mathcal{F}_{t_\ell}$-measurable random variable, $\tilde{x}_\ell$ is also $\mathcal{F}_{t_\ell}$-measurable. Furthermore, $\tilde{x}_\ell$ has a p-th order moment. The latter fact is a consequence of the estimates

$$\begin{aligned} (\mathbb{E}|\tilde{x}_\ell|^p)^{\frac{1}{p}} &\leq (\mathbb{E}|\tilde{x}_\ell - x_\ell^0|^p)^{\frac{1}{p}} + |x_\ell^0| \\ &\leq L_4\{(\mathbb{E}|\tilde{x}_{\ell-1}|^p)^{\frac{1}{p}} + (\mathbb{E}|b_\ell|^p)^{\frac{1}{p}}\} + |x_\ell^0|, \end{aligned}$$

where x_ℓ^0 is the unique solution of the equation $x = \varphi(0, x; t_{\ell-1}, h_\ell)$, and

$$\begin{aligned} (\mathbb{E}|b_\ell|^p)^{\frac{1}{p}} &\leq (\mathbb{E}|\psi(\tilde{x}_{\ell-1}; t_{\ell-1}, h_\ell, I_{t_{\ell-1},h_\ell})|^p)^{\frac{1}{p}} + (\mathbb{E}|\tilde{d}_\ell|^p)^{\frac{1}{p}} \\ &\leq h^{\frac{1}{2}} L_3\{(\mathbb{E}|\tilde{x}_{\ell-1}|^p)^{\frac{1}{p}} + (\mathbb{E}|\psi(0; t_{\ell-1}, h_\ell, I_{t_{\ell-1},h_\ell})|^p)^{\frac{1}{p}}\} + (\mathbb{E}|\tilde{d}_\ell|^p)^{\frac{1}{p}} \end{aligned}$$

and of condition (A4).

Next we derive the stability estimate (3.4). The estimate (3.5) was shown in [23] for $p = 2$, but its proof carries over to the more general situation $p \geq 1$.

Let d_ℓ^* and $\tilde{d}_\ell$ for $\ell = 1, \ldots, N$ be perturbations of the discrete system and let x_ℓ^* and $\tilde{x}_\ell$, $\ell = 1, \ldots, N$, be their unique solutions. We introduce the following notations for $i = 1, \ldots, N$

$$\begin{aligned} e_i &:= x_i^* - \tilde{x}_i, \quad \triangle\varphi_i := \varphi(x_{i-1}^*, x_i^*; t_{i-1}, h_i) - \varphi(\tilde{x}_{i-1}, \tilde{x}_i; t_{i-1}, h_i), \\ d_i &:= d_i^* - \tilde{d}_i, \quad \triangle\psi_i := \psi(x_{i-1}^*; t_{i-1}, h_i, I_{t_{i-1},h_i}) - \psi(\tilde{x}_{i-1}; t_{i-1}, h_i, I_{t_{i-1},h_i}), \end{aligned}$$

and obtain from (3.3) and Hölder's inequality that

$$\begin{aligned} e_i &= e_{i-1} + \triangle\varphi_i + \triangle\psi_i + d_i = e_0 + \sum_{k=1}^{i} \triangle\varphi_k + \sum_{k=1}^{i} \triangle\psi_k + \sum_{k=1}^{i} d_k, \\ a_\ell &:= \mathbb{E}(\max_{i=1,\ldots,\ell} |e_i|^p) \leq 4^{p-1}\Big\{\mathbb{E}(|e_0|^p) + \mathbb{E}\Big(\max_{i=1,\ldots,\ell} |\sum_{k=1}^{i} \triangle\varphi_k|^p\Big) \\ &\qquad + \mathbb{E}\Big(\max_{i=1,\ldots,\ell} |\sum_{k=1}^{i} \triangle\psi_k|^p\Big) + \mathbb{E}\Big(\max_{i=1,\ldots,\ell} |\sum_{k=1}^{i} d_k|^p\Big)\Big\} \end{aligned}$$

holds for each $i, \ell = 1, \ldots, N$. For the second summand in the right-hand side of the latter estimate we continue by using (A1) and $ih \leq a(T - t_0)$

$$|\sum_{k=1}^{i} \triangle\varphi_k|^p \leq i^{p-1} \sum_{k=1}^{i} |\triangle\varphi_k|^p \leq (2i)^{p-1} h^p \sum_{k=1}^{i} \{L_1^p |e_{k-1}|^p + L_2^p |e_k|^p\} \leq \hat{L}_2 \frac{1}{N} \sum_{k=0}^{i} |e_k|^p.$$

where $\hat{L}_2 := 2^p (a(T - t_0))^p \max\{L_1^p, L_2^p\}$. Hence, we obtain the estimate

$$\mathbb{E}\Big(\max_{i=1,\ldots,\ell} |\sum_{k=1}^{i} \triangle\varphi_k|^p\Big) \leq \hat{L}_2 \frac{1}{N} \mathbb{E}\Big(\sum_{k=0}^{\ell} |e_k|^p\Big) \leq \hat{L}_2 \frac{1}{N} \Big(\sum_{k=0}^{\ell-1} a_k + a_\ell\Big).$$

For estimating the third summand we observe that the discrete parameter process $\{M_i := \sum_{k=1}^{i} \triangle\psi_k, \mathcal{F}_{t_{i-1}}\}_{i=1}^{N}$ is a martingale due to (A2) and has finite p-th order moments. Then Burkholder's and Davis's inequalities [19, Chapter VII.3] yield the estimate

$$\mathbb{E}\Big(\max_{i=1,\ldots,\ell}|\sum_{k=1}^{i}\triangle\psi_k|^p\Big) \le B_p\mathbb{E}\Big(\sum_{k=1}^{\ell}|\triangle\psi_k|^2\Big)^{\frac{p}{2}}$$

for each $\ell = 1,\ldots,N$ with some universal constant $B_p > 0$. For instance, $B_p := \frac{18p^{\frac{5}{2}}}{(p-1)^{\frac{3}{2}}}$ is such a constant if $p > 1$. Hence, we obtain

$$\begin{aligned}
\mathbb{E}\Big(\max_{i=1,\ldots,\ell}|\sum_{k=1}^{i}\triangle\psi_k|^p\Big) &\le B_p\ell^{\frac{p}{2}-1}\sum_{k=1}^{\ell}\mathbb{E}|\triangle\psi_k|^p \le B_p\ell^{\frac{p}{2}-1}\sum_{k=0}^{\ell-1}h_k^{\frac{p}{2}}L_3^p\mathbb{E}|e_k|^p \\
&\le B_pL_3^p\ell^{\frac{p}{2}-1}h^{\frac{p}{2}}\sum_{k=1}^{\ell}\mathbb{E}|e_k|^p \\
&\le B_pL_3^p(a(T-t_0))^{\frac{p}{2}}\frac{1}{N}\sum_{k=0}^{\ell-1}a_k.
\end{aligned}$$

for $\ell = 1,\ldots,N$ by using Hölder's inequality and condition (A3). Setting $\hat{L}_3 := B_pL_3^p(a(T-t_0))^{\frac{p}{2}}$ we arrive, altogether, at the estimate

$$a_\ell \le 4^{p-1}\Big\{\mathbb{E}|e_0|^p + \hat{L}_2\frac{1}{N}\Big(\sum_{k=0}^{\ell-1}a_k + a_\ell\Big) + \hat{L}_3\frac{1}{N}\sum_{k=0}^{\ell-1}a_k + \mathbb{E}\Big(\max_{i=1,\ldots,\ell}|\sum_{k=1}^{i}d_k|^p\Big)\Big\}$$

for $\ell = 1,\ldots,N$. If necessary, we choose h^0 smaller such that $4^{p-1}\hat{L}_2\frac{1}{N} \le \frac{3}{4}$ holds if $h < h^0$. We conclude that

$$a_\ell \le 4^p\Big\{\mathbb{E}(|e_0|^p) + (\hat{L}_2+\hat{L}_3)\frac{1}{N}\sum_{k=0}^{\ell-1}a_k + \mathbb{E}\Big(\max_{i=1,\ldots,\ell}|\sum_{k=1}^{i}d_k|^p\Big)\Big\}$$

holds for $\ell = 1,\ldots,N$. By applying the lemma this leads to the semifinal estimate

$$\max_{\ell=1,\ldots,N}a_\ell = \mathbb{E}(\max_{i=1,\ldots,N}|e_i|^p) \le 4^p\exp(4^p(\hat{L}_2+\hat{L}_3))\Big\{\mathbb{E}|e_0|^p + \mathbb{E}\max_{i=1,\ldots,N}|\sum_{k=1}^{i}d_k|^p\Big\}.$$

It remains to decompose the perturbation difference d_k into $d_k = r_k + s_k$ with $\mathbb{E}(s_k|\mathcal{F}_{t_{k-1}}) = 0$ for $k = 1,\ldots,N$. Then $\{\sum_{k=1}^{i}s_k, \mathcal{F}_{t_{i-1}}\}_{i=1}^{N}$ is a martingale having finite p-th order moments. Appealing again to Burkholder's and Davis's inequalities provides

$$\begin{aligned}\mathbb{E}\Big(\max_{i=1,\dots,N}|\sum_{k=1}^{i}s_k|^p\Big) &\le B_p\mathbb{E}\Big(\sum_{k=1}^{N}|s_k|^2\Big)^{\frac{p}{2}} \le B_pN^{\frac{p}{2}}\max_{k=1,\dots,N}\mathbb{E}|s_k|^p \\ &\le B_p(a(T-t_0))^{\frac{p}{2}}\frac{\max_{k=1,\dots,N}\mathbb{E}|s_k|^p}{h^{\frac{p}{2}}}.\end{aligned}$$

Summarizing we obtain the final estimate

$$\begin{aligned}\mathbb{E}(\max_{i=1,\dots,N}|e_i|^p) &\le \hat{S}^p\Big\{\mathbb{E}|e_0|^p+\mathbb{E}\max_{i=1,\dots,N}|\sum_{k=1}^{i}s_k|^p+\mathbb{E}\max_{i=1,\dots,N}|\sum_{k=1}^{i}r_k|^p\Big\} \\ &\le \hat{S}^p\Big\{\mathbb{E}|e_0|^p+B_p(a(T-t_0))^{\frac{p}{2}}\frac{\max_{k=1,\dots,N}\mathbb{E}|s_k|^p}{h^{\frac{p}{2}}} \\ &\qquad +(a(T-t_0))^p\frac{\mathbb{E}\max_{k=1,\dots,N}|r_k|^p}{h^p}\Big\}\end{aligned}$$

where $\hat{S}^p := 4^p2^{p-1}\exp(4^p(\hat{L}_2+\hat{L}_3))$. This completes the proof. □

Acknowledgment

The authors wish to thank Uwe Feldmann and Georg Denk (Infineon Technologies) for invaluable discussions and the excellent cooperation. Further thanks are due to Roswitha März (Humboldt University Berlin) for beneficial discussions.

References

[1] U. Ascher and L. Petzold. *Computer methods for ordinary differential equations and differential-algebraic equations.* SIAM, Philadelphia 1998.

[2] A. Blum. *Elektronisches Rauschen.* Teubner, 1996.

[3] K. Burrage and P.M. Burrage. Numerical methods for stochastic differential equations with applications. Manuscript, University of Queensland, March 2002.

[4] A. Demir and A. Sangiovanni-Vincentelli. *Analysis and simulation of noise in nonlinear electronic circuits and systems.* Kluwer, Boston 1998.

[5] G. Denk and R. Winkler. Modeling and simulation of transient noise in circuit simulation. To appear in: Proceedings of 4th MATHMOD, Vienna, February 2003.

[6] D. Estevez Schwarz and C. Tischendorf. Structural analysis for electronic circuits and consequences for MNA. *Int. J. Circ. Theor. Appl.* 28 (2000), 131–162.

[7] M. Gelbrich. Simultaneous time and chance discretization for stochastic differential equations. *J. Comp. Appl. Math.* 58 (1995), 255–289.

[8] M. Günther and U. Feldmann. CAD-based electric-circuit modeling in industry I. Mathematical structure and index of network equations. *Surv. Math. Ind.* 8 (1999), 97–129.

[9] P.E. Kloeden and E. Platen. *Numerical solution of stochastic differential equations.* Springer, Berlin 1992.

[10] R. März. On initial value problems in differential-algebraic equations and their numerical treatment. *Computing* 35 (1985), 13–37.

[11] R. März. Numerical methods for differential-algebraic equations. *Acta Numerica 1992*, 141–198.

[12] G.N. Milstein. *Numerical integration of stochastic differential equations.* Kluwer, Dordrecht 1995.

[13] G.N. Milstein and M.V. Tretyakov. Mean-square numerical methods for stochastic differential equations with small noise. *SIAM J. Sci. Comput.* 18 (1997), 1067–1087.

[14] C. Penski. A new numerical method for SDEs and its application in circuit simulation. *J. Comp. Appl. Math.* 115 (2000), 461–470.

[15] E. Platen. An introduction to numerical methods for stochastic differential equations. *Acta Numerica 1999*, 197–246.

[16] W. Römisch and R. Winkler. Stepsize control for mean-square numerical methods for SDEs with small noise. In preparation.

[17] O. Schein. *Stochastic differential algebraic equations in circuit simulation.* PhD thesis, Technische Universität Darmstadt, 1999.

[18] O. Schein and G. Denk. Numerical solution of stochastic differential-algebraic equations with applications to transient noise simulation of microelectronic circuits. *J. Comp. Appl. Math.* 100 (1998), 77–92.

[19] A.N. Shiryaev. Probability (Second Edition). Springer, New York 1996.

[20] C. Tischendorf. Topological index calculation of DAEs in circuit simulation. *Surv. Math. Ind.* 8 (1999), 187–199.

[21] L. Weiß. *Rauschen in nichtlinearen elektronischen Schaltungen und Bauelementen – ein thermodynamischer Zugang.* PhD thesis, Otto-von-Guericke Universität Magdeburg, 1999.

[22] L. Weiß and W. Mathis. A thermodynamical approach to noise in nonlinear networks. *Int. J. Circ. Theor. Appl.* 26 (1998), 147–165.

[23] R. Winkler. Stochastic differential algebraic equations of index 1 and applications in circuit simulation, *J. Comp. Appl. Math.* (to appear).

Werner Römisch
Humboldt-Universität zu Berlin
Institut für Mathematik
Unter den Linden 6
D-10099 Berlin
email: romisch@mathematik.hu-berlin.de

Renate Winkler
Humboldt-Universität zu Berlin
Institut für Mathematik
Unter den Linden 6
D-10099 Berlin
email: winkler@mathematik.hu-berlin.de

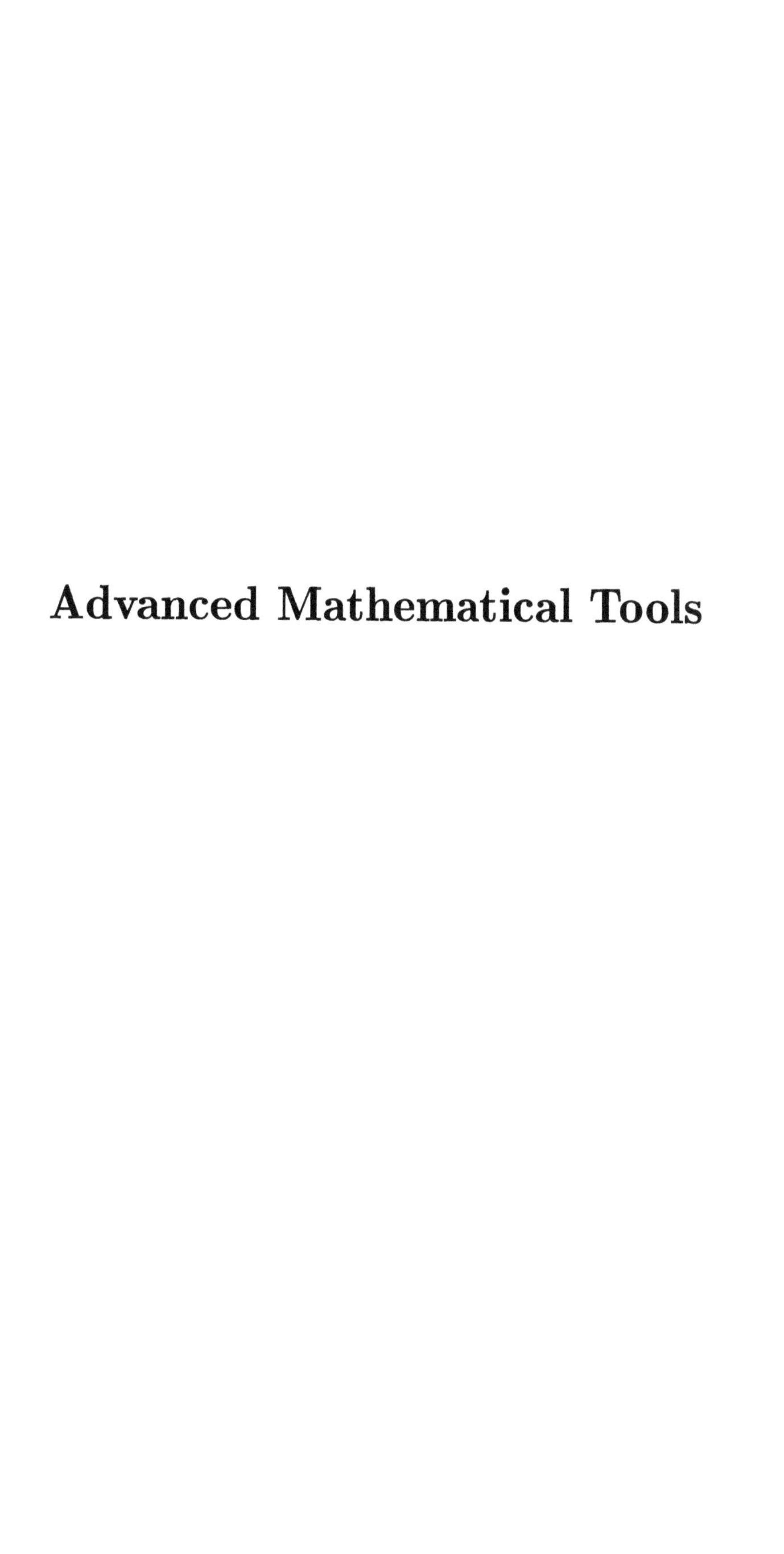

Advanced Mathematical Tools

Modeling, Simulation and Optimization of Integrated Circuits
(Eds.) K. Antreich, R. Bulirsch, A. Gilg and P. Rentrop

International Series of Numerical Mathematics Vol. 146, 321–332

An Efficient Way to Avoid the Order Reduction of Linearly Implicit Runge-Kutta Methods for Nonlinear IBVP's

M. P. Calvo, J. de Frutos, and J. Novo

Abstract. In the present paper we show that the strategy proposed in [7] to avoid the order reduction of Runge-Kutta methods when integrating linear initial boundary value problems can be extended to also avoid the order reduction in nonlinear cases. Furthermore, we see that if the Runge-Kutta method is replaced by an appropriate linearly implicit Runge-Kutta method, this strategy is interesting not only from a theoretical point of view, but also provides an efficient procedure to time integrate initial boundary value problems.

1. Introduction

Let us consider an abstract initial boundary value problem

$$\begin{cases} u'(t) = \widetilde{A}u(t) + f(t), & 0 \leq t \leq T, \\ u(0) = u_0, & \\ \partial u(t) = g(t), & 0 \leq t \leq T, \end{cases} \tag{1.1}$$

where $\widetilde{A} : D(\widetilde{A}) \subset X \to X$ and $\partial : D(\widetilde{A}) \subset X \to Y$ are linear operators, X and Y are Banach spaces, $u_0 \in X$, $f : [0,T] \to X$ and $g : [0,T] \to Y$. Many evolutionary partial differential equations of practical interest can be written in this abstract form [14]. It is well known that when (1.1) is integrated in time with a Runge-Kutta method of classical order p and stage order q [10], the order of convergence is $\min\{p, q+\theta\}$ with $0 \leq \theta < 2$ [5]. This is the so-called *order reduction phenomenom.*

Different remedies to avoid the order reduction of Runge-Kutta methods when applied to the time integration of initial boundary value problems have been proposed in the literature [1, 2, 3, 4, 5, 7, 9, 11, 12, 15, 16, 18]. The aim of the

---

2000 *Mathematics Subject Classification.* Primary 65M12, 65M20.
Key words and phrases. Runge-Kutta, order reduction, linearly implicit methods.
This research has been supported by DGI-MCYT under project BFM2001-2138 cofinanciated by FEDER funds, and by Junta de Castilla y León under project VA112/02. Research completed while the second author was visiting professor at GERAD and HEC (Montréal), under grant SEEU PR2002-0223.

present paper is to show that the strategy proposed in [7] to deal with linear problems can be applied to also avoid the order reduction of linearly implicit Runge-Kutta methods when time integrating nonlinear problems.

The basic idea of the strategy proposed in [7] is as follows. The solution of (1.1) is decomposed as $u(t) = v(t) + u^*(t)$, where $v(t)$ can be computed directly in terms of the data and $u^*(t)$ is the solution of a suitable initial value problem for which there is no order reduction. In order to extend this idea to deal with nonlinear problems, we need to describe more precisely how these functions are constructed in [7].

For the sake of simplicity, throughout the paper we assume that the problem

$$\begin{cases} \widetilde{A}x = 0, \\ \partial x = y \end{cases} \tag{1.2}$$

has a unique solution and we denote it by Ey. The technical hypothesis to be fulfilled by $\widetilde{A}$ and ∂ are described in Section 2 of [7]. Set $D(A) = \ker(\partial) = \{ x \in D(\widetilde{A}) \,/\, \partial x = 0 \}$ and let us consider the restriction $A = \widetilde{A}|_{D(A)} : D(A) \subset X \to X$ which, under these hypothesis, is the infinitesimal generator of a C_0-semigroup in X. Let us also assume that the solution u of (1.1) belongs to $C^{p+1}([0,T], D(\widetilde{A}^r))$ for some $1 \le r \le p-q$. (Here p and q denote the order and the stage order of the Runge-Kutta method, respectively.) We want to stress that, in practice, this is a natural requirement since it only demands regularity of u both, in time and space, but it does not impose any restriction on the boundary values of u. The correcting term is defined in [7] as

$$v_r(t) = \sum_{s=0}^{r-1} A^{-s} E \partial \widetilde{A}^s u(t), \qquad 0 \le t \le T. \tag{1.3}$$

Then $u^{*,r}(t) = u(t) - v_r(t)$ is the solution of the initial value problem

$$\begin{cases} (u^{*,r})'(t) = Au^{*,r}(t) + f^{*,r}(t), \quad 0 \le t \le T, \\ u^{*,r}(0) = u_0 - v_r(0), \end{cases} \tag{1.4}$$

where

$$f^{*,r}(t) = f(t) - \sum_{s=0}^{r-1} A^{-s} E \partial \widetilde{A}^s f(t) - A^{-(r-1)} E \left[g^{(r)}(t) - \sum_{j=0}^{r-1} \partial \widetilde{A}^{r-j-1} f^{(j)}(t) \right].$$

We want to stress that, as shown in [7],

$$\partial \widetilde{A}^s u(t) = g^{(s)}(t) - \sum_{j=0}^{s-1} \partial \widetilde{A}^{s-j-1} f^{(j)}(t), \qquad 1 \le s \le r-1.$$

Therefore, $v_r(t)$ and $f^{*,r}(t)$ can be computed in terms of f and g and their derivatives. This will not happen when dealing with nonlinear problems.

Denoting by $u_n^{*,r}$ the Runge-Kutta approximation to the solution of (1.4) at time level t_n, $0 \le n \le N$ and setting $\hat{u}_n = u_n^{*,r} + v_r(t_n)$ (then $\hat{u}_n$ is the numerical approximation to the solution of (1.1) at time level t_n), in [7] the following theorem is proved:

Theorem 1.1. *Let $u \in C^{p+1}([0,T], D(\widetilde{A}^r))$, $r = p - q$, be the solution of the initial boundary value problem (1.1). Assume also that $f \in C^p([0,T], X)$ and $g \in C^{p+1}([0,T], Y)$. Let $v_r : [0,T] \to X$ be the correcting term defined in (1.3). Then*

$$\|u(t_n) - \hat{u}_n\| = \|u^{*,r}(t_n) - u_n^{*,r}\| \leq C\rho_n \sum_{m=1}^{n} k_{m-1}^{p+1} I_m(u),$$

where $C > 0$ is a constant and

$$\rho_n = \|\prod_{m=1}^{n} r(k_{m-1}A)\|,$$

$$I_m(u) = \sup_{q+1 \leq l \leq p+1} \|u^{(l)}\|_{L^\infty([t_{m-1},t_m], D(\widetilde{A}^r))}.$$

Here $r(z)$ is the stability function of the Runge-Kutta method, and the step sizes k_m, $0 \leq m \leq n-1$, are small enough to ensure that the operators $r(k_m A)$ are well defined in X. The constant C depends on T, A and the Runge-Kutta method but it is independent of u, f and g.

Observe that $r = 1$ in (1.3) corresponds to the usual correction $v_1(t) = Eg(t)$ mentioned in [16] to convert problem (1.1) into an initial boundary value problem with homogeneous boundary conditions, which is less affected than (1.1) by the order reduction [5], [13]. For $r = 2$,

$$v_2(t) = Eg(t) + A^{-1}E[g'(t) - \partial f(t)] \tag{1.5}$$

and

$$f^{*,2}(t) = f(t) - E\partial f(t) - A^{-1}E[g''(t) - \partial f'(t)]. \tag{1.6}$$

In Section 2 we present a nonlinear initial boundary value problem which exhibits order reduction when is time integrated using a third-order linearly implicit Runge-Kutta method. Section 3 is devoted to adapt the strategy of [7] to deal with nonlinear problems. In Section 4 we show how this strategy can be efficiently implemented and, finally, in Section 5 numerical results are provided.

2. A nonlinear problem with order reduction

Let us consider the one-dimensional forced Burgers' equation

$$u_t = u_{xx} - uu_x + f(t,x), \qquad -1 \leq x \leq 1, \qquad t > 0, \tag{2.1}$$

subject to homogeneous Dirichlet boundary conditions and with initial data

$$u_0(x) = \begin{cases} 1 + x + \dfrac{x^2}{2} + \dfrac{x^3}{6} - \dfrac{x^4}{3} & \text{if } x \in [-1,0], \\ 1 + x + \dfrac{x^2}{2} + \dfrac{x^3}{6} - \dfrac{8x^4}{3} & \text{if } x \in [0,1]. \end{cases}$$

The forcing term f is defined in order to have $u(t,x) = u_0(x)/(1+t)$ as the exact solution of the problem. We chose homogeneous boundary conditions since, in practice, before starting the numerical integration, the original problem is transformed into an initial boundary value problem satisfying homogeneous boundary conditions.

Let us numerically integrate equation (2.1). For the spatial discretization we consider a Chebyshev spectral Galerkin method [8] using the basis functions proposed in [17] and evaluating the nonlinear terms using collocation and Fast Fourier Transform techniques. The resulting semidiscretized system has the form

$$\mathbf{y}' = L\mathbf{y} + N(t,\mathbf{y}), \tag{2.2}$$

where matrix L stems from the spatial discretization of the diffusive term and $N(t,\mathbf{y})$ arises from the discretization of nonlinear and source terms. The main difficulty in dealing with (2.2) is that the use of explicit integrators is usually inefficient because the system becomes stiffer as the spatial mesh is refined. On the other hand, if a stiffly accurate integrator is chosen, one has to solve nonlinear equations that are difficult to handle with, especially in connection with spectral methods [8]. In order to avoid these difficulties, the linearly implicit Runge-Kutta methods proposed in [6] combine an L-stable singly diagonally implicit Runge-Kutta method to treat the linear part with and explicit Runge-Kutta scheme for the nonlinear terms, in such a way that the overall method has good stability properties for spectral discretizations of advection-reaction-diffusion equations (see [6] for more details). These methods have Butcher tableaux

$$\begin{array}{c|c|c} \mathbf{c} & \mathcal{A} & \hat{\mathcal{A}} \\ \hline & \mathbf{b}^T & \mathbf{b}^T \end{array} \tag{2.3}$$

with $\mathcal{A} = (a_{ij})_{i,j=1}^{s+1}$, $\hat{\mathcal{A}} = (\hat{a}_{ij})_{i,j=1}^{s+1}$, $\mathbf{c} = [0, c_2, \dots, c_{s+1}]^T$ and $\mathbf{b}^T = [0, b_2, \dots, b_{s+1}]$. Matrix $\mathcal{A}$ is lower triangular with constant diagonal elements and has null first column, while $\hat{\mathcal{A}}$ is strictly lower triangular. The equations to advance a step of length k in the integration of (2.2) from t_n to $t_{n+1} = t_n + k$ take the form

$$\begin{aligned}
\mathbf{Y}_1 &= \mathbf{y}_n, \\
\mathbf{Y}_i &= \mathbf{y}_n + k\left(\sum_{j=2}^{i} a_{ij} L\mathbf{Y}_j + \sum_{j=1}^{i-1} \hat{a}_{ij}\mathbf{N}(t_n + c_j k, \mathbf{Y}_j)\right), \quad 2 \le i \le s+1, \\
\mathbf{y}_{n+1} &= \mathbf{y}_n + k\left(\sum_{i=2}^{s+1} b_i L\mathbf{Y}_i + \sum_{i=2}^{s+1} b_i \mathbf{N}(t_n + c_i k, \mathbf{Y}_i)\right),
\end{aligned}$$

where $\mathbf{Y}_i$ denote the internal stages. In our numerical experiments we use the linearly implicit Runge-Kutta method with classical order $p = 3$, stage order $q = 1$ and $s + 1 = 4$ stages proposed in [6]. Due to the special structure of matrices $\mathcal{A}$ and $\hat{\mathcal{A}}$, the method requires 4 evaluations of the nonlinear term per step, 3 matrix-vector multiplications per step and 3 linear systems with the same matrix to be solved at each step.

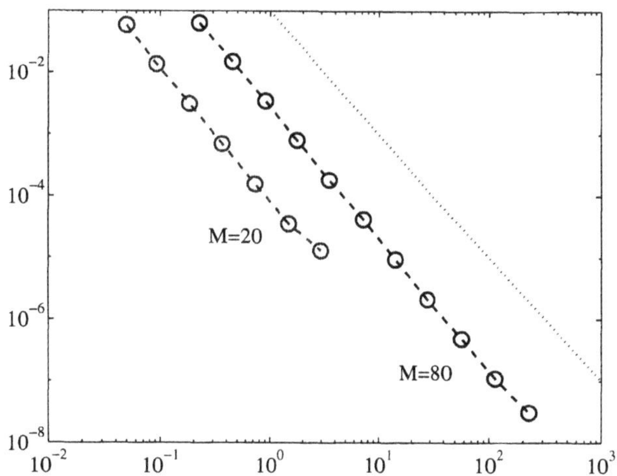

FIGURE 1. Error at $T = 10$ against CPU time for Burgers' equation

In order to observe the errors due to the time integrator, we fix the number of degrees of freedom for the spatial discretization and we consider time steps $k = 0.4, 0.2, \ldots$. In Figure 1 we have plotted error at $T = 10$ (measured in the L^2 norm) against CPU time for two different values of the number of degrees of freedom $M = 80$ and $M = 20$. Only data for which the error due to the spatial discretization is smaller than the error of the time integration have been plotted. We observe that the time integrator behaves as a second-order method as corresponds to the slope -2 of both lines. Then, the order reduction is present because the classical order of the method is 3. We want to stress that although the stage order of the method is only 1, the errors shown in Figure 1 exhibit second-order. This is due to the homogeneous boundary conditions satisfied by the exact solution of (2.1), which imply that the first correction (1.3) with $r = 1$ is already done (see [5], [13] for details).

3. How to avoid the order reduction in the nonlinear case

In this section we show how the strategy of [7] can be adapted to deal with certain families of nonlinear problems, including Burgers' equation.

Let us now consider an abstract initial boundary value problem of the form

$$\begin{cases} u'(t) = \widetilde{A}u(t) + R(u(t)) + f(t), & 0 \le t \le T, \\ u(0) = u_0, & \\ \partial u(t) = g(t), & 0 \le t \le T, \end{cases} \tag{3.1}$$

with $\widetilde{A}$ and ∂ as in (1.1) and $R : D(R) \subset X \to X$ a nonlinear operator satisfying $D(\widetilde{A}) \subset D(R)$. For instance, Burgers' equation can be written in this abstract format as follows: $X = L^2([-1,1])$, $Y = \mathbb{R} \times \mathbb{R}$, $D(\widetilde{A}) = W^{2,2}([-1,1]) = \{\varphi \in X \,/\, \varphi'' \in X\}$, $\widetilde{A}\varphi = \varphi''$, $R(\varphi) = \varphi\varphi'$ and $\partial\varphi = [\varphi(-1), \varphi(1)]^T$, $\varphi \in D(\widetilde{A})$. In this example $D(A) = W^{2,2}([-1,1]) \cap W_0^{1,2}([-1,1]) = \{\varphi \in W^{2,2}([-1,1]) \,/\, \varphi(-1) = \varphi(1) = 0\}$ and for $\xi = [\xi_0, \xi_1]^T \in Y$, $E\xi$ is the mapping sending $x \in [-1,1]$ to $\xi_0 + ((\xi_1 - \xi_0)/2)(x+1)$. Notice that in practice, R collects spatial lower-order derivatives and nonlinear terms.

According to (1.3), if $u : [0,T] \to X$ is a solution of (3.1), let us define

$$\begin{aligned} v_2(t) &= E\partial u(t) + A^{-1}E\partial\widetilde{A}u(t) \\ &= E\partial u(t) + A^{-1}E\partial[u'(t) - R(u(t)) - f(t)]. \end{aligned}$$

Then $u^{*,2}(t) = u(t) - v_2(t)$ is the solution of the initial value problem

$$\begin{cases} (u^{*,2})'(t) = Au^{*,2}(t) + f^{*,2}(t), \quad 0 \le t \le T, \\ u^{*,2}(0) = u_0 + A^{-1}E\partial f(0), \end{cases} \tag{3.2}$$

for an appropriate $f^{*,2}(t)$.

Let us now assume that $\partial u = 0$ and $\partial R(u) = 0$. This includes initial boundary value problems subject to either Dirichlet or Neumann homogeneous boundary conditions and with nonlinearities including reaction terms of the form $R(u)$ with R a smooth real function satisfying $R(0) = 0$, or other nonlinear terms such as $R(u) = (u \cdot \nabla)u$ appearing in Navier-Stokes equations or Burgers' type equations. Under these assumptions,

$$v_2(t) = -A^{-1}E\partial f(t)$$

and

$$f^{*,2}(t) = R(u(t)) + f(t) - \overbrace{\left[E\partial f(t) - A^{-1}E\partial f'(t)\right]}^{H(t)}. \tag{3.3}$$

Notice that $v_2(t)$ and $H(t)$ are computable in terms of the data but the nonlinear term depends on the solution $u(t)$ of the original problem (3.1). This is the main difference with respect to the linear case, where $f^{*,2}(t)$ was computable in terms of f and g and their derivatives (see (1.6)).

From the theoretical point of view this is not a serious drawback since $u(t) = u^{*,2}(t) + v_2(t)$ and $v_2(t)$ is computable in terms of $f(t)$. However, the computation of $v_2(t)$ during the time integration requires the solution of several additional stationary problems per step, which can be expensive in practice. In Section 4 we propose an alternative way of providing approximations to u and $u^{*,2}$ by time integrating (3.1) and (3.2) simultaneously using a linearly implicit Runge-Kutta method as those proposed in [6].

4. An efficient implementation of this strategy

For the time integration of (3.2), let us consider a linearly implicit Runge-Kutta method with Butcher tableaux (2.3). The equations to time integrate (3.2) from $t = 0$ to $t_N = Nk$ take the form:

For $n = 0, 1, 2, \ldots, N-1$,

$$\begin{aligned}
U_1^{*,2} &= u_n^{*,2}, \\
U_i^{*,2} &= u_n^{*,2} + k\sum_{j=2}^{i} a_{ij} AU_j^{*,2} \\
&+ k\sum_{j=1}^{i-1} \hat{a}_{ij}\left[R(u(t_n + c_j k)) + f(t_n + c_j k) - H(t_n + c_j k)\right], \quad 2 \le i \le s+1, \\
u_{n+1}^{*,2} &= u_n^{*,2} + k\sum_{i=2}^{s+1} b_i \left[AU_i^{*,2} + R(u(t_n + c_i k)) + f(t_n + c_i k) - H(t_n + c_i k)\right], \\
t_{n+1} &= t_n + k,
\end{aligned}$$

with $t_0 = 0$ and $u_0^{*,2} = u_0 - v_2(0)$. Then, the approximation to the solution of (3.1) at t_N is given by

$$\hat{u}_N = u_N^{*,2} - A^{-1}E\partial f(t_N). \tag{4.1}$$

However, the nonlinear terms are not known because they depend on the exact solution of (3.1). What we propose in order to overcome this difficulty is to time integrate (3.1) and (3.2) simultaneously. More precisely,
For $n = 0, 1, 2, \ldots, N-1$,

$$\begin{aligned}
U_1 &= u_n, \\
U_1^* &= u_n^*, \\
U_i &= u_n + k\sum_{j=2}^{i} a_{ij} \widetilde{A} U_j \\
&+ k\sum_{j=1}^{i-1} \hat{a}_{ij}\left[R(U_j) + f(t_n + c_j k)\right], \quad 2 \le i \le s+1, \\
U_i^* &= u_n^* + k\sum_{j=2}^{i} a_{ij} AU_j^* \\
&+ k\sum_{j=1}^{i-1} \hat{a}_{ij}\left[R(U_j)) + f(t_n + c_j k) - H(t_n + c_j k)\right], \quad 2 \le i \le s+1, \\
u_{n+1} &= u_n + k\sum_{i=2}^{s+1} b_i \left[\widetilde{A} U_i + R(U_i) + f(t_n + c_i k)\right], \\
u_{n+1}^* &= u_n^* + k\sum_{i=2}^{s+1} b_i \left[AU_i^* + R(U_i) + f(t_n + c_i k) - H(t_n + c_i k)\right], \\
t_{n+1} &= t_n + k,
\end{aligned}$$

with $t_0 = 0$ and $u_0^* = u_0 - v_2(0)$. Now, the approximation to the solution of (3.1) at t_N is given by

$$\hat{\hat{u}}_N = u_N^* - A^{-1}E\partial f(t_N). \tag{4.2}$$

In this way, for $1 \le j \le s+1$, the term $R(u(t_n + c_j k))$ is approximated by $R(U_j)$, being U_j, $1 \le j \le s+1$, the internal stages of the linearly implicit Runge-Kutta method when applied to the original problem (3.1). Notice that the additional computational cost of this implementation, compared with the standard implementation of the linearly implicit method applied to (3.1), is just the computational cost of the solution of 3 linear systems per step, but with the same matrix used in the standard procedure. It is also necessary to evaluate the function $H(t)$ at three different time levels per step, but this can be efficiently done in parallel at the beginning of each step. Furthermore, although each evaluation of $H(t)$ requires the solution of 3 stationary problems, in the one-dimensional case the computational cost of this task is almost negligible since (1.2) can be solved exactly.

On the other hand, in the implementation above, the term $R(u(t_n + c_j k))$ is approximated by $R(U_j)$, $1 \le j \le s+1$, which is only a second-order approximation since the stage order of the linearly implicit Runge-Kutta method is 1. However, the following result holds:

Theorem 4.1. *Let $\hat{u}_N$ and $\hat{\hat{u}}_N$ be the numerical approximations to $u(t_N)$ defined by* (4.1) *and* (4.2) *respectively. If $\|\hat{u}_N - u(t_N)\| = O(k^3)$, then $\|\hat{\hat{u}}_N - u(t_N)\| = O(k^3)$.*

Proof. We will prove that $\|\hat{u}_N - \hat{\hat{u}}_N\| = O(k^3)$. First we observe that substracting (4.2) from (4.1) it is enough to prove that $\|u_N^{*,2} - u_N^*\| = O(k^3)$. To this end, we set $F^{*,2}(t_n) = \{f^{*,2}(t_n + c_i k)\}_{i=1}^{s+1}$, where $f^{*,2}(t)$ is defined by (3.3) and $F^*(t_n) = \{R(U_i) + f(t_n + c_i k) - H(t_n + c_i k)\}_{i=1}^{s+1}$, where $U_1, \ldots, U_{s+1}$ are the internal stages of the linearly implicit Runge-Kutta method when applied to (3.1) to advance from t_n to t_{n+1}. We also set $U^{*,2} = \{U_i^{*,2}\}_{i=1}^{s+1}$ and $U^* = \{U_i^*\}_{i=1}^{s+1}$. According to (3.3), $F^{*,2}(t_n) - F^*(t_n) = \{R(u(t_n + c_i k)) - R(U_i)\}_{i=1}^{s+1}$, being $u(t)$ the exact solution of (3.1). Since the stage order of the linearly implicit Runge-Kutta method is 1,

$$u(t_n + c_i k) - U_i = k^2 a_i(t_n) + O(k^3)$$

and then if R is smooth enough

$$F^{*,2}(t_n) - F^*(t_n) = k^2 \mathbf{a}^*(t_n) + O(k^3). \tag{4.3}$$

The vectors collecting the internal stages $U^{*,2}$ and U^* satisfy

$$U^{*,2} = (\mathbf{e} \otimes I)u_n^{*,2} + k(\mathcal{A} \otimes A)U^{*,2} + k(\hat{\mathcal{A}} \otimes I)F^{*,2}(t_n) \tag{4.4}$$

and

$$U^* = (\mathbf{e} \otimes I)u_n^* + k(\mathcal{A} \otimes A)U^* + k(\hat{\mathcal{A}} \otimes I)F^*(t_n), \tag{4.5}$$

respectively. Here $\mathcal{A}$ and $\hat{\mathcal{A}}$ denote the matrices of the Runge-Kutta tableaux in (2.3), $\mathbf{e} = [1, \ldots, 1]^T \in \mathbb{R}^{s+1}$ and I stands for the identity matrix. On the other hand, the approximations to the solution at t_{n+1} satisfy

$$u_{n+1}^{*,2} = u_n^{*,2} + k(\mathbf{b}^T \otimes A)U^{*,2} + k(\mathbf{b}^T \otimes I)F^{*,2}(t_n) \tag{4.6}$$

and

$$u_{n+1}^{*} = u_n^{*} + k(\mathbf{b}^T \otimes A)U^{*} + k(\mathbf{b}^T \otimes I)F^{*}(t_n), \tag{4.7}$$

respectively, where $\mathbf{b}^T$ are the weights of the linearly implicit Runge-Kutta method (2.3). Substracting (4.5) from (4.4) and using (4.3) we get

$$U^{*,2} - U^* = (\mathbf{e} \otimes I)[u_n^{*,2} - u_n^*] + k(\mathcal{A} \otimes A)[U^{*,2} - U^*] + k(\hat{\mathcal{A}} \otimes I)[k^2 \mathbf{a}^*(t_n) + O(k^3)],$$

which leads to

$$\begin{aligned} U^{*,2} - U^* &= (I - k(\mathcal{A} \otimes A))^{-1}(\mathbf{e} \otimes I)[u_n^{*,2} - u_n^*] \\ &+ k^3 (I - k(\mathcal{A} \otimes A))^{-1}(\hat{\mathcal{A}} \otimes I)\mathbf{a}^*(t_n) + O(k^4). \end{aligned} \tag{4.8}$$

On the other hand, substracting (4.7) from (4.6) and using again (4.3) we get

$$u_{n+1}^{*,2} - u_{n+1}^* = [u_n^{*,2} - u_n^*] + k(\mathbf{b}^T \otimes A)[U^{*,2} - U^*] + k(\mathbf{b}^T \otimes I)[k^2 \mathbf{a}^*(t_n) + O(k^3)],$$

which after using (4.8) and some manipulations leads to

$$\begin{aligned} u_{n+1}^{*,2} - u_{n+1}^* &= \left(I + k\left(\mathbf{b}^T \otimes A\right)\left(I - k(\mathcal{A} \otimes A)\right)^{-1}(\mathbf{e} \otimes I)\right)\left[u_n^{*,2} - u_n^*\right] \\ &+ k^3 (\mathbf{b}^T \otimes I)\mathbf{a}^*(t_n) + O(k^4). \end{aligned}$$

Now, the variation of constants formula applied to the previous equality shows that

$$\begin{aligned} u_{n+1}^{*,2} - u_{n+1}^* &= (r(kA))^{n+1}\left[u_0^{*,2} - u_0^*\right] \\ &+ k^3 \sum_{j=0}^{n} (r(kA))^{n-j}\left[(\mathbf{b}^T \otimes I)\mathbf{a}^*(t_j) + O(k)\right], \end{aligned}$$

where $r(kA) = I + k\left(\mathbf{b}^T \otimes A\right)\left(I - k(\mathcal{A} \otimes A)\right)^{-1}(\mathbf{e} \otimes I)$. Notice that $r(z)$ is nothing but the stability function of the implicit part of the linearly implicit Runge-Kutta method (2.3). As long as $u_0^{*,2} = u_0^*$, and using summation by parts, we get

$$\begin{aligned} u_{n+1}^{*,2} - u_{n+1}^* &= k^3 R_n(kA)(\mathbf{b}^T \otimes I)\mathbf{a}^*(0) \\ &+ k^3 \sum_{j=1}^{n} R_{n-j}(kA)(\mathbf{b}^T \otimes I)\left[\mathbf{a}^*(t_j) - \mathbf{a}^*(t_{j-1})\right] \\ &+ O(k^3), \end{aligned}$$

with $R_j(kA) = \sum_{l=0}^{j} (r(kA))^l$ for $0 \le j \le n$. Since $[\mathbf{a}^*(t_j) - \mathbf{a}^*(t_{j-1})] = O(k)$, it is clear that for $0 \le n \le N-1$,

$$\|u_{n+1}^{*,2} - u_{n+1}^*\| = O(k^3),$$

as we wanted to prove. □

5. Numerical results

In order to illustrate the performance of the strategy described in the previous section, we consider again the forced Burgers' equation (2.1) with the same initial data and forcing term as in Section 2. We repeat the experiment of Section 2 but with the new implementation of the time integrator. In Figure 2 we have plotted

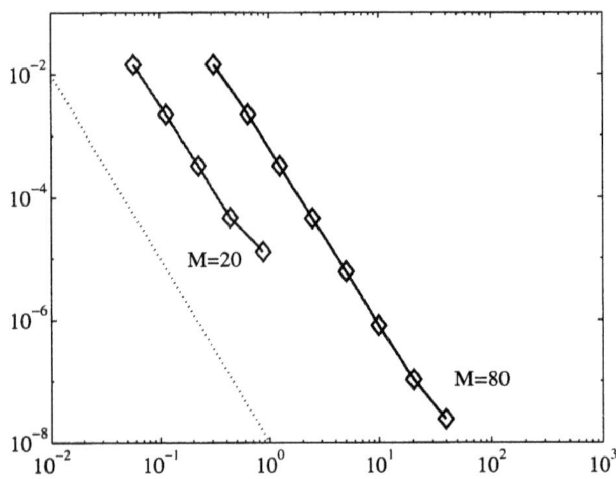

FIGURE 2. Error at $T = 10$ against CPU when integrating (3.2) for Burgers' equation

error at $T = 10$ (measured in the L^2 norm) against CPU time for $M = 80$ and $M = 20$ degrees of freedom and step-sizes $k = 0.4, 0.2, \ldots, 0.003125$ for $M = 80$ and $k = 0.4, 0.2, \ldots, 0.025$ for $M = 20$. Only data for which the error due to the spatial discretization is smaller than the error caused by the time integrator have been plotted. We observe that the time integrator now behaves as a third-order method as corresponds to the slope -3 of both lines. Then, the order reduction has been avoided.

In Figure 3 we have plotted the results obtained using both procedures, in order to see which of them is the most efficient. We use diamonds joined by a solid line for the data obtained with the implementation without order reduction and circles joined by a dashed line for the data generated with the linearly implicit method applied to Burgers' equation. We see that independently of the number of degrees of freedom used in the spatial discretization, the new implementation of the time integrator is more efficient than the standard integration of the original problem. We observe not only the third-order of the new implementation, as expected, but we also observe that the cheapest way to get errors under a prescribed tolerance always corresponds to the time integration using the new procedure. We conclude that the strategy proposed in [7] to avoid the order reduction of

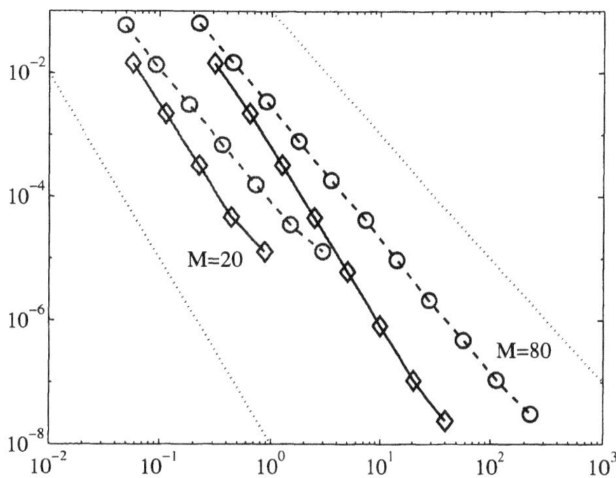

FIGURE 3. Efficiency diagram comparing both procedures

Runge-Kutta methods and adapted in the present paper to also deal with nonlinear problems is interesting not only from a theoretical point of view but also as an efficient procedure to time integrate one-dimensional nonlinear initial boundary value problems.

References

[1] S. Abarbanel, D. Gottlieb & M.H. Carpenter, *On the removal of boundary errors caused by Runge-Kutta integration of nonlinear partial differential equations*, SIAM J. Sci. Comput. **17** (1996), 777–782.

[2] I. Alonso-Mallo, *Explicit single step methods with optimal order of convergence for partial differential equations*, Appl. Numer. Math. **31** (1999), 117–131.

[3] I. Alonso-Mallo, *Rational methods with optimal order of convergence for partial differential equations*, Appl. Numer. Math. **35** (2000), 265–292.

[4] I. Alonso-Mallo & B. Cano, *Spectral/Rosenbrock discretizations without order reduction for linear parabolic problems*, Appl. Numer. Math. **41** (2002), 247–268.

[5] I. Alonso-Mallo & C. Palencia, *Optimal orders of convergence for Runge-Kutta methods and linear, initial boundary value problems*, Appl. Numer. Math. **44** (2003), 1–19.

[6] M.P. Calvo, J. de Frutos & J. Novo, *Linearly implicit Runge-Kutta methods for advection-reaction-diffusion equations*, Appl. Numer. Math. **37** (2001), 539–549.

[7] M.P. Calvo & C. Palencia, *Avoiding the order reduction of Runge-Kutta methods for linear initial boundary value problems*, Math. Comput. **71** (2002), 1529–1543.

[8] C. Canuto, M. Hussaini, A. Quarteroni & T. Zang, *Spectral methods in fluid dynamics*, Springer-Verlag, 1987.

[9] M.H. Carpenter, D. Gottlieb, S. Abarbanel & W.S. Don, *The theoretical accuracy of Runge-Kutta time discretizations for the initial boundary value problem: a study of the boundary error*, SIAM J. Sci. Comput. **16** (1995), 1241–1252.

[10] E. Hairer & G. Wanner, *Solving Ordinary Differential Equations II, Stiff and Differential-Algebraic Problems*, 2nd Edition, Springer, 1996.

[11] S.L. Keeling, *Galerkin/Runge-Kutta discretizations for parabolic equations with time-dependent coefficients*, Math. Comput. **52** (1989), 561–586.

[12] S.L. Keeling, *Galerkin/Runge-Kutta discretizations for semilinear parabolic equations*, SIAM J. Numer. Anal. **27** (1990), 394–418.

[13] A. Ostermann & M. Roche, *Runge-Kutta methods for partial differential equations and fractional orders of convergence*, Math. Comput. **64** (1995), 591–599.

[14] C. Palencia & I. Alonso-Mallo, *Abstract initial boundary value problems*, Proc. Royal Soc. Edinburgh **124A** (1994), 879–908.

[15] D. Pathria, *The correct formulation of intermediate boundary conditions for Runge-Kutta time integration of initial boundary value problems*, SIAM J. Sci. Comput. **18** (1997), 1255–1266.

[16] J.M. Sanz-Serna, J.G. Verwer & W.H. Hundsdorfer, *Convergence and order reduction of Runge-Kutta schemes applied to evolutionary problems in partial differential equations*, Numer. Math. **50** (1986), 405–418.

[17] J. Shen, *Efficient spectral-Galerkin method II. Direct solvers of second- and fourth-order equations using Chebyshev polynomials*, SIAM J. Sci. Comput. **16** (1995), 74–87.

[18] J.G. Verwer & J. Lang, *ROS3P – An accurate third-order Rosenbrock solver designed for parabolic problems*, BIT **41** (2001), 731–738.

Acknowledgment

The authors are grateful to Prof. C. Palencia for helpful discussions and suggestions.

Departamento de Matemática Aplicada y Computación
Universidad de Valladolid
Valladolid, Spain
email: `maripaz@mac.cie.uva.es`

Departamento de Matemática Aplicada Fundamental
Universidad de Valladolid
Valladolid, Spain
email: `frutos@mac.cie.uva.es`

Departamento de Matemática
Universidad Autónoma de Madrid
Madrid, Spain
email: `julia.novo@uam.es`

Modeling, Simulation and Optimization of Integrated Circuits
(Eds.) K. Antreich, R. Bulirsch, A. Gilg and P. Rentrop

International Series of Numerical Mathematics Vol. 146, 333–347

Solution of Linear Systems with Sparse Matrices

Friedrich Grund

To my grandchildren Anna, Janina and Paula

Abstract. For large scale problems in electric circuit simulation as well as in chemical process simulation, the linear solver often needs about 50–80% of the total amount of computing time. For that purpose, we consider direct methods for the numerical solution of linear systems of equations with unsymmetric sparse coefficient matrices. The Gaussian elimination method

$$PAQ = LU,$$
$$Ly = Pb, \qquad UQ^{-1}x = y$$

is applied to solve the linear system $Ax = b$. Here, the row permutation matrix P is used to provide numerical stability and the column permutation matrix Q is chosen to control sparsity. In a new approach, implemented in the solver GSPAR2, the determination of the pivot columns is done with a modified algorithm, which has only a complexity of $O(n)$. A partial pivoting technique is used to maintain numerical stability.

For solving several linear systems with the same pattern structure of the coefficient matrix efficiently, we generate a list of pseudo-code instructions for the factorization of the matrices. With it, the solver GSPAR2 has been proven successful within the simulation of several real life problems. For a number of linear systems arising from different technical problems, the computing times of GSPAR2 are compared to that of some recently released linear solvers.

1. Introduction

For the dynamic simulation of large scale applications in electric circuit analysis [2] as well as in chemical process engineering [3], initial value problems for large systems of nonlinear differential-algebraic equations (DAEs) have to be solved.

2000 *Mathematics Subject Classification.* Primary 65F05; Secondary 65F50, 65Y05.
Key words and phrases. Direct Linear Solver, Sparse–Matrix–Techniques, Parallel Computation, Circuit Simulation, Chemical Process Simulation.
This work was supported by the Federal Ministry of Education and Research of the Federal Republic of Germany under grants GA7FVB and GR7FV1.

For solving such problems, implicit integration methods like, e.g., backward differentiation formulas [11] are used and the resulting systems of nonlinear equations are then solved with Newton-type methods [6]. Thus, for the dynamic simulation of real life problems, large linear systems have to be solved repeatedly. In this applications, the linear solver often needs about 50–80% of the total amount of computing time [18].

The sparse, unsymmetric, and high-dimensional Jacobian matrices, appearing as coefficient matrices of the linear systems, usualy maintain their sparsity structure during the integration over many time steps. For these steps, the Gaussian elimination method can be used with the same ordering of the pivots, in general. Thus, the following approach can be used for the numerical solution of the linear systems in this context. At the beginning of the simulation and at some later time points, e.g., after a discontinuity in the functions have occured, the pivot ordering is determined with the Gaussian elimination method. This step is called *first factorization.* Then a list of pseudo-code instructions is generated to perform repeated factorizations of matrices with the same pattern structure, using the same order of pivots. This code contains only the necessary operations for the factorization and thus it works very efficiently. It is defined independently of a computer and can be adapted to exploit the features of vector as well as parallel computers. The factorization of a matrix by using the pseudo-code instructions is called *fast second factorization.* This fast second factorization can now be used within the numerical integration process until, e.g., a discontinuity in the function appears, the condition number of the matrix becomes too large, or the Newton method does not converge. In the later cases a new first factorization is performed and a new pseudo-code is generated. Then the fast second factorization can be used again.

This algorithms are realized in the linear solver GSPAR [12]. By using a new method for the determination of the pivot ordering, the performance of the first factorization can be improved considerably compared to that of GSPAR. This and other methods have recently been implemented in the solver GSPAR2. The solver has been proven successful for the dynamic process simulation of large real life chemical production plants and for the electric circuit simulation as well. Computing times for complete dynamic simulation runs of industrial applications are given. A modification of the solver is used in the new process simulator BOP (*B*lock *O*riented *P*rocess Simulator) [5].

2. The Gaussian elimination method

The Gaussian elimination method

$$PAQ = LU, \tag{2.1}$$

$$Ly = Pb, \qquad UQ^{-1}x = y \tag{2.2}$$

is used for solving the linear system

$$Ax = b. \tag{2.3}$$

The nonzero elements of the matrix A are stored in compressed sparse row format, also known as sparse row wise format. L is a lower triangular and U an upper triangular matrix. The row permutation matrix P is used to provide numerical stability and the column permutation matrix Q is used to control sparsity.

In GSPAR we consider the following approach for the determination of the matrices P and Q. At each elimination step, the algorithm searches for the first column with a minimal number of nonzero elements. This column becomes the pivot column. Then the columns are reordered. The method can be implemented easily, but the order of the numerical complexity is $O(n^2)$. For keeping the method numerically stable, the pivot $a_{i,j}$ of each elimination step is selected among those candidates satisfying the numerical threshold criterion

$$|a_{i,j}| \geq \beta \max_{l \geq k} |a_{l,j}|,$$

with a given threshold parameter $\beta \in (0, 1]$. This process is called partial pivoting. In our applications we usually choose $\beta = 0.01$ or $\beta = 0.001$.

In GSPAR2 we consider another approach. Here, we first reorder the columns according to the number of nonzeros in ascending order. Therefore, the first column becomes the pivot column of the first elimination step. At the end of each elimination step, the number of nonzeros in the remaining columns is corrected. The algorithm uses linked list techniques and is more complicated as the algorithm for GSPAR. The advantage of the algorithm of GSPAR2 over the algorithm of GSPAR is its minor numerical complexity. The complexity is here only $O(n)$. For the numerical stability, partial pivoting is used as described above.

Matrix	discipline	n	$\|A\|$
bayer01	chemical	57 735	277 774
bayer02	engineering	13 935	63 679
bayer03		6 747	56 196
bayer04		20 545	159 082
bayer06		3 008	27 576
bayer10		13 436	94 926
advice3388	circuit	3 388	40 545
meg1	simulation	2 904	58 142
meg4		5 860	46 842
rlxADC_tr		5 355	32 251
zy3315		3 315	15 985
poli	account of	4 008	8 188
poli_large	capital links	15 575	33 074

TABLE 1. Test matrices

In Table 2 on page 336 the performance of the first factorization with GSPAR2 is compared to that of GSPAR for some test matrices characterized in Table 1. The given CPU times are for a DEC AlphaServer with a processor 21164A operating at 400 MHZ. The number of columns and rows is denoted by n and $|A|$ identifies the number of nonzero elements in the matrix. Many of test matrices can be found in [7].

Matrix	GSPAR	GSPAR2
bayer01	34.92	2.35
bayer02	2.20	0.55
bayer03	0.67	0.30
bayer04	5.18	1.82
bayer06	0.82	0.83
bayer10	3.07	1.27

TABLE 2. First factorization with GSPAR and GSPAR2, CPU times in seconds

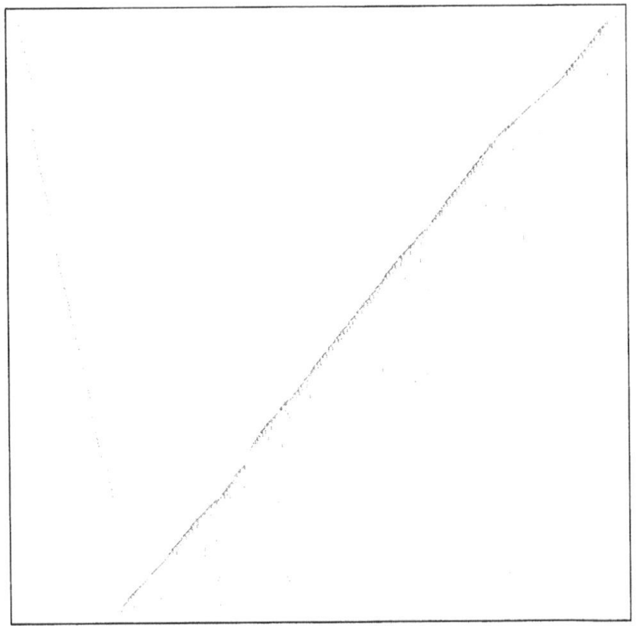

FIGURE 1. Matrix structure for bayer04

In Figure 1 the structure of the matrix bayer04 is shown. For the creation of the figure, a tool from T. Davis [7] has been used. The matrix bayer04 as well as the other bayer matrices in Table 1 are very unstructured and unsymmetric.

3. Fast second factorization

If initial value problems for large systems of differential-algebraic equations have to be treated numerically, e.g., in the dynamic simulation of large scale applications in electric circuit analysis or chemical process engineering, numerous linear systems have to be solved. In this cases it mostly happens, that the Gaussian elimination method has to be used repeatedly for the factorization of a number of matrices with the same pattern structure. To exploit this circumstance for an efficient implementation, a list of pseudo-code instructions for the computation of the factorization is generated.

For the generation of these pseudo-code instructions, the factorization of the Gaussian elimination method is used as shown in Figure 2. The algorithm in this form needs exactly n divisions. Only a few different types of pseudo-code instructions are needed in this case. The types of the pseudo-code instructions and the corresponding lists of indices are stored in integer arrays. For the factorization of matrices with a very large number of nonzero elements, it can happen that the number of integer array elements needed for the storage of the pseudo-code instructions exceed main memory limits [12]. In this cases the pseudo-code technique can not be used and may be replaced by using FORTRAN subroutines.

for $i = 2, n$ do
    $a_{i-1,i-1} = 1/a_{i-1,i-1}$
    for $j = i, n$ do
        $a_{j,i-1} = (a_{j,i-1} - \sum_{k=1}^{i-2} a_{j,k}a_{k,i-1})a_{i-1,i-1}$
    enddo
    for $j = i, n$ do
        $a_{i,j} = a_{i,j} - \sum_{k=1}^{i-1} a_{i,k}a_{k,j}$
    enddo
enddo
$a_{n,n} = 1/a_{n,n}$.

FIGURE 2. Gaussian elimination method

4. Vectorization and parallelization

The pseudo-code instructions can be used for both vectorization as well as parallelization. In both cases, elements of the factorized matrix have to be found which can be computed independently of each other.

For the factorization, a matrix

$$M = (m_{i,j}), \quad m_{i,j} \in \mathbb{N} \cup \{0, 1, 2, \ldots, n^2\}$$

is assigned to the matrix

$$LU = PAQ,$$

where $m_{i,j}$ denotes the level of independence.

The elements with the assigned level zero do not need any operations. All elements with the same level in the factorized matrix (2.1) can be computed independently. First all elements with level one are computed, then all elements with level two and so on. The levels of independence for the matrix elements in (2.1) can be computed with the algorithm of Yamamoto and Takahashi [17]. The algorithm for the determination of the levels of independence $m_{i,j}$ is shown in Figure 3.

For an application on a vector computer, we have to find vector instructions at the different levels of independence [12]. For parallelization, it needs to be distinguished between parallel computers with shared memory and with distributed memory. In the case of parallel computers with shared memory and p processors,

```
M = 0
for i = 1, n − 1 do
    for all {j : a_{j,i} ≠ 0 & j > i}
        m_{j,i} = 1 + max(m_{j,i}, m_{i,i})
        for all {k : a_{i,k} ≠ 0 & k > i}
            m_{j,k} = 1 + max(m_{j,k}, m_{j,i}, m_{i,k})
        enddo
    enddo
enddo
```

FIGURE 3. Algorithm of Yamamoto and Takahashi

we assign the pseudo-code for each level of independence in p parts of approximately same size to the processors. After the processors have executed their part of the pseudo-code instructions of a level concurrently, a synchronization among the processors is needed. Then the execution of the next level can be started. If the processors are vector processors then this property can be used analogously. The parallel computer Compaq AlphaServer GS80 6/731 and the IBM 7040-681, pSeries 690 are examples for such computers.

In the case of parallel computers with distributed memory and q processors, the pseudo-code for each level of independence is again partitioned into q parts of approximately same size. But in this case, the parts of the pseudo-code are moved to the memory of each individual processor. The transfer of parts of the code to the memories of the individual processors is done only once. A synchronization is carried out analogously to the shared memory case. The partitioning and the storage of the matrix as well as of the vectors is implemented in the following way. For small problems the elements of the matrix, right-hand side, and solution vector are located in the memory of one processor, whereas for large problems, they have to be distributed over the memories of several processors. We assume that the data communication between the processors for the exchange of data concerning elements of the matrix, right-hand side, and solution vector is supported by the operating system. The massive parallel computer Cray T3E is an example for such a computer.

This approach can be illustrated by small artificial example. We assume that the permutation matrices P and Q are given, so that

$$PAQ = \begin{pmatrix} 2 & 4 & & & \\ 5 & 7 & & & 9 \\ & 2 & 9 & & 1 \\ & & 1 & 7 & 8 \\ & & 1 & 3 & 5 \end{pmatrix}. \tag{4.1}$$

The nonzero elements of the matrix A are stored in sparse row format in the vector a. Let $\boxed{\text{i}}$ denote the index of the i-th element in the vector a, then the elements of the matrix PAQ are stored in the following way

$$\begin{pmatrix} \boxed{7} & \boxed{8} & & & \\ \boxed{12} & \boxed{13} & & & \boxed{14} \\ & \boxed{2} & \boxed{1} & & \boxed{3} \\ & & \boxed{9} & \boxed{10} & \boxed{11} \\ & & \boxed{4} & \boxed{5} & \boxed{6} \end{pmatrix}. \tag{4.2}$$

The matrix M assigned to the matrix PAQ is found to be

$$M = \begin{pmatrix} 0 & 0 & & & \\ 1 & 2 & & & 0 \\ & 3 & 0 & & 4 \\ & & 1 & 0 & 5 \\ & & 1 & 1 & 6 \end{pmatrix}. \tag{4.3}$$

We can see from (4.3), that six independent levels exist for the factorization. The detailed instructions for the factorization of the matrix A resulting from (4.1)–(4.3) are shown in Table 3 on page 340.

Let us now consider, for example, the instructions of level one in Table 3 only. Then, on one hand on a vector computer, one vector instruction of the length four can be generated. And on the other hand on a parallel computer with distributed or shared memory, a allocation of the four instructions to two or four processors can be done [12].

From our experiments with many different matrices arising from the process simulation of chemical plants and the circuit simulation respectively, it was found that the number of levels of independence is in general small. The number of instructions in the first two levels is very large, in the next four to six levels it is large and finally it becomes smaller and smaller.

Level	Instructions
1	a(12) = a(12)/a(7)
	a(9) = a(9)/a(1)
	a(4) = a(4)/a(1)
	a(5) = a(5)/a(10)
2	a(13) = a(13) – a(12) ⋆ a(8)
3	a(2) = a(2)/a(13)
4	a(3) = a(3) – a(2) ⋆ a(14)
5	a(11) = a(11) – a(5) ⋆ a(3)
6	a(6) = a(6) – a(4) ⋆ a(3) – a(5) ⋆ a(11)

TABLE 3. Instructions for the factorization

5. Numerical results

The developed numerical methods are realized in the program packages GSPAR and GSPAR2. The latest version GSPAR2 has been released in 2002. Whereas GSPAR is written in Fortran 77 and implemented on workstations (Digital AlphaStation, IBM RS/6000, SGI, Sun UltraSparc 1 and 2), vector computers (Cray J90, C90), parallel computers with shared memory (Cray J90, C90, SGI Origin2000, Digital AlphaServer) and parallel computers with distributed memory (Cray T3D), its successor GSPAR2 is written in Fortran 90 and currently implemented on Windows PC's, workstations (Compaq Workstation, SGI, IBM) and parallel computers (Compaq AlphaServer GS80 6/731, IBM 7040-681, pSeries 690).

Numerical results will be given for GSPAR, GSAPAR2, and some other lately released linear solvers. GSPAR has been applied to the test matrices from Table 1 on page 335. In Table 4 on page 341 results are shown using the solver GSPAR on a DEC AlphaServer with an alpha EV5.6 (21164A) processor. Here *# op LU* is the number of operations (only multiplications and divisions) and *fill-in* is the number of fill-ins during the factorization. The CPU time (in seconds) for the first factorization, presented in *First factor.*, includes the times for the analysis as well as for the numerical factorization. The CPU time for the generation of the pseudo-code is given in *Pseudo code.*

In Table 5 on page 341, CPU times (in seconds) for the second factorization are shown for the linear solvers UMFPACK V1.0 [9], SuperLU V1.0 with minimum degree ordering of A^TA (upper index *) or of $A^T + A$ (upper index $^+$) [16, 10], and GSPAR, using a DEC AlphaStation with an alpha EV4.5 (21064) processor.

In many applications, mainly in the numerical simulation of physical and chemical problems, the analysis step including ordering and first factorization is performed only a few times, but the second factorization is performed very often.

Therefore, the CPU time for the second factorization is essential for the overall simulation time.

Matrix	# op LU	fill-in	First factor.	Pseudo code
bayer01	10 032 621	643 898	35.18	12.72
bayer02	2 095 207	134 546	2.28	1.30
bayer03	1 000 325	64 130	0.68	0.47
bayer04	5 954 718	268 006	5.33	3.93
bayer05	119 740	11 024	0.15	0.03
bayer06	3 042 620	73 773	0.85	1.00
bayer09	364 731	23 145	0.18	0.15
bayer10	5 992 500	227 675	3.05	2.55
advice3388	310 348	9 297	0.38	0.65
meg1	796 797	40 436	0.32	0.40
meg4	420 799	38 784	0.68	0.62
rlxADC_dc	73 612	5 404	0.38	0.13
rlxADC_tr	988 759	47 366	0.85	1.13
zy3315	47 326	8 218	0.12	0.03
poli	4 620	206	0.15	0
poli_large	43 310	10 318	2.38	0.25

TABLE 4. GSPAR first factorization and generation pseudo-code

Matrix	UMFPACK V1.0	SuperLU V1.0	GSPAR
bayer01	5.02	6.70 *	3.20
bayer02	1.13	1.47 *	0.55
bayer03	0.72	0.70 *	0.27
bayer04	3.37	2.77 *	1.70
bayer05	0.13	0.75 *	0.05
bayer06	0.83	0.90 *	0.82
bayer09	0.23	0.23 *	0.10
bayer10	1.60	1.57 *	1.65
advice3388	0.25	0.28 +	0.10
meg1	0.58	1.43 +	0.22
meg4	0.37	0.75 +	0.13
rlxADC_dc	0.15	0.18 +	0.03
rlxADC_tr	0.40	0.90 +	0.30
zy3315	0.15	0.18 +	0.02
poli	0.03	0.07 +	0.00
poli_large	0.13	0.27 +	0.03

TABLE 5. CPU times for second factorization

GSPAR achieves a good performance for second factorization for all linear systems in Table 5. For systems with a large number of equations, GSPAR is at least two times faster than UMFPACK V1.0 and SuperLU V1.0 respectively.

In Table 6, wall-clock times (in seconds) are shown for the second factorization, using GSPAR with different pivoting on a DEC AlphaServer with four alpha EV5.6 (21164A) processors. The parallelization technique is based on OpenMP [15]. The wall-clock times have been determined with the system routine *gettimeofday*.

processors	bayer01	bayer04	bayer01	bayer04
1	0.71	0.39	1.08	0.43
2	0.54	0.27	0.75	0.29
3	0.45	0.23	0.63	0.25
4	0.49	0.24	0.70	0.30

TABLE 6. Wall-clock times for second factorization

In the following, we compare the performance of some of the new software packages for solving linear systems with sparse matrices on a single CPU of a Compaq AlphaServer GS80 6/731 with alpha EV6.7 (21264A) processors, operating at 731 MHz and on an IBM 7040-681, pSeries 690 with 64-bit POWER4 processors, operating at 1.3 GHz. The used packages are now GSPAR2, UMFPACK V3.0, and WSMP Version 1.7 (Watson Sparse Matrix Packages) [13]. We compare GSPAR2 and UMFPACK V3.0 on the Compaq and GSPAR2 and WSMP Version 1.7 on the IBM for the test matrices in Table 7 respectively.

Matrix	discipline	n	$\|A\|$	nnc
bayer01	Process simul.	57 735	277 774	33
big	Device simul.	13 209	91 465	12
circuit_3	Circuit simul.	12 127	48 137	5 682
bayer02	Process simul.	13 935	63 679	28
lhr34c	Process simul.	35 152	764 014	36
meg4	Circuit simul.	5 860	46 842	1 194
bayer04	Process simul.	20 545	159 082	43
nopoly	Device simul.	10 774	70 842	11
circuit_4	Circuit simul.	80 209	307 604	8 900
pesa	Device simul.	11 738	79 566	10
shermanACb	Circuit simul.	18 510	145 149	10 405
bayer10	Process simul.	13 436	94 926	32

TABLE 7. Test matrices

Here, $|A|$ identifies the number of nonzero elements in the matrix and nnc the maximum number of nonzeros in columns.

Matrix	First Fac.	Gen. Code	Sec. Fac.	Solv.	Fast Fac.	NNZ-LU
bayer01	1.583	1.248	1.134	0.094	0.328	932 190
big	3.371	2.658	2.656	0.076	0.727	685 441
circuit_3	0.126	0.013	0.014	0.004	0.005	64 608
bayer02	0.243	0.182	0.155	0.015	0.049	192 752
lhr34c	11.713	12.255	11.579	0.252	2.735	3 065 587
meg4	0.106	0.050	0.042	0.003	0.011	84 694
bayer04	0.887	0.799	0.695	0.038	0.181	458 086
nopoly	0.707	0.516	0.409	0.028	0.126	321 350
circuit_4	2.736	1.005	1.155	0.057	0.098	420 875
pesa	1.326	0.997	0.797	0.043	0.263	476 560
shermanACb	6.846	3.509	3.136	0.053	2.011	544 953
bayer10	0.473	0.394	0.305	0.022	0.102	306 865

TABLE 8. GSPAR2 on the Compaq GS80 6/731

In Table 8 the results are shown for GSPAR2 on the Compaq GS80 6/731. The CPU times (in seconds) are given for the first factorization in *First Fac.*, for the generation of the pseudo-codes in *Gen. Code*, for the second factorization using a Fortran routine in *Sec. fac.*, for the solving in *Solv.* and for the fast second factorization in *Fast Fac.*. The number of nonzero elements in the L- and U-matrix is denoted by *NNZ-LU*.

Matrix	Symb Fac.	Num. Fac.	Solv.	Red. Fac.	NNZ-LU
bayer01	0.583	1.017	0.050	0.917	1 080 365
big	0.117	0.367	0.017	0.300	828 480
circuit_3	0.083	0.450	0.000	0.400	106 286
bayer02	0.117	0.200	0.000	0.167	224 979
lhr34c	1.200	1.933	0.083	1.833	2 975 184
meg4	0.117	0.383	0.000	0.350	55 215
bayer04	0.267	0.383	0.017	0.367	362 366
nopoly	0.083	0.167	0.000	0.150	368 545
circuit_4	15.266	934.746	1.217	930.063	43 569 250
pesa	0.100	0.233	0.000	0.200	491 854
shermanACb	0.833	24.666	0.050	23.949	1 266 626
bayer10	0.133	0.233	0.017	0.200	245 986

TABLE 9. UMFPACK V3.0 on Compaq GS80 6/731

In Table 9 on page 343 results are given for UMFPACK V3.0 on the GS80 6/731. The solver UMFPACK V3.0 is written in ANSI/ISO C and contains different new algorithms [8]. The CPU times (in seconds) are identified for the symbolic factorization in *Symb. Fac.*, for the numeric factorization in *Num. Fac.*, for solving in *Solv.* and for the redo numeric factorization in *Red. Fac.*

Matrix	First Fac.	Gen. Code	Sec. Fac.	Solv.	Fast Fac.
bayer01	0.840	0.740	0.450	0.020	0.160
big	1.850	1.320	0.820	0.020	0.360
circuit_3	0.080	0.010	0.010	0.010	0.010
bayer02	0.160	0.130	0.070	0.010	0.020
lhr34c	6.260	6.540	3.830	0.070	1.460
meg4	0.070	0.030	0.020	0.000	0.000
bayer04	0.510	0.460	0.250	0.010	0.100
nopoly	0.390	0.270	0.140	0.010	0.050
circuit_4	1.430	0.570	0.480	0.020	0.050
pesa	0.730	0.590	0.290	0.010	0.140
shermanACb	3.020	1.900	0.970	0.010	0.730
bayer10	0.290	0.230	0.120	0.000	0.050

TABLE 10. GSPAR2 on IBM pSeries 690

CPU times (in seconds) are shown for GSPAR2 on an IBM pSeries 690 in Table 10. The description of the headline is the same as in Table 8.

Matrix	Analysis	Factorization	Back substitution	NNZ-LU
bayer01	2.640	0.460	0.180	1 562 405
big	0.290	0.090	0.020	701 524
circuit_3	0.130	0.030	0.010	91 060
bayer02	0.600	0.100	0.040	376 018
lhr34c	2.710	0.470	0.110	2 955 105
meg4	0.080	0.030	0.010	111 676
bayer04	0.860	0.120	0.040	536 935
nopoly	0.190	0.060	0.020	390 502
circuit_4	1.980	0.300	0.120	603 556
pesa	0.210	0.080	0.020	509 346
shermanACb	0.460	0.110	0.040	430 300
bayer10	0.680	0.100	0.030	455 494

TABLE 11. WSMP Version 1.7 on an IBM pSeries 690

In Table 11 on page 344 results are given for the linear solver WSMP Version 1.7 [14] on an IBM pSeries 690. The CPU times (in seconds) are given for the analysis step in *Analysis*, for the factorization step in *Factorization* and for the back substitution step in *Back substitution*.

From the results in the Tables 8–11 one can see that the second factorization with GSPAR2 is fast. The symbolic factorization with UMFPACK V3.0 is fast too and the time difference between the factorization and the redo factorization is marginal. The package WSMP has a fast factorization step also for large matrices.

6. Application in process engineering

The more and more integrated modeling in process engineering leads to large scale problems in static and dynamic process simulation. The complex real world process models used in this field usually depend on numerous parameters and are in general highly nonlinear. Using concentrated physical models, high-dimensional systems of nonlinear and differential-algebraic equations have to be solved in static and dynamic process simulation respectively. For that purpose robust and efficient numerical simulation tools are needed. They are urgently necessary to improve process design, analysis, as well as operation in todays process industries.

The hierarchical modular simulation concept developed at WIAS [3, 4, 5] is based on divide and conquer techniques and exploits the modular structure of the process, which in most cases is defined by the hierarchical unit structure of the underlying plant. With it the corresponding system of equations is structured into subsystems according to the units and can be partitioned into m blocks

$$F_i(t, y(t), \dot{y}(t), u(t)) = 0, \; i = 1(1)m,$$

$$F_i : \mathbb{R} \times \mathbb{R}^n \times \mathbb{R}^n \times \mathbb{R}^q \rightarrow \mathbb{R}^{n_i}, \; \sum_{i=1}^{m} n_i = n, \; t \in [t_0, t_{end}]$$

which can then be treated almost concurrently within appropriately modified numerical methods. The approach has been implemented in the Block Oriented Process simulator *WIAS-BOP* that uses an own compiler to generate a hierarchically structured data interface from a process description with its modeling language MLPE (Modeling Language for Process Engineering). The numerical solution within *BOP* require a repeated solution of linear systems with the same pattern structure of sparse, unsymmetric coefficient matrices, and with multiple right-hand sides. A slightly modified version of the direct solver GSPAR is used within *BOP* to solve these linear systems. For several real life problems, GSPAR has proven to be a robust and reliable linear solver for this application on parallel computers Cray J90 and SGI Origin 2000. For some distillation processes of Bayer AG Leverkusen, resulting into DAE system with several ten thousand equations, the dynamic simulation with *BOP* has achieved speedup factors of up to 10 in wall-clock time on a Cray J90 with 24 processors. For detailed results in this case, we refer to [5].

Another application of GSPAR in process simulation has been performed using the commercial process simulator SPEEDUP [1]. In SPEEDUP the vector versions of the linear solvers FAMP and GSPAR have been used alternatively. FAMP is the default linear solver in SPEEDUP and optimized on the Cray computer architecture. Table 12 shows two large scale industrial problems of the Bayer AG Leverkusen. The number of differential-algebraic equations are given.

name	chemical plants	equations
bayer04	nitration plant	3 268
bayer10	distillation column	13 436

TABLE 12. Large scale industrial problems

The problems have been solved on a vector computer Cray C90. The CPU times (in seconds) for complete dynamic simulation runs are shown in Table 13.

name	FAMP	GSPAR	in %
bayer04	451.7	283.7	62.8
bayer10	380.9	254.7	66.9

TABLE 13. CPU time for complete dynamic simulation

References

[1] SPEEDUP, *User Manual, Library Manual.* Aspen Technology, Inc., Cambridge, Massachusetts, USA, 1995.

[2] J. Borchardt, F. Grund, D. Horn, M. Uhle, *MAGNUS – Mehrstufige Analyse großer Netzwerke und Systeme.* Weierstraß-Institut für Angewandte Analysis und Stochastik, Report No. 9, Berlin, 1994.

[3] J. Borchardt, F. Grund, D. Horn, *Parallelized Methods for Large Nonlinear and Linear Systems in the Dynamic Simulation of Industrial Applications.* Surv. Math. Ind. **8** (1999), 201–211.

[4] J. Borchardt, K. Ehrhardt, F. Grund, D. Horn, *Parallel Modular Dynamic Process Simulation.* In F. Keil, W. Mackens, H. Voss, J. Werther, editors, *Scientific Computing in Chemical Engineering II*, volume 2, pages 152–159, Springer-Verlag Berlin Heidelberg New York, 1999.

[5] J. Borchardt, *Newton-Type Decomposition Methods in Large-Scale Dynamic Process Simulation.* Comput. Chem. Engng. **25** (2001), 951–961.

[6] K.E. Brenann, S.L. Campbell, L.R. Petzold, *Numerical Solution of Initial-Value Problems in Differential-Algebraic Equations.* North-Holland, New York, 1996.

[7] T.A. Davis, *University of Florida Sparse Matrix Collection.* http://www.cise.ufl.edu/~davis/sparse

[8] T.A. Davis, *UMFPACK V3.0.* http://www.cise.ufl.edu/research/sparse/umfpack/

[9] T.A. Davis, I.S. Duff, *An unsymmetric-pattern multifrontal method for sparse LU factorization.* Tech. Report **TR-94-038**, CIS Dept., Univ. of Florida, Gainsville, FL, 1994.

[10] J.W. Demmel, S.C. Eisenstat, J.R. Gilbert, X.S. Li and J.W. H. Liu, *A Supernodal Approach to Sparse Partial Pivoting.* SIAM J. Matrix Anal. Appl. **20** (1999), 720–755.

[11] C.W. Gear, *Numerical Initial Value Problems in Ordinary Differential Equations.* Prentice-Hall, Inc., Englewood Cliffs, New Jersey, 1971.

[12] F. Grund, *Direct Linear Solvers for Vector and Parallel Computers.* In José M. L. M. Palma and Jack Dongarra and Vicente Hernández, editors, *Vector and Parallel Processing – VECPAR'98*, Lecture Notes in Computer Science **1573**, Springer-Verlag Berlin Heidelberg New York, 1999.

[13] A. Gupta, *Recent Advances in Direct Methods for Solving Unsymmetric Sparse Systems of Linear Equations.* ACM Trans. Math. Softw. **28** (2002), 301–324.

[14] A. Gupta, *WSMP: Watson Sparse Matrix Package.* http://www-users.cs.umn.edu/~agupta/wsmp.html

[15] *OpenMP.* http://www.openmp.org.

[16] X.S. Li, *Sparse Gaussian elimination on high performance computers.* Technical Reports **UCB//CSD-96-919**, Computer Science Division, U.C. Berkeley, Ph.D. dissertation, 1996.

[17] F. Yamamoto, S. Takahashi, *Vectorized LU decomposition algorithms for large-scale nonlinear circuits in the time domain.* IEEE Trans. on Computer-Aided Design **CAD-4** (1985), 232–239.

[18] S.E. Zitney, L. Brüll, L. Lang, R. Zeller, *Plantwide dynamic simulation on supercomputers: modeling a Bayer distillation process.* AIChE Symposium Series **91** (1995), 313–316.

Acknowledgment

The author thanks his coworkers Jürgen Borchardt and Dietmar Horn for useful discussions. The valuable assistance and the technical support from the Bayer AG Leverkusen, the Cray Research Munich and Aspen Technology, Inc., Cambridge, MA, USA are gratefully acknowledged.

Friedrich Grund
Weierstrass Institute for Applied Analysis and Stochastics
Mohrenstrasse 39
10117 Berlin, Germany
email: `grund@wias-berlin.de`

Modeling, Simulation and Optimization of Integrated Circuits
(Eds.) K. Antreich, R. Bulirsch, A. Gilg and P. Rentrop

International Series of Numerical Mathematics Vol. 146, 349–360

The Performances of the Code TOM on the Holt Problem

Lidia Aceto, Francesca Mazzia, and Donato Trigiante

Abstract. The code TOM, for the solution of boundary value problems, is based on linear multistep methods used as BVMs [5, 6, 16, 17]. Among the peculiar features of this code, the mesh selection strategy, based on two measures of conditioning of the problem, seems the most interesting. In this paper the application of the code to the classical Holt problem, one of the most famous and difficult BVP, will permit to stress the effectiveness of such approach along with the utility of the additional information provided by the two measures.

1. Introduction

General purpose codes for initial value problems (IVPs) for ordinary differential equations have reached a very high level of performance. They have by now an history more then thirty years long. Many of them, easily available in the public domain or as part of commercial packages, have become an essential tool in researches of almost all fields of science and technology.

On the contrary, codes for boundary value problems (BVPs) have a shorter history and a lower level of performance. The reason, obviously, is that they are conceptually more difficult. To be more clear, we shall refer to the case of linear constant coefficients. A well-conditioned IVP is one having non positive eigenvalues and a relatively short interval of integration. This implies that, in principle, one can always solve a well-conditioned sequence of problems by reducing appropriately the successive intervals of integration.

On the other side, a well-conditioned BVP is one having the so-called dichotomy, i.e., negative and positive eigenvalues and appropriate boundary conditions. The traditional and popular shooting technique transforms well-conditioned BVPs into ill-conditioned IVPs because of the positive eigenvalues. The literature is full of examples of problems for which the shooting technique does not work.

The direct approach, i.e., the approach transforming the continuous BVP in a discrete BVP has been taken into account relatively later. The most recent codes,

---

2000 *Mathematics Subject Classification.* 65L10; 65L50.
Key words and phrases. boundary value problems; mesh selection; conditioning; Holt problem.
Work supported by G.N.C.S..

however, are based on such approach. Of course, the properties of the numerical methods for BVPs should be different from those tailored for IVPs. The study of such properties are by now satisfactory and codes having very good performances have been proposed [2, 3, 7, 8, 10, 13].

The code TOM, written in MATLAB and not yet released, is in the phase of testing. Some of its parts have been presented in preliminary papers [5, 6, 16, 17].

After a brief description of it, we will show the performances on the classical Holt problem [11].

2. Description of the code

In this section we shall briefly describe the requirements needed by a robust code for BVPs and the way they have been satisfied by the code TOM. As before, in order to be more clear, we shall continue to refer to the constant coefficients problem

$$\begin{cases} \dfrac{dy(t)}{dt} = L\,y(t), & t_0 \le t \le T \\ \\ G_0\,y(t_0) + G_1\,y(T) = \eta, \end{cases} \tag{2.1}$$

although the code has been designed to work in the general nonlinear case. In (2.1) $y \in \mathbb{R}^m$, L, G_0 and G_1 are square matrices of size m and η is a known vector. The main problems to be solved when designing a code are:

1. choice of the appropriate methods;
2. mesh selection;
3. solutions of linear and nonlinear algebraic equations;
4. estimates and control of the global error.

There is a variety of solutions for all of them, some well-exploited, some less. Each of the existing codes makes its own choices. It is then important to compare the results provided by the codes on the most difficult problems in order to classify their effectiveness.

2.1. Choice of the methods

Usually a code uses more than one method having different orders. It is important that such methods share similar stability properties, in the sense specified below.

It has been widely recognized that a good method for BVP should be symmetric, i.e., should treat similarly both the decreasing and the increasing components of the solution. As already said, both of them generally exist because of the existence of eigenvalues with positive and negative real parts. Methods having such property are called *perfect stable* and have the imaginary axis as boundary between the stability and instability regions. In principle, for BVPs, unlike the IVPs case, it is not important to have very large stability regions since they usually work in the relative stability ones [6]. However, it is known that methods having the imaginary axis as required above, also have unbounded stability regions. In practice, such property is important when the mesh is not yet the correct one and the method is not yet working in the relative stability region over the entire interval.

All the most recent codes utilize symmetric methods, although of different kinds. The code COLNEW [3] is based on truncated powers collocation, while the codes TWPBVP, MIRCDC and BVP4c use Mono Implicit Runge Kutta Methods (MIRK) [8, 10, 13].

The code TOM utilizes a special class of methods called Top Order Methods (TOMs) in the family of boundary value methods (BVMs) (see [6] p. 171). In few words, they are multistep methods having the number of steps k odd and the highest order ($2k$). In the past they were considered useless because unstable when used as discrete IVP. It has been shown (see [6] p. 173) that when used as discrete BVP, i.e., when the $k-1$ *additional conditions* are appropriately split by imposing $\nu = (k-1)/2$ of them at the beginning of the interval and an equal number at the end, they are perfectly stable.

Considering that the coefficients characterizing each TOM satisfy the conditions [6]

$$\alpha_i = -\alpha_{k-i}, \qquad \beta_i = \beta_{k-i}, \qquad i = 0, 1, \ldots, k,$$

the equations describing the method, said *main method*, when applied to problem (2.1) can be written as follows

$$\sum_{i=0}^{\nu} \alpha_i (y_{n+i} - y_{n+k-i}) = h \sum_{i=0}^{\nu} \beta_i L(y_{n+i} + y_{n+k-i}). \tag{2.2}$$

The additional conditions are imposed by inserting ν equations before (2.2) and the same number of equations after (2.2). They describe methods (*additional methods*) taken from the class of BVMs known as ETRs (Extended Trapezoidal Rules) ([6], p. 162). All together the discrete problem, using constant stepsize, looks like

$$\begin{cases} G_0\, y_0 + G_1\, y_N = \eta \\ \displaystyle\sum_{i=0}^{k} \alpha_i^{(n)} y_i - h \sum_{i=0}^{k} \beta_i^{(n)} L y_i = 0, \quad n = 1, \ldots, \nu \\ \displaystyle\sum_{i=0}^{\nu-1} \alpha_i (y_{n+i} - y_{n+k-i}) - h \sum_{i=0}^{\nu-1} \beta_i L(y_{n+i} + y_{n+k-i}) = 0, \quad n = \nu+1, \ldots, N-\nu \\ \displaystyle\sum_{i=0}^{k} \alpha_i^{(n-N+k)} y_{N-k+i} - h \sum_{i=0}^{k} \beta_i^{(n-N+k)} L y_{N-k+i} = 0, \quad n = N-\nu+1, \ldots, N, \end{cases}$$

where N is the number of steps.

The coefficients of the initial and final methods verify the following relations:

$$\alpha_i^{(j)} = -\alpha_{k-i}^{(k-j+1)}, \qquad \beta_i^{(j)} = \beta_{k-i}^{(k-j+1)}, \qquad j = 1, 2, \ldots, \nu, \;\; i = 0, 1, \ldots, k.$$

By introducing the all-inclusive square matrix M_h of size $m(N+1)$, the overall process can be stated in a more compact form as:

$$M_h \mathbf{y} = e_1 \otimes \eta$$

where $\mathbf{y} = (y_0^T, y_1^T, \dots, y_N^T)^T$ and $e_1 = (1, 0, \dots, 0)^T$ is of length $N + 1$. It is straightforward to show that the resulting scheme is symmetric.

In general, when difficult problems are treated, the stepsize h cannot remain constant. In such a case the methods can be represented by $M_{\mathbf{h}}$ where $\mathbf{h} = (0, h_1, \dots h_N)$. This permits to change the conditioning of the matrix $M_{\mathbf{h}}$ if the sequence $\{h_i\}$ is chosen appropriately (optimal mesh).

2.2. Mesh selection

The mesh selection is then a crucial step because from it depend, in the case of difficult problems, the subsequent steps (convergence of Newton iterations, error control, etc.). Most of the existing codes base the mesh selection on estimates of the global error. This approach is not completely satisfying in the case of multiscale problems.

In the present code, this step has then been developed with special care. The details are described in [5, 17]. Essentially, two parameters $\kappa_d(\mathbf{h})$ and $\gamma_d(\mathbf{h})$, describing the conditioning of the discrete problem in two different norms, are defined. As the code provides better approximations of the solution, the sequences of the two parameters are requested to converge to the corresponding parameters κ_c and γ_c defined for the continuous problem, i.e., it should be true that

$$\lim_{\mathbf{h} \to 0} \kappa_d(\mathbf{h}) = \kappa_c, \quad \lim_{\mathbf{h} \to 0} \gamma_d(\mathbf{h}) = \gamma_c.$$

Of course, κ_c and γ_c are unknown, but still the behaviors of the two sequences $\{\kappa_d(\mathbf{h})\}$ and $\{\gamma_d(\mathbf{h})\}$ give important information about the approximating process. Such information is used dynamically by the code. It may also be of interest to the user and, for this reason, we briefly describe them as follows:

1. both $\kappa_d(\mathbf{h})$ and $\gamma_d(\mathbf{h})$ are of moderate size and do not change sensibly with $\mathbf{h}$. The problem is well conditioned and then an easy one;
2. $\kappa_d(\mathbf{h}) >> \gamma_d(\mathbf{h})$. The problem is stiff. It can be solved with a relatively small amount of points essentially concentrated in small intervals;
3. both $\kappa_d(\mathbf{h})$ and $\gamma_d(\mathbf{h})$ are large. The problem is ill-conditioned and a large amount of points is needed to obtain its solution;
4. both sequences $\kappa_d(\mathbf{h})$ and $\gamma_d(\mathbf{h})$ are unbounded as $\mathbf{h} \to 0$. It is very likely possible that the continuous problem has no solution or, more generally, is not well posed;
5. the continuous problem may depend on a parameter μ. In this case the sequences $\{\kappa_d^{(\mu)}(\mathbf{h})\}$ and $\{\gamma_d^{(\mu)}(\mathbf{h})\}$, evaluated for different values of the parameter μ, may provide useful information. Suppose, for example, that there is the suspect that for $\mu \to \bar{\mu}$ the problem has no solutions. If the sequence of values of both parameters in correspondence of a sequence of $\mu_j \to \bar{\mu}$ is diverging, then this could confirm the suspect.

All such possibilities will arise in the example below (Holt problem).

The information on the two discrete conditioning parameters are taken from the first block column of the matrix $M_{\mathbf{h}}^{-1}$, as described in [17]. At each step, the new mesh is obtained by minimizing, with respect to $\mathbf{h}$, a quantity related to one of the two parameters. This is obtained by a process of equidistribution of a suitable monitor function which takes into account such parameters.

As soon as the global error becomes small, it also enters in the definition of the monitor function. This permits to make more reliable the required conditions on the tolerance. For a more accurate description see [6, 17].

2.3. Solutions of linear and nonlinear algebraic equations

In order to efficiently exploit the mesh selection strategy described above for linear problems, the code first applies Newton method to the continuous problem and then discretizes. This leads to solve sequences of linear continuous problems. Such approach is known in literature as *quasilinearization* [9, 16]. The linear boundary value problems are then solved by using TOM on a sequence of meshes, until the exit tolerances are satisfied. Each mesh generates a linear system that is solved using the sparse MATLAB LU solver.

2.4. Estimates of the global error

The global error and the local truncation error are approximated using a method of higher order in the same class, as described in [17]. In short, let $F_p(\mathbf{y}) = 0$ be the discrete problem obtained when the continuous BVP is solved using a method of order p and M_p the associated Jacobian matrix. Then, using a classical deferred correction technique, the local truncation error τ is approximated by $F_q(\mathbf{y})$, i.e., $\tau = -F_q(\mathbf{y}) + O(h^{p+2})$, where now q is a higher-order method, and consequently the global error by $-M_p^{-1}F_q(\mathbf{y})$.

3. Performance on a difficult problem

The code has been tested on some of the most difficult problems that can be found in literature. For example, in [17] it has been used to solve second order singular perturbation problems and in [15] the rendez-vous problem. In [16] it has been tested on a difficult problem of elasticity. Here we show the performances on a classical problem chosen because it presents different kinds of difficulties.

In all our test we shall consider $Reltol = Abstol = 10^{-6}$, i.e., the relative and the absolute tolerance for the residual must be less than such value. Moreover, the maximum number of mesh points allowed is equal to 2000.

In the paper [11] Holt presents three problems as examples for which the shooting method does not work appropriately. Holt himself presents a semi-analytical approach and is able to give many information, although not complete, on the solution. Many authors have, later on, referred to the first two of them as good test for numerical methods [1, 4, 12, 18, 19, 20]. It seems however that nobody has provided so far more information than Holt did, especially on the second example. Such problem, that we shall consider here, is more difficult, nonlinear and it is described by the following equations

$$\Psi''' + \frac{3-n}{2}\,\Psi\,\Psi'' + n(\Psi')^2 - 1 + G^2 - s\Psi' = 0 \tag{3.1}$$
$$G'' + \frac{3-n}{2}\,\Psi\,G' + (n-1)\,\Psi'\,G - s\,(G-1) = 0$$

subject to boundary conditions

$$\Psi(0) = \Psi'(0) = G(0) = 0, \quad \Psi'(\infty) = 0,\ G(\infty) = 1, \tag{3.2}$$

with n and s given parameters.

When $s = 0$, it is essentially (apart a sign correction) taken from [14] where the underlying physical problem is also described. Such problem represents the laminar boundary layer produced by the rotating flow of a viscous incompressible fluid over a stationary infinite disk, without magnetic interaction.

In vector form the BVP (3.1)–(3.2) becomes

$$\begin{cases} \dfrac{d\mathbf{y}}{dt} = \mathbf{f}(t,\mathbf{y}), & t \geq 0 \\ \\ B_I\,\mathbf{y}(0) + B_F\,\mathbf{y}(\infty) = \mathbf{e}_5, \end{cases} \tag{3.3}$$

where $\mathbf{y} = (\Psi, \Psi', \Psi'', G, G')^T$,

$$B_I = \begin{pmatrix} 1 & 0 & 0 & 0 & 0 \\ 0 & 1 & 0 & 0 & 0 \\ 0 & 0 & 0 & 1 & 0 \\ 0 & 0 & 0 & 0 & 0 \\ 0 & 0 & 0 & 0 & 0 \end{pmatrix}, \quad B_F = \begin{pmatrix} 0 & 0 & 0 & 0 & 0 \\ 0 & 0 & 0 & 0 & 0 \\ 0 & 0 & 0 & 0 & 0 \\ 0 & 1 & 0 & 0 & 0 \\ 0 & 0 & 0 & 1 & 0 \end{pmatrix},$$

and $\mathbf{e}_5 = (0,0,0,0,1)^T$.

First of all, the interval of integration is substituted by a finite one, i.e., $t \in [0,T]$. Thus, the right-hand side boundary conditions are imposed at T. From now on, we shall refer to the problem on the infinite interval as IH and to the problem on the finite interval as FH. The problem FH is a good approximation of the problem IH only if the value of its solution at $[0,T]$ differs from the asymptotic one less than a prefixed tolerance. As an example, we consider the pictures in Figure 1 obtained by the code and reproducing the Figures 3–4 of Holt's paper. Starting from the initial mesh of stepsize $h = T/15$, i.e., $t_j = jh, j = 0,\dots,15$, and the approximation of the solution $\mathbf{y}^{(0)}(t_j) = (0,0,0,1,0)^T$, we have solved the problems obtained in correspondence of the parameters reported in Figure 1. For the interval $[0,20]$ the results provided by the code TOM are reported in Table 1.

It may happen, however, that the problem FH has solutions which do not exist if $T \to \infty$ (*spurious solutions*). As we shall see, we will be able to recognize them by looking the behavior of the parameters $\kappa_d(\mathbf{h})$ and $\gamma_d(\mathbf{h})$, provided by the code (see Section 2.2).

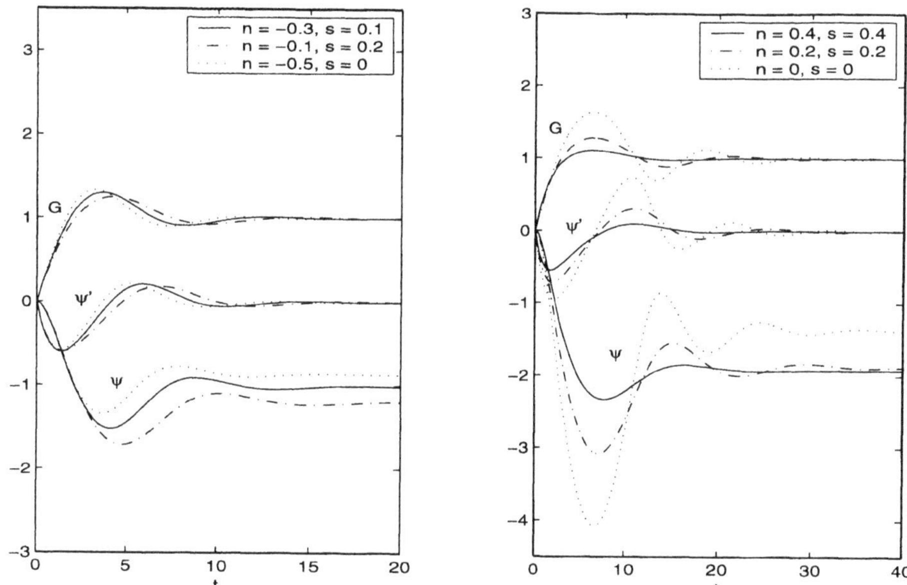

FIGURE 1. Solutions of the Holt problem in correspondence of some values of the parameters.

TABLE 1. The performances of the code TOM on problems on the left-hand side of Figure 1 ($t \in [0, 20]$).

n	s	N	function eval.	b. c. eval.	$\kappa_d(\mathbf{h})$	$\gamma_d(\mathbf{h})$
-0.3	0.1	125	576	11	9.41755	4.48684
-0.1	0.2	120	582	12	9.73223	5.15327
-0.5	0	125	576	11	8.52988	3.75967

Holt provides a picture in the (n, s)-plane, here reproduced in Figure 2, representing the difficulty of the numerical problem. In the region denoted by A, Holt was able to obtain the solution of the problem, while it was not obtained in the region B. For the parameters in the region A the code gives the solution (some of them are reported in Figure 1). It is worth to note that in correspondence of parameters $n = s = 0$, depending on the initial guess, the code provides two different solutions for the FH problem for all values of T considered (T ranging from 10 to 100). For example, choosing $T = 90$ and starting from $\mathbf{y}^{(0)}(t_j) = (-4, 0, 0, 0, 0)^T$, the code provides the solution on the right of Figure 3, while using the solution $\mathbf{y}^{(0)}(t_j) = (0, 0, 0, 1, 0)^T$ as starting guess, the picture on the left is obtained.

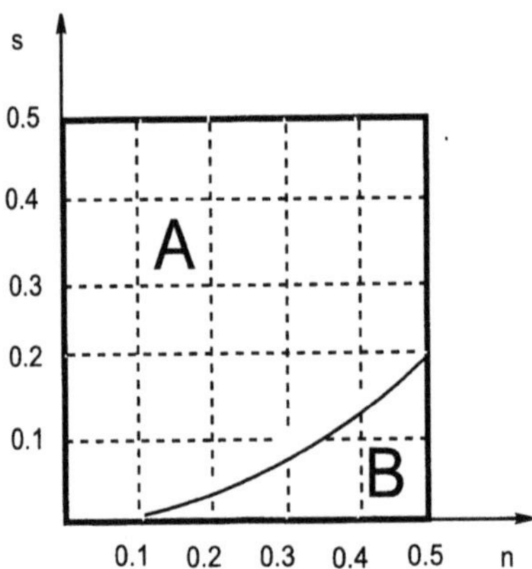

FIGURE 2. Regions A and B.

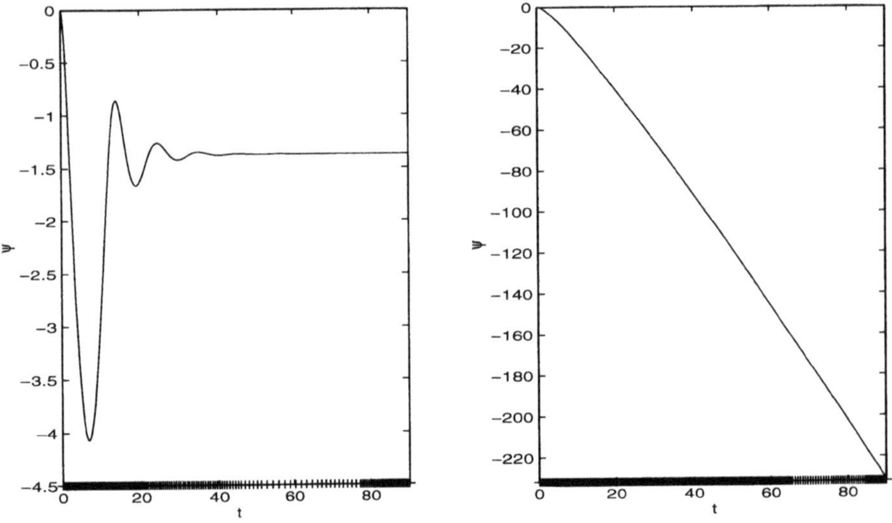

FIGURE 3. Solutions of the Holt problem at $n = s = 0$ (first component only).

It would be interesting to recognize which one of the two is candidate to be solution of the problem IH. According to what has been said in Section 2.2, the sequences of the parameters $\kappa_d^{(T)}(\mathbf{h})$ and $\gamma_d^{(T)}(\mathbf{h})$, for increasing values of T, give the information that the first solution survives for $T \to \infty$, while the second one does not. In fact, in the first case $\{\kappa_d^{(T)}(\mathbf{h})\}$ and $\{\gamma_d^{(T)}(\mathbf{h})\}$ do not increase as $T \to \infty$, as shown on the left-hand side of Figure 4. On the contrary, such parameters increase with T, as we can see on the right of the same picture. This permits to recognize the utility of such additional information, generally not provided by other codes.

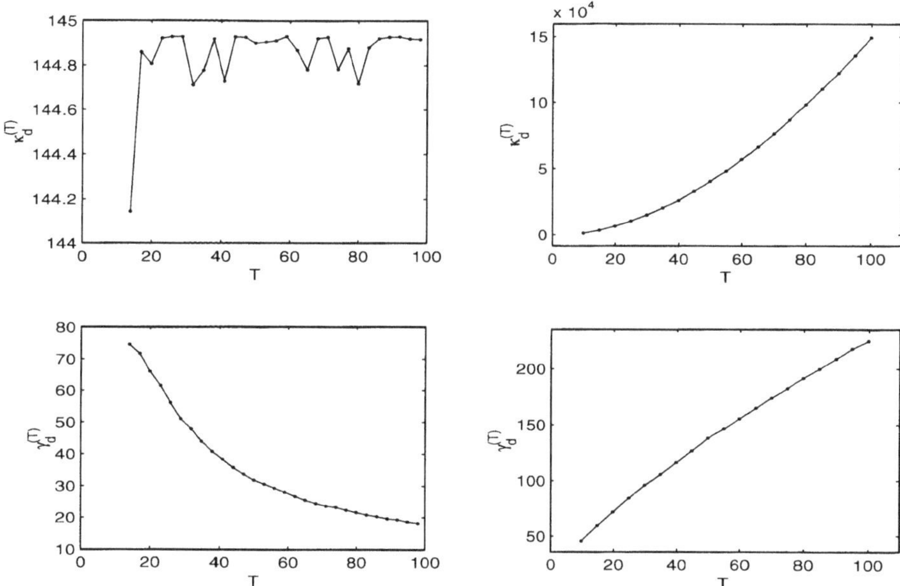

FIGURE 4. Values of $\{\kappa_d^{(T)}(\mathbf{h})\}$ and $\{\gamma_d^{(T)}(\mathbf{h})\}$ for the problem considered in Figure 3.

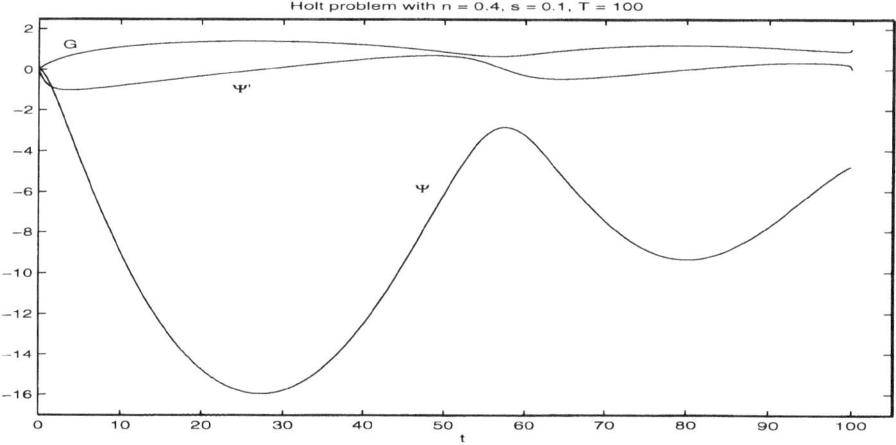

FIGURE 5. Example of solution in the region B.

Concerning the parameters n and s in the region B, the results show that, for $s > 0$, the oscillating solution exists both for FH and IH. Such solution, however, has much larger wavelength (see, for example, Figure 5). This implies that much larger values of T are needed in order to have FH a good approximation for IH.

TABLE 2. The Holt problem with $n = 0.4, s = 0.1, T = 100$.

n	s	N	function eval.	b.c. eval.	$\kappa_d(\mathbf{h})$	$\gamma_d(\mathbf{h})$
0.5	0.3	235	1361	16	23.7637	11.3657
0.5	0.2	275	2676	11	86.7574	36.7855
0.4	0.2	275	2208	8	61.5143	22.9938
0.4	0.15	275	2484	9	145.398	61.896
0.4	0.14	275	1380	5	178.582	80.1947
0.4	0.13	315	2288	8	222.601	101.757
0.4	0.12	315	1896	6	282.248	135.839
0.4	0.11	315	2212	7	365.196	186.425
0.4	0.1	315	2528	8	687.061	255.614

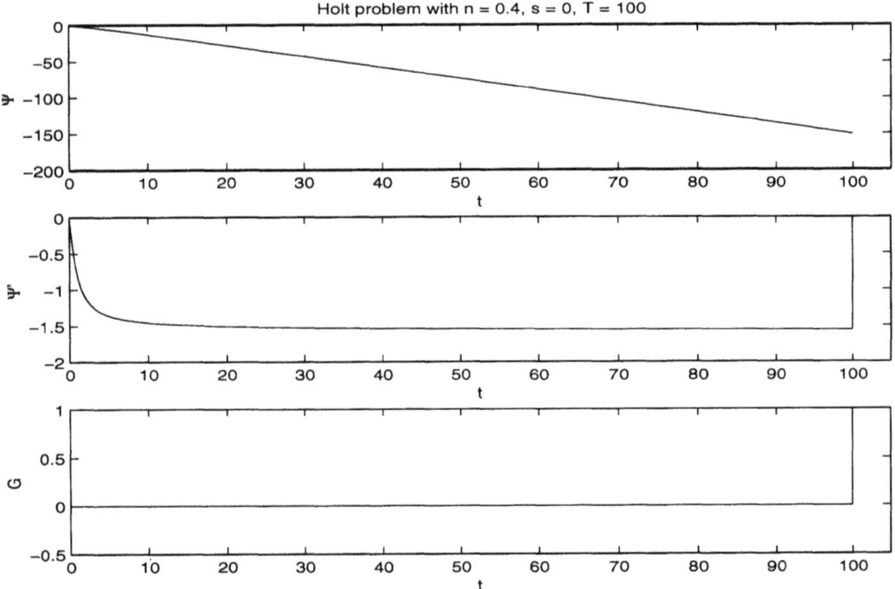

FIGURE 6. Example of a solution in the region B.

In addition, the initial guess must be chosen with more care. We take the latter as the solution obtained in correspondence of a sequence of couples (n, s). Since $\kappa_d(\mathbf{h}), \gamma_d(\mathbf{h})$ become large, the problem tends to be ill-conditioned (see Table 2).

For $s = 0, n \neq 0$, the oscillating solution does not appear, while the other solution, considered spurious for $n = 0$, does (see the case $n = 0.4, s = 0$ in Figure 6). For finite values of T some components of such solution have a layer at T. The problem FH turns out to be a stiff problem and both the parameters $\kappa_d^{(T)}(\mathbf{h})$ and

$\gamma_d^{(T)}(\mathbf{h})$ grow. We deduce then that such a solution will not survive for $T \to \infty$, i.e., the IH problem will not have a solution. As matter of fact, it is known that for $n = 1$ and $s = 0$, IH problem has no solution [14].

References

[1] R.P. Agarwal and R.C. Gupta, *On the solution of the Holt's problem.* BIT **24** (1984), 342–346.

[2] U. Ascher, J. Christiansen, and R.D. Russell, *Colsys – a collocation code for boundary value problems.* Lecture Notes Comp. Sc. (B. Childs et al., ed.), vol. 76, Springer Verlag, 1979, pp. 164–185.

[3] U. Ascher, R. Mattheij, and R.D. Russell, *Numerical solution of boundary value problems for odes.* Prentice-Hall, Englewood CliffsNJ, 1988.

[4] K. Balla and M. Vicsek, *On the reduction of Holt's problem to a finite interval.* Numer. Math. **51** (1987), 291–302.

[5] L. Brugnano and D. Trigiante, *A new mesh selection strategy for odes.* Appl. Numer. Math. **24** (1997), 1–21.

[6] ———, *Solving differential problems by multistep initial and boundary value methods.* Gordon & Breach, Amsterdam, 1998.

[7] J. Cash, G. Moore, and R. Wright, *An automatic continuation strategy for the solution of singularly perturbed nonlinear boundary value problems.* ACM Transaction of Mathematical Software **27** (2001), no. 2, 245–266.

[8] J.R. Cash and M.H. Wright, *A deferred correction method for nonlinear two-point boundary value problems: implementation and numerical evaluation.* SIAM J. Sci. Statist. Comput. **12** (1991), no. 4, 971–989.

[9] E.J. Dean, *An inexact Newton method for nonlinear two-point boundary value problems.* J. Optim. Theory Appl. **75** (1992), no. 3, 471–486.

[10] W.H. Enright and P.H. Muir, *Runge-Kutta software with defect control for boundary value odes.* SIAM J. Sci. Comput. **17** (1996), 479–497.

[11] J.F. Holt, *Numerical solution of nonlinear two-point boundary problems by finite difference methods.* Comm. ACM **7** (1964), 366–373.

[12] D.J. Jones, *Solution of Troesch's, and other, two-point boundary problems by shooting techniques.* J. Comput. Phys. **12** (1973), 429–434.

[13] J. Kierzenka and L.F. Shampine, *A BVP solver based on residual control and the* MATLAB *pse.* ACM Transaction of Mathematical Software **27** (2001), no. 3, 299–316.

[14] W.S. King and W.S. Lewellen, *Boundary-layer similarity solutions for rotating flows with and without magnetic interaction.* Phys. Fluids **7** (1964), 1674–1680.

[15] A. Labianca, F. Mazzia, and D. Trigiante, *Soluzione numerica di problemi ai valori al contorno: applicazione al modello differenziale del rendez-vous.* CAPI2002 – 6° Workshop sul Calcolo ad Alte Prestazioni in Italia http://www.cilea.it/convegni/CAPI2002/index.htm.

[16] F. Mazzia and I. Sgura, *Numerical approximation of nonlinear bvps by means of bvms.* Appl. Numer. Math. **42** (2002), no. 1–3, 337–352.

[17] F. Mazzia and D. Trigiante, *Mesh selection strategy for Boundary Value Problems* (to be published on Numerical Algorithms).

[18] A. Miele, A.K. Agarwal, and J.L. Tietze, *Solution of two-point boundary-value problems with Jacobian matrix characterized by large positive eigenvalues.* J. Comput. Phys. **15** (1974), 117–133.

[19] M.R. Osborne, *On shooting methods for boundary value problems.* J. Math. Anal. Appl. **27** (1969), 417–433.

[20] S.M. Roberts and J.S. Shipman, *Multipoint solution of two-point boundary-value problems.* J. Optim. Theory Applic. **7** (1971), 301–318.

Lidia Aceto
Dipartimento di Energetica
Via C. Lombroso 6/17
50134 Firenze, Italy
email: `aceto@ingfi1.ing.unifi.it`

Francesca Mazzia
Dipartimento di Matematica
Via Orabona 4
70125 Bari, Italy
email: `mazzia@dm.uniba.it`

Donato Trigiante
Dipartimento di Energetica
Via C. Lombroso 6/17
50134 Firenze, Italy
email: `trigiant@unifi.it`

GPSR Compliance
The European Union's (EU) General Product Safety Regulation (GPSR) is a set of rules that requires consumer products to be safe and our obligations to ensure this.

If you have any concerns about our products, you can contact us on

ProductSafety@springernature.com

In case Publisher is established outside the EU, the EU authorized representative is:

Springer Nature Customer Service Center GmbH
Europaplatz 3
69115 Heidelberg, Germany

www.ingramcontent.com/pod-product-compliance
Ingram Content Group UK Ltd.
Pitfield, Milton Keynes, MK11 3LW, UK
UKHW021444280726
14060UKWH00001BA/229

* 9 7 8 3 7 6 4 3 2 1 9 2 5 *